Applied Mathematical Sciences

EDITORIAL STATEMENT

The mathematization of all sciences, the fading of traditional scientific boundaries, the impact of computer technology, the growing importance of mathematical-computer modelling and the necessity of scientific planning all create the need both in education and research for books that are introductory to and abreast of these developments.

The purpose of this series is to provide such books, suitable for the user of mathematics, the mathematician interested in applications, and the student scientist. In particular, this series will provide an outlet for material less formally presented and more anticipatory of needs than finished texts or monographs, yet of immediate interest because of the novelty of its treatment of an application or of mathematics being applied or lying close to applications.

The aim of the series is, through rapid publication in an attractive but inexpensive format, to make material of current interest widely accessible. This implies the absence of excessive generality and abstraction, and unrealistic idealization, but with quality of exposition as a goal.

Many of the books will originate out of and will stimulate the development of new undergraduate and graduate courses in the applications of mathematics. Some of the books will present introductions to new areas of research, new applications and act as signposts for new directions in the mathematical sciences. This series will often serve as an intermediate stage of the publication of material which, through exposure here, will be further developed and refined. These will appear in conventional format and in hard cover.

MANUSCRIPTS

The Editors welcome all inquiries regarding the submission of manuscripts for the series. Final preparation of all manuscripts will take place in the editorial offices of the series in the Division of Applied Mathematics, Brown University, Providence, Rhode Island.

SPRINGER-VERLAG NEW YORK INC., 175 Fifth Avenue, New York, N. Y. 10010

Applied Mathematical Sciences | Volume 66

Applied Mathematical Sciences

1. John: **Partial Differential Equations,** 4th ed.
2. Sirovich: **Techniques of Asymptotic Analysis.**
3. Hale: **Theory of Functional Differential Equations,** 2nd ed.
4. Percus: **Combinatorial methods.**
5. von Mises/Friedrichs: **Flid Dynamics.**
6. Freiberger/Grenander: **A Short Course in Computational Probability and Statistics.**
7. Pipkin: **Lectures on Viscoelasticity Theory.**
9. Friedrichs: **Spectral Theory of Operators in Hilbert Space.**
11. Wolovich: **Linear Multivariable Systems.**
12. Berkovitz: **Optimal Control Theory.**
13. Bluman/Cole: **Similarity Methods for Differential Equations.**
14. Yoshizawa: **Stability Theory and the Existence of Periodic Solution and Almost Periodic Solutions.**
15. Braun: **Differential Equations and Their Applications,** 3rd ed.
16. Lefschetz: **Applications of Algebraic Topology.**
17. Collatz/Wetterling: **Optimization Problems.**
18. Grenander: **Pattern Synthesis: Lectures in Pattern Theory, Vol I.**
20. Driver: **Ordinary and Delay Differential Equations.**
21. Courant/Friedrichs: **Supersonic Flow and Shock Waves.**
22. Rouche/Habets/Laloy: **Stability Theory by Liapunov's Direct Method.**
23. Lamperti: **Stochastic Processes: A Survey of the Mathematical Theory.**
24. Grenander: **Pattern Analysis: Lectures in Pattern Theory, Vol. II.**
25. Davies: **Integral Transforms and Their Applications,** 2nd ed.
26. Kushner/Clark: **Stochastic Approximation Methods for Constrained and Unconstrained Systems**
27. de Boor: **A Practical Guide to Splines.**
28. Keilson: **Markov Chain Models—Rarity and Exponentiality.**
29. de Veubeke: **A Course in Elasticity.**
30. Sniatycki: **Geometric Quantization and Quantum Mechanics.**
31. Reid: **Sturmian Theory for Ordinary Differential Equations.**
32. Meis/Markowitz? **Numerical Solution of Partial Differential Equations.**
33. Grenander: **Regular Structures: Lectures in Pattern Theory, Vol. III.**
34. Kevorkian/Cole: **Perturbation methods in Applied Mathematics.**
35. Carr: **Applications of Centre Manifold Theory.**
36. Bengtsson/Ghil/Källén: **Dynamic Meterology: Data Assimilation Methods.**
37. Saperstone: **Semidynamical Systems in Infinite Dimensional Spaces.**
38. Lichtenberg/Lieberman: **Regular and Stochastic Motion.**
39. Piccini/Stampacchia/Vidossich: **Ordinary Differential Equations in $\mathbf{R}^n$.**
40. Naylor/Sell: **Linear Operator Theory in Engineering and Science.**
41. Sparrow: **The Lorenz Equations: Bifurcations, Chaos, and Strange Attractors.**
42. Guckenheimer/Holmes: **Nonlinear Oscillations, Dynamical Systems and Bifurcations of Vector Fields.**
43. Ockendon/Tayler: **Inviscid Fluid Flows.**

(continued on inside back cover)

I. Hlaváček J. Haslinger
J. Nečas J. Lovíšek

Solution of Variational Inequalities in Mechanics

Springer-Verlag
New York Berlin Heidelberg
London Paris Tokyo

I. Hlaváček
Mathematical Institute of the
Czechoslovak Academy of Sciences
Žitná 25
115 67 Praha 1
Czechoslovakia

J. Haslinger, J. Nečas
Faculty of Mathematics and Physics
of the Charles University
Prague
Czechoslovakia

J. Lovíšek
Faculty of Civil Engineering
Slovak Technical University
Bratislava
Czechoslovakia

Mathematics Subject Classification (1980): 73K25

Library of Congress Cataloging-in-Publication Data
Solution of variational inequalities in mechanics.
(Applied mathematical sciences; v. 66)
Translated from the Slovak.
Bibliography: p.
Includes index.
1. Continuum mechanics. 2. Variational
inequalities (Mathematics) I. Hlaváček, Ivan.
II. Series: Applied mathematical sciences (Springer-Verlag New York Inc.); v.66.
QA1.A647 vol. 66 510 s 87-20767
[QA808.2] [531]

This book is a translation of the original edition: *Riešenie variačných nerovností v mechanike.*

Printed and bound by R.R. Donnelley & Sons, Harrisonburg, Virginia.
Printed in the United States of America.

9 8 7 6 5 4 3 2 1

ISBN 0-387-96597-1 Springer-Verlag New York Berlin Heidelberg
ISBN 3-540-96597-1 Springer-Verlag Berlin Heidelberg New York

Preface

The idea for this book was developed in the seminar on problems of continuum mechanics, which has been active for more than twelve years at the Faculty of Mathematics and Physics, Charles University, Prague. This seminar has been pursuing recent directions in the development of mathematical applications in physics; especially in continuum mechanics, and in technology. It has regularly been attended by upper division and graduate students, faculty, and scientists and researchers from various institutions from Prague and elsewhere. These seminar participants decided to publish in a self-contained monograph the results of their individual and collective efforts in developing applications for the theory of variational inequalities, which is currently a rapidly growing branch of modern analysis.

The theory of variational inequalities is a relatively young mathematical discipline. Apparently, one of the main bases for its development was the paper by G. Fichera (1964) on the solution of the Signorini problem in the theory of elasticity. Later, J. L. Lions and G. Stampacchia (1967) laid the foundations of the theory itself.

Time-dependent inequalities have primarily been treated in works of J. L. Lions and H. Brézis. The diverse applications of the variational inequalities theory are the topics of the well-known monograph by G. Duvaut and J. L. Lions, *Les inéquations en mécanique et en physique* (1972). *Analyse numérique des inéquations variationnelles* (1976), by R. Glowinski, J. L. Lions, and R. Trémolières, deals with the numerical analysis of various problems formulated in terms of variational inequalities. 1980 heralded the appearance of the excellent book *An Introduction to Variational Inequalities and Their Applications*, by D. Kinderlehrer and G. Stampacchia.

Our intention was to loosely follow the ideas of these and other works, especially those of some chapters of the book *Mathematical Theory of Elas-*

tic and Elasto-Plastic Bodies (J. Nečas and I. Hlaváček, 1981). Thus, the second and the third chapter develop, both theoretically and from the point of view of numerical analysis, the solution of the Signorini generalized problem and two models of the theory of plastic flow, respectively.

In both the engineering and physics literature we can find many problems solved by intuitive, ad hoc methods, regardless of the fact that there exists a possibility of formulating and solving these problems in the framework of the theory of variational inequalities, and thus penetrating more profoundly to the very core of the problems. The general purpose of our book is to acquaint the wider reading public with the progress of modern mathematics in this field, and to help this readership find an adequate formulation of the problem as well as an economic method of computation. Our specific desire when writing this book was to share experience gained in the course of solving individual problems for some industrial organizations.

Our thanks are due to our colleagues Dr. O. John and L. Trávníček for their careful reading of the original version of the book, and to Dr. J. Jarník for the translation into English.

I. Hlaváček
J. Haslinger
J. Nečas
J. Lovíšek

Contents

Chapter 1

Unilateral Problems for Scalar Functions

In this chapter we will consider boundary-value problems with one-sided conditions in the form of inequalities, where the unknown is a single real scalar function in a given domain of an n-dimensional space $\mathbf{R}^n$, $n = 2, 3$. First we pay attention to problems in which the inequalities are prescribed only on the boundary of the domain considered. Then we deal with problems which involve inequalities inside the domain, i.e., with an obstacle inside the domain. For the sake of simplicity we restrict our exposition to the second order elliptic operators.

We will show how to formulate these problems in terms of variational inequalities, which in the case of a symmetric elliptic operator are equivalent to the variational problem of finding the minimum of a quadratic functional on a certain closed convex set. Thus, we will in fact obtain a generalization of two dual variational principles of the minimum of the potential energy or the minimum of the complementary energy, well known in mechanics. Moreover, we will also present the mixed variational formulations—a generalization of the Hellinger–Reissner-type canonical variational principles.

Each of the above-mentioned three variational formulations may serve as the basis for deriving approximate variational methods of solution. We will show here applications of one of the simplest variants of the finite element methods, namely, the elements with linear polynomials on a triangular net. We will establish some a priori error bounds provided the exact solution is sufficiently regular. We will also prove that the finite element method converges even if the solution fails to be regular.

The use of dual analysis—that is, the simultaneous solution by means of both dual variational formulations—makes it possible to find even a posteriori error bounds, as well as two-sided bounds for the energy of the exact solution.

1.1 Unilateral Boundary Value Problems for Second Order Equations

Let us first consider elliptic equations of the second order with boundary conditions in the form of inequalities. Such problems describe steady-state phenomena in some branches of mathematical physics, as for example, thermics, fluid mechanics, and electrostatics (see Duvaut and Lions (1972)).

In order to grasp the crucial points of the solution of this class of problems as easily as possible, let us first choose several representative model problems of a relatively simple character.

Thus, let $\Omega \subset \mathbf{R}^n$ be a bounded domain with a Lipschitzian boundary (for the definition, see Nečas (1967) or Hlaváček and Nečas (1981)), and let us consider "the equation with an absolute term"

$$-\Delta u + u = f \quad \text{in } \Omega, \tag{1.1}$$

with boundary conditions of the so-called Signorini type

$$u \geq 0, \quad \frac{\partial u}{\partial \nu} \geq 0, \quad u\frac{\partial u}{\partial \nu} = 0 \quad \text{on } \partial\Omega = \Gamma, \tag{1.2}$$

where $\partial u/\partial \nu$ stands for the derivative with respect to the outer normal ν to the boundary Γ.

Problem (1.1) and (1.2) will briefly be called the "problem $\mathcal{P}_1$".

Secondly, let us consider "the equation without an absolute term"

$$-\Delta u = f \quad in \ \Omega, \tag{1.3}$$

with boundary conditions of two types:

$$u = 0 \quad \text{on } \Gamma_u \subset \Gamma, \tag{1.4}$$

$$u \geq 0, \quad \partial u/\partial \nu \geq 0 \text{ and } u\, \partial u/\partial \nu = 0 \quad \text{on } \Gamma_a = \Gamma \dot{-} \Gamma_u. \tag{1.5}$$

Problem (1.3), (1.4), and (1.5) will be briefly called "problem $\mathcal{P}_2$".

The results which will be established in what follows can be easily extended to equations of the form

$$-\sum_{i,j=1}^{n} \frac{\partial}{\partial x_i}\left(a_{ij}(x)\frac{\partial u}{\partial x_j}\right) + a_0(x)u = f,$$

where $a_{ij} \in L^\infty(\Omega), a_0 \in L^\infty(\Omega)$, a_{ij} is a symmetric matrix that is positive definite in Ω, and

$$a_0(x) \geq c > 0 \quad \text{or} \quad a_0(x) = 0,$$

hold almost everywhere in Ω for the class of problems which generalize $\mathcal{P}_1$ or $\mathcal{P}_2$, respectively.

In what follows we shall make use of the Sobolev spaces $W^{k,p}(\Omega)$ of functions which possess generalized derivatives integrable with the p-th power up to and including the k-th order. For $p = 2$, we shall write $W^{k,2}(\Omega) = H^k(\Omega), H^0(\Omega) = L_2(\Omega)$. Further, we introduce the scalar product in $L_2(\Omega)$ by

$$(f,g)_0 = \int_\Omega f(x)g(x)dx$$

$(x = x_1 \dots x_n, dx = dx_1 \dots dx_n)$. The norm in $H^k(\Omega)$ will be denoted by $\|\cdot\|_{k,\Omega}$, with the subscript Ω omitted if no misunderstanding can occur. The symbol $|u|_{k,\Omega}$ will stand for the seminorm,

$$|u|_{k,\Omega} = \left(\sum_{|\alpha|=k} \|D^\alpha u\|^2_{0,\Omega}\right)^{1/2}, \quad \|u\|_{k,\Omega} = \left(\sum_{j=0}^{k} |u|^2_{j,\Omega}\right)^{1/2}.$$

In $H^1(\Omega)$ we introduce the scalar product

$$(u,v)_1 = (u,v)_0 + (\nabla u, \nabla v)_0,$$

where

$$(\nabla u, \nabla v)_0 = \sum_{i=1}^{n} \left(\frac{\partial u}{\partial x_i}, \frac{\partial v}{\partial x_i}\right)_0.$$

The norm fulfills

$$\|u\|_{1,\Omega} = (u,u)_1^{1/2}.$$

If $u \in [H^k(\Omega)]^n$, then $\|u\|_{k,\Omega}$ means the usual norm

$$\|u\|_{k,\Omega} = \|u\|_k = \left(\sum_{i=1}^{n} \|u_i\|^2_{k,\Omega}\right)^{1/2}, \quad k = 1,2,\dots;$$

similarly for the seminorm.

Let us always assume that the right-hand side of equations (1.1) and (1.3) fulfills $f \in L_2(\Omega)$.

1.1.1 Primal and Dual Variational Problems

Problem $\mathcal{P}_1$ may be formulated as a variational problem. To this end, let us define the set

$$K_1 = \{v \mid v \in H^1(\Omega),\ \gamma v \geq 0\},$$

where γv denotes the trace of the function v on the boundary Γ (e.g., Nečas (1967)), and the functional of potential energy

$$\mathcal{L}_1(v) = \frac{1}{2}\|v\|_1^2 - (f, v)_0.$$

Then the problem: find $u \in K_1$ such that

$$\mathcal{L}_1(u) \leq \mathcal{L}_1(v) \quad \forall v \in K_1, \tag{1.6}$$

is a variational formulation of problem $\mathcal{P}_1$.

A function $u \in K_1$, which is a solution of (1.6), is called a generalized solution of problem $\mathcal{P}_1$. The relation between problem $\mathcal{P}_1$ and problem (1.6) is fundamentally seen from:

Theorem 1.1. *Let V be a Banach space, let a functional $\mathcal{F} : V \to \mathbf{R}$ possess the Gateaux differential in V and let K be a convex subset of V. Then u minimizes $\mathcal{F}$ on the set K only if*

$$D\mathcal{F}(u, v - u) \geq 0 \quad \forall v \in K. \tag{1.7}$$

If, moreover, $\mathcal{F}$ is convex, then (1.7) is a necessary and sufficient condition for u to minimize $\mathcal{F}$ on K.

Proof. See Céa (1971). □

As K_1 is a convex set (indeed, $\gamma(\lambda u + (1-\lambda)v) = \lambda\gamma u + (1-\lambda)\gamma v \geq 0$, $\forall \lambda \in (0,1)$, $\forall u, v \in K_1$), and $\mathcal{L}_1$ is a convex functional in $V = H^1(\Omega)$, the problem (1.6) is equivalent to the variational inequality

$$D\mathcal{L}_1(u, v - u) \geq 0 \quad \forall v \in K_1,$$

which may be written in the form

$$(u, v - u)_1 \geq (f, v - u)_0 \quad \forall v \in K_1. \tag{1.8}$$

Choosing here

$$v = u \pm \varphi \quad \varphi \in C_0^\infty(\Omega),$$

we obtain

$$(u, \varphi)_1 = (f, \varphi)_0 \quad \forall \varphi \in C_0^\infty(\Omega),$$

which implies that the equation (1.1) is satisfied in the sense of distributions and

$$\Delta u = u - f \in L_2(\Omega).$$

In order to find out in what sense the boundary conditions are satisfied, we first introduce the space of traces of functions from $H^1(\Omega)$ on the boundary Γ.

Definition 1.1. If Ω is a domain with a Lipschitzian boundary Γ, then the image of the space $H^1(\Omega)$ is denoted by

$$\gamma(H^1(\Omega)) = H^{1/2}(\Gamma),$$

and the norm in $H^{1/2}(\Gamma)$ is defined as

$$\|w\|_{1/2,\Gamma} = \inf_{\substack{v \in H^1(\Omega) \\ \gamma v = w}} \|v\|_{1,\Omega}. \tag{1.9}$$

Remark 1.1. In Nečas (1967), the reader will find another definition of $H^{1/2}(\Gamma)$ together with a norm which is equivalent to the norm (1.9). $H^{1/2}(\Gamma)$ is a linear subspace of $L_2(\Gamma)$.

Theorem 1.2. *Let us denote by*

$$H^{-1/2}(\Gamma) = [H^{1/2}(\Gamma)]'$$

the space of linear continuous functionals over the space $H^{1/2}(\Gamma)$*. Further, let* $u \in H^1(\Omega)$ *satisfy* $\Delta u \in L_2(\Omega)$ *and let* $w \in H^{1/2}(\Gamma), v \in H^1(\Omega), \gamma v = w$*. Then the formula*

$$\langle \frac{\partial u}{\partial \nu}, w \rangle = (\nabla u, \nabla v)_0 + (v, \Delta u)_0 \tag{1.10}$$

determines an element from $H^{-1/2}(\Gamma)$*, which will be denoted by* $\partial u / \partial \nu$*.*

Remark 1.2. The reader will easily verify that the left-hand side of formula (1.10) for functions $u \in C^\infty(\bar{\Omega})$ reduces to the integral

$$\langle \frac{\partial u}{\partial \nu}, w \rangle = \int_\Gamma \frac{\partial u}{\partial \nu} \, w d\Gamma.$$

Consequently, theorem 1.2 extends the notion of the derivative with respect to the normal to a more general class of functions.

Proof of Theorem 1.2. 1^0 The right-hand side of (1.10) is independent of the choice of $v \in H^1(\Omega)$. Indeed, the identity

$$(\nabla u, \nabla \varphi)_0 + (\varphi, \Delta u)_0 = 0 \quad \forall \varphi \in C_0^\infty(\Omega)$$

directly follows from the definition of Δu in the sense of distributions.

2^0 As the mapping γ is additive and homogeneous, the functional $\partial u/\partial \nu$ is linear.

3^0

$$\left|\langle\frac{\partial u}{\partial \nu}, w\rangle\right| \leq (\|u\|_1^2 + \|\Delta u\|_0^2)^{1/2}\|v\|_1 \quad \forall u \in H^1(\Omega), \ \gamma v = w.$$

Passing to the infimum we find that $\partial u/\partial \nu$ is bounded in $H^{1/2}(\Gamma)$. □

Now let us substitute $v = 0$ and $v = 2u$ into inequality (1.8). We obtain

$$(u, u)_1 = (f, u)_0 \tag{1.11}$$

and, since $\Delta u = u - f$, we have

$$\begin{aligned} \langle\frac{\partial u}{\partial \nu}, \gamma u\rangle &= (\nabla u, \nabla u)_0 + (u, \Delta u)_0 \\ &= (\nabla u, \nabla u)_0 + (u, u)_0 - (f, u)_0 = 0. \end{aligned} \tag{1.12}$$

Substituting (1.11) into (1.8), we find

$$\begin{aligned} 0 &\leq (u, v)_1 - (f, v)_0 = (\nabla u, \nabla v)_0 + (u - f, v)_0 \\ &= (\nabla u, \nabla v)_0 + (\Delta u, v)_0 = \langle\frac{\partial u}{\partial \nu}, \ \gamma v\rangle \quad \forall v \in K_1. \end{aligned} \tag{1.13}$$

Inequality (1.13) states that the functional $\partial u/\partial \nu$ is nonnegative, as will be seen later on. Thus, we conclude that a solution u of problem $\mathcal{P}_1$ fulfills the second and last condition from the triplet of boundary conditions (1.2) in the functional sense.

Conversely, each sufficiently smooth solution of problem $\mathcal{P}_1$ solves problem (1.6). Indeed, it suffices to multiply (1.1) by a function $v \in K_1$ and to integrate by parts, thus deriving (1.12) and (1.13). Hence, the variational inequality (1.8) follows.

The same method may also be applied to problem $\mathcal{P}_2$. Let us define the set

$$K_2 = \{v \mid v \in H^1(\Omega), \gamma v = 0 \ \text{ on } \ \Gamma_u, \gamma v \geq 0 \text{ on } \Gamma_a\},$$

and the functional of potential energy

$$\mathcal{L}_2(v) = \frac{1}{2}|v|_1^2 - (f, v)_0,$$

where

$$|v_1|^2 = (\nabla v, \nabla v)_0.$$

Then the problem: find $u \in K_2$ such that

$$\mathcal{L}_2(u) \leq \mathcal{L}_2(v) \quad \forall v \in K_2 \tag{1.14}$$

is a variational formulation of problem $\mathcal{P}_2$.

The relation between the solution of problem $\mathcal{P}_2$ and the variational problem (1.14) is quite analogous to the case of problem $\mathcal{P}_1$.

Problems (1.6) and (1.14) are called the primal variational formulations of the one-sided problems $\mathcal{P}_1$ and $\mathcal{P}_2$, respectively. Although the methods of solution for both of these primal problems differ only insignificantly, we shall see that the difference between the corresponding dual variational formulations is essential.

1.1.11. Dual Variational Formulation. We can formulate problems $\mathcal{P}_1$ and $\mathcal{P}_2$ in such a way that the unknown is not the function u but rather its gradient, which is often an important and interesting quantity from the physical point of view. However, instead of the direct transformation of problems $\mathcal{P}_1$ and $\mathcal{P}_2$, we will derive variational formulations which are dual with respect to the primal variational formulations (1.6) and (1.14).

Let us introduce the set

$$H(\mathrm{div}, \Omega) = \{q \mid q \in [L_2(\Omega)]^n, \ \mathrm{div}\, q \in L_2(\Omega)\},$$

where the divergence operator is understood in the sense of distributions:

$$\int_\Omega q \cdot \nabla\varphi \, dx = -\int_\Omega \varphi \, \mathrm{div}\, q \, dx \quad \forall \varphi \in C_0^\infty(\Omega).$$

Theorem 1.3. *Let $q \in H(\mathrm{div}, \Omega)$, $w \in H^{1/2}(\Gamma)$. Then the formula*

$$\langle q \cdot \nu, w \rangle = \int_\Omega (q \cdot \nabla v + v \, \mathrm{div}\, q) dx,$$

with $v \in H^1(\Omega)$ and $\gamma v = w$, defines a functional $q \cdot \nu \in H^{-1/2}(\Gamma)$.

Proof. Almost identical with that of theorem 1.2. □

We shall write $s \geq 0$ on Γ (or Γ_a) for a functional $s \in H^{-1/2}(\Gamma)$ if

$$\langle s, \gamma v \rangle \geq 0 \quad \forall v \in K_1 \quad (K_2, \text{respectively}).$$

Let us introduce the class of admissible functions

$$\mathcal{U}_1 = \{q \mid q \in [L_2(\Omega)]^{n+1}, \quad q = [\bar{q}, q_{n+1}], \quad \bar{q} \in H(\mathrm{div}, \Omega),$$

$$q_{n+1} = f + \mathrm{div}\, \bar{q}, \ \bar{q} \cdot \nu \geq 0 \ \text{ on } \Gamma\},$$

and the functional of the complementary energy

$$\mathcal{S}_1(q) = \frac{1}{2}\sum_{i=1}^{n+1} \|q_i\|_0^2.$$

The problem: find $q^0 \in \mathcal{U}_1$ such that

$$\mathcal{S}_1(q^0) \leq \mathcal{S}_1(q) \quad \forall q \in \mathcal{U}_1, \tag{1.15}$$

will be called *the dual variational formulation* with respect to the problem (1.6).

Let us consider problem $\mathcal{P}_2$. Introduce the set

$$\mathcal{U}_2 = \{q \mid q \in H(\text{div}, \Omega), \ \ \text{div } q + f = 0, \ \ q \cdot \nu \geq 0 \ \text{ on } \Gamma_a\}$$

and the functional of complementary energy

$$\mathcal{S}_2(q) = \frac{1}{2}\sum_{i=1}^{n} \|q_i\|_0^2.$$

The problem: find $q^0 \in \mathcal{U}_2$ such that

$$\mathcal{S}_2(q^0) \leq \mathcal{S}_2(q) \quad \forall q \in \mathcal{U}_2 \tag{1.16}$$

will be called *the dual variational formulation* with respect to the problem (1.14).

Theorem 1.4. *Both of the primal problems* (1.6) *and* (1.14), *as well as both of the dual problems* (1.15) *and* (1.16), *has a unique solution.*

Proof. Based on the following general result. □

Theorem 1.5. *Let $\mathcal{F}$ be a strictly convex, continuous functional defined on a reflexive Banach space X. Let $K \subset X$ be a closed convex set and let $\mathcal{F}$ be coercive on K, i.e.,*

$$v \in K, \ \ \|v\| \to \infty \Rightarrow \mathcal{F}(v) \to +\infty. \tag{1.17}$$

Then there is one and only one solution of the problem:

$$\mathcal{F}(u) = \min \ \ on \ K.$$

(Proof is found in the book by Fučík and Kufner (1978). See theorem 26.8.)

The functionals $\mathcal{L}_i, i = 1,2$ are continuous, strictly convex, and coercive in the spaces $V_1 = H^1(\Omega)$ and $V_2 = \{v \in H^1(\Omega) \mid \gamma v = 0 \text{ on } \Gamma_u\}$, respectively.

Indeed, take for example $\mathcal{L}_2$. It has a second differential which satisfies

$$D^2\mathcal{L}_2(u,v,v) = |v|_1^2 \geq C\|v\|_1^2 \quad \forall v \in V_2, \ (\text{mes } \Gamma_u > 0),$$

where $C = \text{const} > 0$. Hence, $\mathcal{L}_2$ is strictly convex and satisfies (1.17). It is not difficult to verify that the sets K_i are closed and convex in V_i.

In order to prove existence and uniqueness for problem (1.15), we first transform it to an equivalent problem, formulated only for "reduced" vector functions $\bar{q} \in H(\text{div},\Omega)$. Namely, after the substitution $q_{n+1} = f + \text{div}\,\bar{q}$ we have a new equivalent problem: find $\bar{q} \in \mathcal{U}_0$ such that

$$I(\bar{q}^0) \leq I(\bar{q}) \quad \forall \bar{q} \in \mathcal{U}_0, \tag{1.18}$$

where

$$\mathcal{U}_0 = \{\bar{q} \in H(\text{div},\Omega) \mid \bar{q}\cdot\nu \geq 0 \ \text{ on } \Gamma\},$$

$$I(\bar{q}) = \frac{1}{2}\left(\sum_{i=1}^{n} \|\bar{q}\|_0^2 + \|\text{div}\,\bar{q}\|_0^2\right) + (f, \text{div}\,\bar{q})_0.$$

If the norm in the space $H(\text{div},\Omega)$ is introduced by

$$\|q\|_{H(\text{div},\Omega)} = \left(\sum_{i=1}^{n} \|q_i\|_0^2 + \|\text{div}\,q\|_0^2\right)^{1/2},$$

then $H(\text{div},\Omega)$ is a Hilbert space. $\mathcal{U}_0$ is closed and convex in $H(\text{div},\Omega)$. To prove for example the closedness of $\mathcal{U}_0$, it is sufficient to realize that the mapping $q \to q\cdot\nu$ of the space $H(\text{div},\Omega)$ into $H^{-1/2}(\Gamma)$ is linear and continuous.

The functional I is continuous on $H(\text{div},\Omega)$, strictly convex, and coercive. Hence, according to theorem 1.5 we obtain the existence and uniqueness of the solution to problem (1.15).

We proceed analogously with problem (1.16). $\mathcal{U}_2$ is closed and convex in the space $H(\text{div},\Omega)$, and the functional S_2 is continuous and strictly convex in $H(\text{div},\Omega)$. Since

$$q \in \mathcal{U}_2, \ \|q\|_{H(\text{div},\Omega)} \to \infty \quad \Rightarrow \quad \left(\sum_{i=1}^{n} \|q_i\|_0^2 + \|f\|_0^2\right) \to \infty \Rightarrow$$
$$\Rightarrow \quad S_2(q) \to +\infty,$$

S_2 is also coercive in $\mathcal{U}_2$.

1.1.12. Relation Between the Primal and Dual Variational Formulations. We will now show how the solutions of the dual problems are related to those of the primal problems. To this end we will use the *saddle-point method* (the min-max method), the sequel of which will also enable us to derive the mixed variational formulation of the original problem $\mathcal{P}_1$ or $\mathcal{P}_2$, corresponding to the canonical variational principles of Hellinger–Reissner type (see Hlaváček and Nečas (1981)).

Let us consider problem $\mathcal{P}_1$ together with its variational formulations.

Theorem 1.6. *If u is a solution of* (1.6) *and q^0 a solution of* (1.15), *then*

$$q_i^0 = \frac{\partial u}{\partial x_i}, \quad i = 1, 2, \dots n, \tag{1.19}$$

$$q_{n+1}^0 = u,$$

$$\mathcal{S}_1(q^0) + \mathcal{L}_1(u) = 0. \tag{1.20}$$

Proof. For $v \in H^1(\Omega)$ let us denote

$$\mathcal{E}(v) = [\nabla v, v] \in [L_2(\Omega)]^{n+1}, \quad M = [L_2(\Omega)]^{n+1},$$

$$\mathcal{W} = K_1 \times M$$

and let us introduce the Lagrangian functional

$$\mathcal{H}(v, \mathcal{N}; q) = \frac{1}{2}\|\mathcal{N}\|^2 - (f, v)_0 + (q, \mathcal{E}(v) - \mathcal{N}),$$

where $v \in K_1$, $\mathcal{N} \in M, q \in M$, and $\|\cdot\|$, $(.,.)$ denote the norm and the scalar product in $[L_2(\Omega)]^{n+1}$, respectively.

We easily find that

$$\sup_{q \in M}(q, \mathcal{E}(v) - \mathcal{N}) = \begin{matrix} 0 & \text{for } \mathcal{N} = \mathcal{E}(v) \\ +\infty & \text{for } \mathcal{N} \neq \mathcal{E}(v), \end{matrix}$$

and thus

$$\inf_{v \in K_1} \mathcal{L}_1(v) = \inf_{[v, \mathcal{N}] \in \mathcal{W}} \sup_{q \in M} \mathcal{H}(v, \mathcal{N}; q). \tag{$\mathcal{P}$}$$

□

In the theory of duality (e.g., Ekeland and Temam (1974)), the problem

$\mathcal{P}^*$

$$\sup_{q \in M} \inf_{[v, \mathcal{N}] \in \mathcal{W}} \mathcal{H}(v, \mathcal{N}; q)$$

is called the dual problem to $\mathcal{P}$. The question arises when both values coincide.

Let us denote

$$S_0(q) = \inf_{[v, \mathcal{N}] \in \mathcal{W}} \mathcal{H}(v, \mathcal{N}; q)$$

and let us try to explicitly calculate this functional.

First of all, we have

$$S_0(q) \leq \inf_{\substack{[v, \mathcal{N}] \in \mathcal{W} \\ \mathcal{N} = \mathcal{E}(v)}} \mathcal{H}(v, \mathcal{N}; q) = \inf_{v \in K_1} \mathcal{L}_1(v) = \mathcal{L}_1(u)$$

for all $q \in M$. This implies

$$\sup_M S_0(q) \leq \mathcal{L}_1(u). \tag{1.21}$$

Furthermore,

$$S_0(q) = \inf_{\mathcal{N} \in M} \mathcal{H}_1(\mathcal{N}, q) + \inf_{v \in K_1} \mathcal{H}_2(v, q)$$

where

$$\mathcal{H}_1(\mathcal{N}, q) = \frac{1}{2}\|\mathcal{N}\|^2 - (q, \mathcal{N}),$$

$$\mathcal{H}_2(v, q) = (q, \mathcal{E}(v)) - (f, v)_0.$$

It is easy to calculate

$$\inf_{\mathcal{N} \in M} \mathcal{H}_1(\mathcal{N}, q) = -\frac{1}{2}\|q\|^2.$$

Lemma 1.1. *We have*

$$\inf_{v \in K_1} \mathcal{H}_2(v, q) = \begin{matrix} 0 & \textit{for } q \in \mathcal{U}_1, \\ -\infty & \textit{for } q \notin \mathcal{U}_1. \end{matrix}$$

Proof. 1^0 $\mathcal{H}_2(\cdot, q)$ is a continuous linear functional in $H^1(\Omega)$. Let there exist $v_0 \in K_1$ such that $\mathcal{H}_2(v_0, q) < 0$. Then if $t \to +\infty$, $tv_0 \in K_1$ and $\lim \mathcal{H}_2(tv_0, q) = -\infty$; hence the infimum is $-\infty$. Consequently, if the infimum is to be greater than $-\infty$, then necessarily

$$\mathcal{H}_2(v, q) \geq 0 \quad \forall v \in K_1,$$

or

$$\sum_{i=1}^{n} \left(\bar{q}_i, \frac{\partial v}{\partial x_i} \right)_0 + (q_{n+1}, v)_0 - (f, v)_0 \geq 0 \quad \forall v \in K_1.$$

Substituting in the inequality $v = \pm\varphi \in C_0^\infty(\Omega)$, we obtain

$$q_{n+1} - f = \text{div}\ \bar{q} \in L_2(\Omega).$$

Hence, $\bar{q} \in H(\text{div}, \Omega)$.

Further, according to theorem 1.3 we have for all $v \in K_1$

$$0 \le \mathcal{H}_2(v, q) = \int_\Omega (q \cdot \nabla v + v\ \text{div}\ \bar{q})dx = \langle \bar{q} \cdot \nu, \gamma\nu \rangle.$$

Hence, $\bar{q} \cdot \nu \ge 0$ on Γ, which means $q \in \mathcal{U}_1$.

2^0 Let $q \in \mathcal{U}_1$, $v \in K_1$. Then

$$\begin{aligned} (q, \mathcal{E}(v)) &= \int_\Omega (\bar{q} \cdot \nabla v + v\ \text{div}\ \bar{q} + vf)dx \\ &= \langle \bar{q} \cdot \nu, \gamma v \rangle + (f, v)_0 \ge (f, v)_0. \end{aligned}$$

Hence, for $q \in \mathcal{U}_1$ we have $\mathcal{H}_2(v, q) \ge 0$ $\forall v \in K_1$. Substituting $v = 0 \in K_1$ we derive that the infimum is equal to zero, which completes the proof of lemma 1.1. □

Altogether, we can write

$$\begin{aligned} \mathcal{S}_0(q) &= -\frac{1}{2}\|q\|^2 = -\mathcal{S}_1(q) \quad \text{for } q \in \mathcal{U}_1, \\ \mathcal{S}_0(q) &= -\infty \quad \text{for } q \notin \mathcal{U}_1. \end{aligned}$$

Thus, by virtue of (1.21) we have

$$\mathcal{L}_1(u) \ge \sup_M \mathcal{S}_0(q) = \sup_{\mathcal{U}_1}[-\mathcal{S}_1(q)] = \inf_{\mathcal{U}_1} \mathcal{S}_1(q) = -\mathcal{S}_1(q^0). \tag{1.22}$$

Setting $q = \hat{q} = \mathcal{E}(u)$ we obtain $\hat{q} \in \mathcal{U}_1$. Indeed, at the beginning of section 1.1.1 we found that $\Delta u = u - f \in L_2(\Omega)$. Hence:

$$\text{div}\ \bar{\hat{q}} = \text{div}\ \nabla u = \Delta u \in L_2(\Omega) \Longrightarrow \bar{\hat{q}} \in H\ (\text{div}, \Omega),$$

$$\text{div}\ \bar{\hat{q}} = \hat{q}_{n+1} - f.$$

Finally, it follows from (1.13) that

$$\bar{\hat{q}} \cdot \nu = \frac{\partial u}{\partial \nu} \ge 0 \text{ on } \Gamma.$$

Substituting $q = \hat{q}$ into $\mathcal{S}_1$ and using equation (1.11), we obtain

$$-\mathcal{S}_1(\hat{q}) = -\frac{1}{2}\|\hat{q}\|^2 = -\frac{1}{2}\|u\|_1^2 = \frac{1}{2}\|u\|_1^2 - (f, u)_0 = \mathcal{L}_1(u). \tag{1.23}$$

Hence, $-\mathcal{S}_1$ assumes its maximum at the point $\hat{q}$, as is seen in (1.22). However, the uniqueness of the solution of problem (1.15)—see theorem 1.4—implies that $\hat{q} = q^0$. This, together with (1.23), yields relation (1.20).

Remark 1.3. The point $\{[u, \mathcal{E}(u)], q^0\} \in \mathcal{W} \times M$ is called a *saddle point* of the functional $\mathcal{H}([v, \mathcal{N}]; q)$ in the Cartesian product $\mathcal{W} \times M$. The following identity holds for this point as a consequence of (1.23) and (1.22):

$$\begin{aligned} \mathcal{H}([u, \mathcal{E}(u)]; q^0) &= \min_{\mathcal{W}} \sup_M \mathcal{H}([v, \mathcal{N}]; q) \\ &= \max_M \inf_{\mathcal{W}} \mathcal{H}([v, \mathcal{N}]; q). \end{aligned} \tag{1.24}$$

An analogous theorem holds for problem $\mathcal{P}_2$ and its variational formulations:

Theorem 1.7. *If u is a solution of* (1.14) *and q^0 a solution of* (1.16), *then*

$$q^0 = \nabla u,$$
$$\mathcal{S}_2(q^0) + \mathcal{L}_2(u) = 0.$$

Proof. Can be carried out analogously to that of theorem 1.6. □

1.1.2 Mixed Variational Formulations

In this section we will derive new, equivalent variational formulations of problems $\mathcal{P}_1$ and $\mathcal{P}_2$. The difference between the primal formulations and the new ones consists of the fact that our new formulations do not require the one-sided boundary conditions to be explicitly fulfilled. Let us restrict our considerations to problem $\mathcal{P}_1$.

The symbol $H_+^{-1/2}(\Gamma)$ $(\Gamma \equiv \partial\Omega)$ denotes the set of all nonnegative linear functionals over $H^{1/2}(\Gamma)$ (in the sense of section 1.1.11). It is evident that

$$v \in K_1 \Longleftrightarrow -\langle \varphi, \gamma v \rangle \le 0 \quad \forall \varphi \in H_+^{-1/2}(\Gamma). \tag{2.1}$$

Moreover,

$$\sup_{H_+^{-1/2}(\Gamma)} \{-\langle \varphi, \gamma v \rangle\} = \begin{array}{ll} 0 & \text{provided } v \in K_1, \\ +\infty & \text{provided } v \notin K_1. \end{array}$$

Indeed, let $v \in K_1$. Then,

$$-\langle \varphi, \gamma v \rangle \le 0 \quad \forall \varphi \in H_+^{-1/2}(\Gamma)$$

and

$$\langle \theta, \gamma v \rangle = 0,$$

where θ is the zero element of $H^{-1/2}(\Gamma)$. On the other hand, if $v \notin K_1$, then there exists $\mu \in H_+^{-1/2}(\Gamma)$ such that

$$-\langle \mu, \gamma v \rangle = c > 0.$$

It is easily verified that the elements of the form $\rho\mu$ belong to $H_+^{-1/2}(\Gamma)$ for every $\rho \geq 0$. Consequently,

$$\sup_{H_+^{-1/2}(\Gamma)} \{-\langle \varphi, \gamma v \rangle\} \geq -\langle \rho\mu, \gamma v \rangle = \begin{matrix} \rho c \to +\infty, \\ c \to +\infty. \end{matrix}$$

Thus, we can formally write

$$\inf_{K_1} \mathcal{L}_1(v) = \inf_{H^1(\Omega)} \sup_{H_+^{-1/2}(\Gamma)} \{\mathcal{L}_1(v) - \langle \mu, \gamma v \rangle\}.$$

Let us write $\mathcal{H}(v, \mu) = \mathcal{L}_1(v) - \langle \mu, \gamma v \rangle$. We shall establish the relation between the solution of $\mathcal{P}_1$ and the saddle point of $\mathcal{H}$ in $H^1(\Omega) \times H_+^{-1/2}(\Gamma)$.

Theorem 2.1. *A pair $(w, \lambda) \in H^1(\Omega) \times H_+^{-1/2}(\Gamma)$ is a saddle point of $\mathcal{H}$ in $H^1(\Omega) \times H_+^{-1/2}(\Gamma)$ if and only if*

$$w = u, \quad \lambda = \partial u / \partial \nu,$$

where $u \in K_1$ is a solution of $\mathcal{P}_1$.

Proof. (i) Let (w, λ) be a saddle point of $\mathcal{H}$ in $H^1(\Omega) \times H_+^{-1/2}(\Gamma)$:

$$(w, \lambda) \in H^1(\Omega) \times H_+^{-1/2}(\Gamma):$$

$$\mathcal{H}(w, \mu) \leq \mathcal{H}(w, \lambda) \leq \mathcal{H}(v, \lambda) \quad \forall (v, \mu) \in H^1(\Omega) \times H_+^{-1/2}(\Gamma). \tag{2.2}$$

Substituting into this inequality first $\mu = \theta$, then $\mu = 2\lambda$, we obtain

$$\langle \lambda, \gamma w \rangle = 0, \tag{2.3$_1$}$$

and consequently

$$-\langle \mu, \gamma w \rangle \leq 0 \quad \forall \mu \in H_+^{-1/2}(\Gamma). \tag{2.3$_2$}$$

Hence, of necessity $w \in K_1$. Taking into account (2.1), the second inequality in (2.2) and $(2.3)_{1,2}$, as well as the definition of $\mathcal{H}$, we conclude

$$\mathcal{L}_1(w) \leq \mathcal{L}_1(v) - \langle \mu, \gamma v \rangle \leq \mathcal{L}_1(v) \quad \forall v \in K_1.$$

This means that $w \in K_1$ is an element which minimizes $\mathcal{L}_1$ in K_1. Such an element, as we know, is unique; hence, $w = u$. The functional $\mathcal{H}(\cdot, \lambda)$ assumes its minimum in $H^1(\Omega)$ at the point $v = w$. Therefore,

$$(w, v)_1 - \langle \lambda, \gamma v \rangle = (f, v)_0 \quad \forall v \in H^1(\Omega).$$

Using Green's formula (1.10) together with (1.1), we finally obtain

$$\langle \partial w / \partial v, \gamma v \rangle - \langle \lambda, \gamma v \rangle = 0 \quad \forall v \in H^1(\Omega),$$

which implies $\partial w / \partial \nu = \lambda$.

(ii) Let u be a solution of $\mathcal{P}_1$. Then, (1.13) implies $(u, \partial u/\partial \nu) \in H^1(\Omega) \times H_+^{-1/2}(\Gamma)$. Further,

$$\begin{aligned} \mathcal{H}(u, \partial u/\partial \nu) &= \mathcal{L}_1(u) - \langle \partial u/\partial \nu, \gamma u \rangle \geq \mathcal{L}_1(u) - \langle \mu, \gamma u \rangle \\ &= \mathcal{H}(u, \mu) \quad \forall \mu \in H_+^{-1/2}(\Gamma). \end{aligned}$$

Similarly,

$$\begin{aligned} \mathcal{H}(u, \partial u/\partial \nu) - \mathcal{H}(v, \partial u/\partial \nu) &= \mathcal{L}_1(u) - \mathcal{L}_1(v) + \langle \partial u/\partial \nu, \gamma v - \gamma u \rangle \\ = -\frac{1}{2}\|u - v\|_1^2 &+ (u, u - v)_1 + (f, v - u)_0 + \langle \partial u/\partial \nu, \gamma v - \gamma u \rangle \\ &= -\frac{1}{2}\|u - v\|_1^2 \leq 0 \quad \forall v \in H^1(\Omega) \end{aligned}$$

by virtue of Green's formula (1.10) and also (1.1). Thus, we have verified that $(u, \partial u/\partial \nu)$ is a saddle point of $\mathcal{H}$ in $H^1(\Omega) \times H_+^{-1/2}(\Gamma)$. □

Remark 2.1. By using partial variations $\delta_v \mathcal{H}$, $\delta_\mu \mathcal{H}$ with respect to the variables v, μ, it is possible to equivalently characterize (2.2) by the following relations (see Ekeland-Temam (1974)):

$$\begin{aligned} &(w, \lambda) \in H^1(\Omega) \times H_+^{-1/2}(\Gamma), \\ &\delta_v \mathcal{H}(w, \lambda; v) = 0 \quad \forall v \in H^1(\Omega) \Longleftrightarrow (w, v)_1 - \langle \lambda, v \rangle \\ &\qquad = (f, v)_0 \quad \forall v \in H^1(\Omega), \end{aligned} \tag{2.2'}$$

$$\begin{aligned} \delta_\mu \mathcal{H}(w, \lambda; \mu - \lambda) \ &\leq \ 0 \quad \forall \mu \in H_+^{-1/2}(\Gamma) \Longleftrightarrow \\ &\langle \mu - \lambda, \gamma w \rangle \geq 0 \quad \forall \mu \in H_+^{-1/2}(\Gamma). \end{aligned}$$

The relations (2.2) or (2.2′) will be called the *mixed formulation* of $\mathcal{P}_1$. Similar formulation is available for $\mathcal{P}_2$.

It is seen from the definition of the mixed variational formulation that the first component ranges over the *whole* space $H^1(\Omega)$, and not only over the set K_1, as was the case with the primal variational formulation. The fact that the first component of the saddle point is eventually an element of K_1 is a consequence of the mixed formulation itself; more precisely of the first inequality in (2.2) or (2.2′).

We introduce one more example, which serves as a simpler model problem for the so-called one-sided boundary value friction problems. Such problems will be studied in more detail in Chapter 2.

Let

$$\mathcal{L}_3(v) = \frac{1}{2}\|v\|_1^2 - (f, v)_0 + \int_\Gamma |v|\, ds, \quad f \in L_2(\Omega),$$

where $\Gamma \equiv \partial\Omega$ is the boundary of a bounded domain Ω with Lipschitz boundary. Let us define the following problem:

$$\text{find } u \in H^1(\Omega) \text{ such that}$$

$$\mathcal{L}_3(u) \leq \mathcal{L}_3(v) \quad \forall v \in H^1(\Omega). \tag{2.4}$$

Since $\mathcal{L}_3$ is strictly convex, weakly lower semicontinuous, and coercive in $H^1(\Omega)$, there exists a unique u satisfying (2.4). However, unlike $\mathcal{L}_1$, $\mathcal{L}_2$, the functional $\mathcal{L}_3$ *is not differentiable* in $H^1(\Omega)$. In order to be able to interpret (2.4), we have to make use of the following generalization concerning theorem 1.1:

Theorem 2.2. *Let V be a Banach space, let $K \subseteq V$ be a nonempty convex closed subset, and let $\mathcal{F} = \mathcal{F}_1 + \mathcal{F}_2 : V \mapsto \mathbf{R}$ where $\mathcal{F}_1, \mathcal{F}_2$ are convex, and, moreover $\mathcal{F}_1$ has its Gâteaux differential in V. Then, the assertions*

(i) $u \in K: \quad \mathcal{F}(u) \leq \mathcal{F}(v) \quad \forall v \in K;$

(ii) $D\mathcal{F}_1(u, v-u) + \mathcal{F}_2(v) - \mathcal{F}_2(u) \geq 0 \quad \forall v \in K$

are equivalent.

Proof. Can be found in Céa (1971). □

This theorem implies that $u \in H^1(\Omega)$ is a solution of (2.4) if and only if

$$(u, v-u)_1 + \int_\Gamma (|v| - |u|)ds \geq (f, v-u)_0 \quad \forall v \in H^1(\Omega).$$

Hence, using Green's formula and proceeding similarly to the case of problem $\mathcal{P}_i$, $i = 1, 2$, we derive that u is a solution of the problem

$$-\Delta u + u = f \quad \text{a.e. in } \Omega,$$

$$|\partial u/\partial \nu| \leq 1, \quad (\partial u/\partial \nu)u + |u| = 0, \quad \text{a.e. on } \Gamma.$$

We omit the proof, since in Chapter 2 we will prove an analogous assertion for one-sided contact problems with friction.

One of the main difficulties, particularly from the practical point of view, is the fact that $\mathcal{L}_3$ is not differentiable in $H^1(\Omega)$. A number of methods have been developed to avoid this difficulty. One of these methods is based on an idea similar to that exploited in the mixed formulation of problem $\mathcal{P}_1$.

We evidently have

$$\int_\Gamma |v|\, ds = \sup_\Lambda \int \mu\, v ds,$$

where

$$\Lambda = \{\mu \in L_2(\Gamma) \mid \ |\mu| \leq 1 \quad \text{a.e. on } \Gamma\}$$

is a convex, closed, and bounded subset of $L_2(\Gamma)$. Therefore, we may formally write

$$\inf_{H^1(\Omega)} \mathcal{L}_3(v) = \inf_{H^1(\Omega)} \sup_\Lambda \tilde{\mathcal{H}}(v, \mu),$$

where $\tilde{\mathcal{H}} : H^1(\Omega) \times \Lambda \longmapsto \mathbf{R}$ is given by the formula

$$\tilde{\mathcal{H}}(v, \mu) = \frac{1}{2}\|v\|_1^2 - (f, v)_0 + \int_\Gamma \mu\, v ds.$$

Instead of problem (2.4), let us now consider the problem of finding a saddle point (w, λ) of the functional $\tilde{\mathcal{H}}$ in $H^1(\Omega) \times \Lambda$:

$$(w, \lambda) \in H^1(\Omega) \times \Lambda,$$
$$\tilde{\mathcal{H}}(w, \mu) \leq \tilde{\mathcal{H}}(w, \lambda) \leq \tilde{\mathcal{H}}(v, \lambda) \quad \forall (v, \mu) \in H^1(\Omega) \times \Lambda. \tag{2.5}$$

It is now possible to prove a result analogous to theorem 2.1. Indeed, the following theorem holds:

Theorem 2.3. *A pair $(w, \lambda) \in H^1(\Omega) \times \Lambda$ is a saddle point of $\tilde{\mathcal{H}}$ in $H^1(\Omega) \times \Lambda$ if and only if*

$$w = u, \quad \lambda = -\partial u / \partial \nu,$$

where u is a solution of problem (2.4).

Problem (2.5) will be called the *mixed variational formulation* of problem (2.4). We see that the problem of minimization of a nondifferentiable functional is reduced to the problem of finding a saddle point of the functional $\tilde{\mathcal{H}}$, which already is differentiable in the variables v, μ. This is one of the main advantages of the mixed formulation (2.5). An analogous approach will be used in Chapter 2 when solving the Signorini problem with friction.

1.1.3 Solution of Primal Problems by the Finite Element Method and Error Bounds

Before we start to study the general theory of approximations of the primal formulations of variational inequalities of elliptic type, we shall present an example illustrating how to solve approximately problems of this type. In doing so, we shall restrict ourselves to problem $\mathcal{P}_1$.

1.1.31. Approximation of Problem $\mathcal{P}_1$ by the Finite Element Method. For the sake of simplicity, let us assume that $\Omega \subset \mathbf{R}^2$ is a bounded domain with a *polygonal* boundary Γ. In what follows, a triangulation of the domain Ω will mean any *finite* collection of closed triangles $\{T_i\}_{i \in J}$ such that

$$\bar{\Omega} = \bigcup_{i \in J} T_i,$$

$$T_i \cap T_j = \begin{matrix} \emptyset \\ A \\ \ell \end{matrix} \quad \forall i, j \in \mathcal{T},\ i \neq j,$$

where A or 1 denote respectively, a common vertex or a *whole* common side of T_i, T_j.

Each triangulation will be characterized by two parameters: *the longest side* h, and *the least inner angle* θ among all the triangles of the given triangulation, which will be denoted by $\{\mathcal{T}_{h,\theta}\}$. Of course, we will consider not a single triangulation but rather a system of triangulations $\{\mathcal{T}_{h,\theta}\}$ for $h \to 0_+$. For our purposes, we shall consider only *regular* systems of triangulations.

We say that a given *system* $\{\mathcal{T}_{h,\theta}\}$, $h \to 0_+$ is *regular* if there exists $\theta_0 > 0$ independent of h such that $\theta \geq \theta_0$. In other words, when refining the partition of $\bar{\Omega}$, the triangles of the given triangulation do not reduce to segments. Thus, any regular system of triangulations is actually characterized by the single parameter $h > 0$. In the case of a regular system of triangulations, therefore, we shall use the simpler symbol $\{\mathcal{T}_h\}$ instead of $\{\mathcal{T}_{h,\theta}\}$.

Remark 3.1. When solving problems with various types of boundary conditions prescribed on certain parts of the boundary Γ, we subject the system of triangulations to the additional assumption of being *compatible* with the partition of Γ. That is, each point of Γ at which the boundary condition changes must be a vertex of a certain triangle $T_i \in \mathcal{T}_h$.

Let $\{\mathcal{T}_h\}$, $h \to 0_+$ be a regular system of triangulations of $\bar{\Omega}$. The nodes of a triangulation (i.e., vertices of T_i's) lying on Γ will be denoted by

$a_1, \dots a_{m(h)}$. Each $\mathcal{T}_h$ will be associated with a finite-dimensional space V_h of piece-wise linear functions:

$$V_h = \{v \mid v \in C(\bar{\Omega}), \quad v|_T \in P_1(T) \quad \forall T \in \mathcal{T}_h\},$$

where $P_k(T)$ ($k \geq 0$ is an integer) denotes the set of all polynomials of degree at most k with the definition domain T.

Let us further define

$$K_h = \{v_h \in V_h \mid v_h(a_i) \geq 0 \quad \forall i = 1, 2, \dots m(h)\}.$$

It is easily seen that K_h is a convex closed subset of $V_h, K_h \subset K \quad \forall h > 0$.

To obtain an approximation of $\mathcal{P}_1$, we will make use of a modification of the Ritz method, with which the reader is familiar as one of the possible methods of numerical solution of variational equations. Problem (1.6) will be replaced by the problem of minimization of $\mathcal{L}_1$ in K_h. That is, we look for $u_h \in K_h$ such that

$$\mathcal{L}_1(u_h) \leq \mathcal{L}_1(v_h) \quad \forall v_h \in K_h. \tag{3.1}$$

A complex natural question arises, which is what is the relation between u and u_h, and more precisely, whether $\|u - u_h\|_1 \to 0, \ h \to 0_+$, or possibly, what is the rate of this convergence if expressed in powers of h. We shall study these issues in the subsequent text.

Remark 3.2. Problem (3.1) is already suitable for computer realization. It is easily seen that dim $V_h = M(h)$, where $M(h)$ is the number of all nodes $\{A_i\}_{i=1}^{M(h)}$ of the given triangulation $\mathcal{T}_h$. Let $\{\varphi_i\}_{i=1}^{M(h)}$ be such elements of V_h that $\varphi_i(A_j) = \delta_{ij}$ (Kronecker's symbol). Then, evidently, $\{\varphi_i\}_{i=1}^{M(h)}$ forms a basis of V_h. Moreover,

$$v_h(x) = \sum_{i=1}^{M(h)} v_h(A_i)\varphi_i(x) \quad \forall v_h \in V_h.$$

Let $\mathcal{T} : V_h \to \mathbf{R}^{M(h)}$ be the isomorphism given by

$$\mathcal{T} v_h = \alpha = \big(\alpha_1, \alpha_2, \dots, \alpha_{M(h)}\big) \in \mathbf{R}^{M(h)} \quad \forall v_h \in V_h,$$

where the components of the vector α are the coordinates of v_h with respect to $\{\varphi_i\}_{i=1}^{M(h)}$. By means of $\mathcal{T}$ we can identify the set K_h with a convex closed subset $\mathcal{K}_{M(h)} \subseteq \mathbf{R}^{M(h)}$:

$$\mathcal{K}_{M(h)} = \mathcal{T}(K_h) = \{\alpha \in \mathrm{R}^{M(h)} \mid \alpha_i \geq 0 \quad \forall i \in I\},$$

where $I \subset \{1, \ldots, M(h)\}$ is the set of indices, which in the given numeration, correspond to the vertices $a_i \in \Gamma$. Problem (3.1) is then equivalent to the problem of finding $\alpha^* = (\alpha_1^*, \ldots, \alpha_{M(h)}^*) \in \mathcal{K}_{M(h)}$ such that

$$\mathcal{Y}(\alpha^*) \leq \mathcal{Y}(\alpha) \quad \forall \alpha \in \mathcal{K}_{M(h)}. \tag{3.2}$$

Here,

$$\mathcal{Y}(\alpha) = \mathcal{L}_1(\mathcal{T}^{-1}\alpha) = \frac{1}{2}(\alpha, A\alpha)_{\mathbf{R}^{M(h)}} - (\mathcal{F}, \alpha)_{\mathbf{R}^{M(h)}},$$

$\mathcal{T}^{-1} : \mathbf{R}^{M(h)} \to V_h$ is the mapping inverse to $\mathcal{T}$, $A = ((\varphi_i, \varphi_j)_1)_{i,j=1}^{M(h)}$ is the stiffness matrix, $\mathcal{F} = ((f, \varphi_1)_0, \ldots, (f, \varphi_{M(h)})_0)$ is the vector resulting by integrating the right-hand side f, and $(\cdot, \cdot)_{\mathbf{R}^{M(h)}}$ denotes the scalar product in $\mathbf{R}^{M(h)}$. The desired solution $u_h \in K_h$ is then determined from the formula

$$u_h = \mathcal{T}^{-1}\alpha^* = \sum_{j=1}^{M(h)} \alpha_j^* \varphi_j.$$

Thus, (3.1) leads to the problem of quadratic programming seen in (3.2). At this point, let us briefly mention the algorithm which facilitates the approximate solution of minimization problems of the type (3.2). The definition of the set $\mathcal{K}_{M(h)}$ implies that we can write

$$\mathcal{K}_{M(h)} = \mathcal{K}^1 \times \mathcal{K}^2 \times \cdots \times \mathcal{K}^{M(h)},$$

where $\mathcal{K}^i = [0, +\infty)$ for $i \in I$, $\mathcal{K}^i = \mathrm{R}^1$ provided $i \notin I$. It follows from this formula for the convex set $\mathcal{K}_{M(h)}$ that each of the variables $\alpha_1, \ldots, \alpha_{M(h)}$ is subjected to at most one constraint (provided $i \in I$), and, moreover, that this constraint involves no other variable. In this sense, the variables are *separated.* In order to find the minimum of the quadratic function $\mathcal{Y}$ on the convex closed set $\mathcal{K}_{M(h)}$ of the above-mentioned type, it is advantageous to adopt the following generalization of the well-known SOR method (see Glowinski, Lions, and Trémolières (1976)): choose $\alpha^0 \in \mathcal{K}_M$[1] arbitrarily; if $\alpha^{(n)} \in \mathcal{K}_M$ is already known, we successively correct its individual components in accordance with the following scheme:

$$\alpha_i^{(n+1/2)} = -\frac{1}{a_{ii}} \left(\sum_{j<i} a_{ij}\alpha_j^{(n+1)} + \sum_{j>i} a_{ij}\alpha_j^{(n)} - \mathcal{F}_i \right),$$

$$\alpha_i^{(n+1)} = P_{\mathcal{K}^i}((1-\omega)\alpha_i^{(n)} + \omega\alpha_i^{(n+1/2)}), \quad i = 1, \ldots, M,$$

[1] Here we write M instead of $M(h)$.

and then set $\alpha^{(n+1)} = (\alpha_1^{(n+1)}, \ldots, \alpha_M^{(n+1)})$. Here, $P_{\mathcal{K}^i}$ stands for the projection of R^1 to $\mathcal{K}^i$ and $\omega > 0$ is the relaxation parameter, whose proper choice increases the rate convergence of $\alpha^{(n)}$ to α^* (the minimum point of $\mathcal{Y}$ in $\mathcal{K}_M$). In the case just considered, we have

$$P_{\mathcal{K}^i}(a) = \begin{array}{ll} a & \text{if } a \in [0, +\infty), \\ 0 & \text{if } a < 0 \end{array}$$

for $i \in I$ and

$$P_{\mathcal{K}^i}(a) = a \quad \forall a \in \mathrm{R}^1$$

provided $i \notin I$. In the look mentioned above, the authors proved that for a quadratic function given by a symmetric, positive definite matrix A, the above algorithm for $\omega \in (0, 2)$ converges for an arbitrary choice $\alpha^{(0)} \in \mathcal{K}_M$. The method just described will be called *the superrelaxation method with an additional projection.*

1.1.32. The General Theory of Approximations for Elliptic Inequalities. This section is devoted to the problems of approximation of variational inequalities from a general point of view.

Let V be a real Hilbert space with a norm $\|\cdot\|$, V' the space of continuous linear functionals over V and $\langle f, v\rangle$ the value of the functional $f \in V'$ at the point $v \in V$. Let $\mathcal{Y} : V \mapsto \mathrm{R}^1$ be the general quadratic functional

$$\mathcal{Y}(v) = \frac{1}{2}a(v, v) - \langle f, v\rangle, \quad f \in V', \tag{3.3}$$

where $a : V \times V \mapsto \mathrm{R}^1$ is a continuous, V-elliptic, and symmetric bilinear form, i.e.,

$$\exists M = \text{const} > 0 : |a(u, v)| \leq M\|u\| \cdot \|v\| \quad \forall u, v \in V \text{ (continuity);} \tag{3.4}$$

$$\exists \alpha = \text{const} > 0 : a(v, v) \geq \alpha\|v\|^2 \quad \forall v \in V \text{ (V-ellipticity);} \tag{3.5}$$

$$a(u, v) = a(v, u) \quad \forall u, v \in V \text{ (symmetry).} \tag{3.6}$$

Let $K \subseteq V$ be a nonempty, convex, closed subset. Let us introduce the problem

$$\text{find } u \in K \text{ such that}$$

$$\mathcal{Y}(u) \leq \mathcal{Y}(v) \quad \forall v \in K. \tag{$\mathcal{P}$}$$

If (3.4)–(3.6) hold, then $\mathcal{P}$ has exactly one solution, since $\mathcal{Y}$ satisfies all the assumptions of theorem 1.5. Moreover, theorem 1.1 implies that $\mathcal{P}$ is equivalent to the variational inequality

$$u \in K : \; a(u, v - u) \geq \langle f, v - u\rangle \quad \forall v \in K. \tag{$\mathcal{P}'$}$$

Remark 3.3. If the bilinear form a is merely continuous and V-elliptic, then problem $\mathcal{P}'$ has exactly one solution (e.g., Glowinski, Lions, and Trémolières (1976)).

If moreover, a is symmetric, then $\mathcal{P}'$ is equivalent to the problem of minimization of the quadratic functional $\mathcal{Y}$ on K.

Problem $\mathcal{P}$ or $\mathcal{P}'$ is hardly solvable in most cases; therefore, it must usually be replaced by another sequence of problems, which we are able to deal with algorithmically. To this end, we will use the same method we have sketched in the previous section for the approximation of problem $\mathcal{P}_1$. Let $\{K_h\}$, $h \in (0,1)$ be a collection of nonempty, convex, closed subsets of V. Also, in all cases to be considered, each K_h will be embedded into its own finite-dimensional space V_h, i.e., K_h is a convex and closed subset of V_h. Now we replace problem $\mathcal{P}$ by the sequence of problems

$$\text{find } u_h \in K_h :$$

$$\mathcal{Y}(u_h) \le \mathcal{Y}(v_h) \quad \forall v_h \in K_h.\text{[2]} \tag{$\mathcal{P}_h$}$$

or equivalently,

$$\text{find } u_h \in K_h :$$

$$a(u_h, v_h - u_h) \ge \langle f, v_h - u_h \rangle \quad \forall v_h \in K_h. \tag{$\mathcal{P}_h'$}$$

Remark 3.4. The method which replaces $\mathcal{P}$ by problems $\mathcal{P}_h$ is known in the literature as the Ritz method. If a fails to be symmetric, then $\mathcal{P}_h'$ is an approximation of $\mathcal{P}'$ and we speak of the Galerkin method. In the subsequent text, if the fact that the form is symmetric or not is immaterial, we will use the term Ritz–Galerkin method, and the solution u_h of problems $\mathcal{P}_h$, $\mathcal{P}_h'$ will be called the Ritz–Galerkin approximation of the solution u in K_h.

Definition 3.1. We say that the Ritz–Galerkin method (applied to $\mathcal{P}$ or $\mathcal{P}'$) converges, if

$$\epsilon(h) \equiv \|u - u_h\| \to 0, \quad h \to 0_+. \tag{3.7}$$

We will formulate sufficient conditions for $\{K_h\}$, $h \in (0,1)$, to guarantee (3.7).

Theorem 3.1. *Let a bilinear form a fulfill* (3.4) *and* (3.5), *and let the following assertions hold*:

$$\forall v \in K \ \exists v_h \in K_h : v_h \to v, \quad h \to 0_+ \ \text{ in } V; \tag{3.8}$$

$$v_h \in K_h, \ v_h \rightharpoonup v, \quad h \to 0_+ \ \text{(weakly) in } V \text{ implies } v \in K. \tag{3.9}$$

[2] Thus, $\mathcal{P}_h$ leads to the problem of generally nonlinear programming in a finite dimension.

The Ritz–Galerkin method then converges.

Proof. For a symmetric bilinear form, the proof is found in Céa (1971). For a general bounded and V-elliptic form, the proof is found in Glowinski, Lions, and Trémolières (1976). □

Remark 3.5. Let $\tilde{\mathcal{K}} \subset K$ be another convex closed subset. If we know in advance that the solution u of problem $\mathcal{P}$ lies in $\tilde{\mathcal{K}}$, then it suffices to consider the condition

$$\forall v \in \tilde{\mathcal{K}}\ \exists v_h \in K_h : v_h \to v, \quad h \to 0_+ \ \text{ in } V. \tag{3.8'}$$

Remark 3.6. Generally, K_h need not be subsets of K. If $K_h \subset K \ \ \forall h \in (0,1)$, we say that K_h are *internal* approximations of K. In the opposite case we speak of *external* approximations of K.

Remark 3.7. If $\{K_h\}$, $h \in (0,1)$ form an internal approximation of K, then (3.9) is automatically fulfilled by virtue of the weak closedness of K, i.e., $x_n \rightharpoonup x$, $n \to \infty$, $x_n \in K \Rightarrow x \in K$.

To get a deeper insight into the matter, it is also useful to know the rate of convergence of u_h to u. There are two approaches which provide this information, on the one hand, there exists the method of so-called one-sided approximations, which will be explained in detail in the next section, and on the other hand there exists the method suggested for the first time by Falk. We will describe it now.

Lemma 3.1. *The inequality*

$$\begin{aligned} \alpha\|u-u_h\|^2 \le a(u-u_h, u-u_h) &\le \langle f, u-v_h\rangle + \langle f, u_h - v\rangle \\ &+ a(u_h - u, v_h - u) + a(u, v-u_h) + a(u, v_h - u) \end{aligned} \tag{3.10}$$

holds for every $v \in K$, $v_h \in K_h$, $h \in (0,1)$.

Proof. $\mathcal{P}$ and $\mathcal{P}_h'$ imply that

$$a(u,u) \le \langle f, u - v\rangle + a(u,v) \quad \forall v \in K,$$

$$a(u_h, u_h) \le \langle f, u_h - v_h\rangle + a(u_h, v_h) \quad \forall v_h \in K_h.$$

Hence,

$$\begin{aligned} a(u-u_h, u-u_h) &= a(u,u) + a(u_h,u_h) - a(u_h,u) - a(u,u_h) \\ &\le \langle f, u-v\rangle + a(u,v) + \langle f, u_h - v_h\rangle + a(u_h, v_h) \\ &- a(u_h, u) - a(u, u_h) \\ &= \langle f, u - v_h\rangle + \langle f, u_h - v\rangle + a(u, v-u_h) \\ &+ a(u_h - u, v_h - u) + a(u, v_h - u), \end{aligned}$$

which is the same as (3.10). □

Remark 3.8. If $K_h \subset K \ \forall h \in (0,1)$, then we can write (3.10) in a simpler form, namely

$$\alpha\|u - u_h\|^2 \le a(u - u_h, u - u_h) \le \langle f, u - v_h\rangle$$

$$+a(u_h - u, v_h - u) + a(u, v_h - u) \quad \forall v_h \in K_h, \tag{3.10'}$$

which follows from (3.10) by setting $v = u_h \in K$.

Remark 3.9. Let $|\cdot|$ be a seminorm in V, and let the bilinear form a merely fulfill

$$\exists\alpha = \text{const} > 0: \ a(v,v) \ge \alpha|v|^2 \quad \forall v \in V \tag{3.5'}$$

instead of (3.5). In this case, a solution of problems $\mathcal{P}$, $\mathcal{P}_h$ generally need not exist, and even if it does exist, it need not be unique (the reason why this is so is concretely illustrated in section 1.1.6. Nevertheless, if a solution of $\mathcal{P}$, $\mathcal{P}_h$ exists, then it suffices to replace the symbol $\|\cdot\|$ in (3.10), (3.10′) by the symbol $|\cdot|$ in order to obtain a bound for the seminorm of the error $u_h - u$.

We can analogously generalize theorem 3.1 on the convergence of the approximate solutions u_h to u. Let us formulate the variant, which will be used in the subsequent text.

Let V, H be two Hilbert spaces. Let $V \subset H$ and let the embedding be *totally continuous*. Let

$$\|v\|^2 = |v|^2 + \|v\|_H^2 \quad \forall v \in V,$$

where $\|\cdot\|_H$ denotes the norm in H. If $\mathcal{Y}$ is coercive on $\cup_{h\in(0,1)}K_h$, that is,

$$\lim_{h\to 0_+} \mathcal{Y}(v_h) = +\infty, \ \|v_h\| \to +\infty, \quad v_h \in K_h,$$

and if (3.5′), (3.8), and (3.9) hold, then

$$\|u_h - u\| \to 0, \quad h \to 0_+,$$

provided u, u_h exist and u is uniquely determined.

1.1.33. A Priori Bound for Problem $\mathcal{P}_1$. Let us now go back to the approximation of $\mathcal{P}_1$ in the form described in section 1.1.31. We will show how to apply the results of the previous section to this particular case.

Theorem 3.2. *Let a solution* $u \in H^2(\Omega) \cap K, u \in W^{1,\infty}(\Gamma)^3$ *satisfy* $\partial u/\partial \nu \in L^\infty(\Gamma)$, *and let the set of points from* Γ *at which* u *changes from* $u = 0$ *to* $u > 0$ *be finite. Then*

$$\|u - u_h\|_1 \leq c(u)h, \quad h \to 0_+, \tag{3.11}$$

where $c(u)$ *is a positive constant that depends only on* u.[4]

Proof. As $K_h \subset K \ \ \forall h \in (0,1)$, we can use the relation (3.10′) to obtain the required bound. Let us set $v_h = r_h u$, where $r_h u$ denotes the piecewise linear Lagrange interpolation of the function u. As

$$r_h u(a_i) = u(a_i) \geq 0 \quad \forall i = 1, \ldots,$$

we have $r_h u \in K_h$. Using Green's formula and (1.1), we obtain

$$\begin{aligned} a(u, r_h u - u) &= (u, r_h u - u)_1 \\ &= (-\Delta u + u, r_h u - u)_0 + \int_\Gamma \frac{\partial u}{\partial \nu}(r_h u - u)ds \\ &= (f, r_h u - u)_0 + \int_\Gamma \frac{\partial u}{\partial \nu}(r_h u - u)ds. \end{aligned}$$

Substituting into the right-hand side of (3.10′) and using classical interpolation properties of the function $r_h u$, we conclude that

$$\begin{aligned} \|u - u_h\|_1^2 &= (u - u_h, u - u_h)_1 \\ &\leq (u_h - u, r_h u - u)_1 + \int_\Gamma \frac{\partial u}{\partial \nu}(r_h u - u)ds \\ &\leq \frac{1}{2}\|u_h - u\|_1^2 + \frac{1}{2}\|r_h u - u\|_1^2 + \int_\Gamma \frac{\partial u}{\partial \nu}(r_h u - u)ds \\ &\leq \frac{1}{2}\|u_h - u\|_1^2 + ch^2|u|_{2,\Omega}^2 + \int_\Gamma \frac{\partial u}{\partial \nu}(r_h u - u)ds. \end{aligned} \tag{3.12}$$

It remains to estimate the integral over Γ. We divide the boundary points into two groups:

$$\Gamma_0 = \{x \in \Gamma : u(x) = 0\},$$

[3]Let $\Gamma = \cup_{i=1}^m \overline{A_i A_{i+1}}$ be the boundary of a polygonal domain Ω. We define: $u \in W^{1,\infty}(\Gamma) \Leftrightarrow u|_{A_i A_i} \in W^{1,\infty}(\overline{A_i A_{i+1}}) \ \forall i = 1, \ldots, m$, i.e., the trace u on $\overline{A_i A_{i+1}}$ and its first derivative in the direction of the side $\overline{A_i A_{i+1}}$ is a bounded measurable function of one variable (the parameter of the side $\overline{A_i A_{i+1}}$).

[4]In the sequel, c denotes a general positive constant, which may assume different values at different points of our exposition. If we want to explicitly express its dependence on parameters $t_1, t_2, \ldots, t_s$, we write $c = c(t_1, t_2, \ldots, t_s)$.

$$\Gamma_+ = \{x \in \Gamma : u(x) > 0\}.$$

Let $a_i a_{i+1} \subset \Gamma_0$. Then, $r_h u(a_i) = u(a_i) = r_h u(a_{i+1}) = u(a_{i+1}) = 0$, and consequently $r_h u \equiv 0$ on $a_i a_{i+1}$. Hence,

$$\int_{a_i a_{i+1}} \frac{\partial u}{\partial \nu}(r_h u - u) ds = 0. \tag{3.13}$$

If $a_i a_{i+1} \subset \bar{\Gamma}_+$, then (1.2) implies that $\partial u / \partial \nu \equiv 0$ on $a_i a_{i+1}$, and (3.13) again holds. Let $\mathcal{T}$ be the set of all $a_i a_{i+1}$ whose interiors contain points from both Γ_+, Γ_0. Making use of the assumptions on the smoothness of u, $\partial u / \partial \nu$ on the boundary, and of the interpolation properties of $r_h u|_\Gamma$ we obtain

$$\begin{aligned} \int_{a_i a_{i+1}} \frac{\partial u}{\partial \nu}(r_h u - u) ds &\le \|\frac{\partial u}{\partial \nu}\|_{L^\infty(a_i a_{i+1})} \|r_h u - u\|_{L^\infty(a_i a_{i+1})} h \\ &\le c h^2 |u|_{2, a_i a_{i+1}} \|\frac{\partial u}{\partial \nu}\|_{L^\infty(a_i a_{i+1})}. \end{aligned}$$

According to the assumptions of the theorem, the number of elements of the set $\mathcal{T}$ is bounded from above independently of h. Hence,

$$\int_\Gamma \frac{\partial u}{\partial \nu}(r_h u - u) ds = \sum_{a_i a_{i+1} \in \mathcal{T}} \int_{a_i a_{i+1}} \frac{\partial u}{\partial \nu}(r_h u - u) ds = 0(h^2).$$

This together with (3.12) completes the proof of the theorem. □

The rate of convergence $0(h)$ is obtained under comparatively strong assumptions on the smoothness of u. However, the actual situation is usually such that assumptions analogous to those formulated in the preceding theorem are unrealistic. Therefore we shall try as far as possible, to prove the convergence of u_h to u without additional assumptions on the smoothness of u. Naturally, we shall pay for it by obtaining no information about the rate of convergence. Let us again use problem $\mathcal{P}_1$ to illustrate.

Theorem 3.3. *For an arbitrary regular system of triangulations* $\{\mathcal{T}_h\}, h \to 0_+$, *we have*

$$\|u - u_h\|_1 \to 0, \quad h \to 0_+.$$

Proof. We now will verify the assumptions of theorem 3.1. Since $K_h \subset K \ \forall h \in (0,1)$, it is sufficient to verify (3.8). To this end we shall use the following auxiliary result, which is proven in Haslinger (1977):

$$\overline{C^\infty(\bar{\Omega}) \cap K} = K, \tag{3.14}$$

where $C^\infty(\bar{\Omega})$ is the set of all infinitely differentiable functions in Ω, which together with their derivatives are continuously extensible to $\bar{\Omega}$. The closure in (3.14) is taken in the norm of $H^1(\Omega)$.

Let $v \in K$ be arbitrary. Then, (3.14) implies that there exists a sequence $v_n \in K \cap C^\infty(\bar{\Omega})$ such that

$$v_n \to v, \; n \to \infty \quad \text{in } H^1(\Omega). \tag{3.15}$$

To every function v_n we can construct its piecewise linear Lagrange interpolation $r_h v_n \in K_h$. Then,

$$\|v_n - r_h v_n\|_1 \le ch|v_n|_{2,\Omega}. \tag{3.16}$$

The triangule inequality

$$\|v - r_h v_n\|_1 \le \|v - v_n\|_1 + \|v_n - r_h v_n\|_1,$$

together with (3.15), (3.16) yields a sequence $v_h \in K_h$ satisfying (3.8) (it suffices to set $v_h = r_h v_n$, where n is sufficiently great while $h > 0$ is sufficiently small). □

1.1.4 Solution of Dual Problems by the Finite Element Method and Error Bounds

In dealing with dual problems we have to distinguish problems $\mathcal{P}_1$ and $\mathcal{P}_2$, since they essentially differ in the construction of approximations of admissible functions. Indeed, problem $\mathcal{P}_1$, for equations with absolute terms, can be dually formulated in terms of the equivalent variational problem (1.18) in the set $\mathcal{U}_0$, in which we do not require—in contradistinction to $\mathcal{U}_2$—any differential equation to be fulfilled.

1.1.41. Problems with Absolute Terms. We shall start with the dual equivalent formulation (1.18). Let the domain $\Omega \subset \mathbf{R}^2$ have a polygonal boundary. Let us consider a triangulation $\mathcal{T}_h$ and a space of piecewise linear finite elements

$$V_h = \{v \,|\, v \in C(\bar{\Omega}), \; v|_T \in P_1(T) \quad \forall T \in \mathcal{T}_h\}.$$

We introduce an approximation of the set $\mathcal{U}_0$ by

$$\mathcal{U}_{0h} = \mathcal{U}_0 \cap [V_h]^2.$$

As $[V_h]^2 \subset [H^1(\Omega)]^2 \subset H(\text{div}, \Omega)$, we evidently have

$$\mathcal{U}_{0h} = \{q \in [V_h]^2 \mid q \cdot \nu \ge 0 \text{ on } \Gamma\}.$$

A vector function $q^h \in \mathcal{U}_{0h}$ will be called an approximation of the dual problem (1.18), provided

$$I(q^h) \leq I(q) \quad \forall q \in \mathcal{U}_{0h}. \tag{4.1}$$

Since $\mathcal{U}_{0h}$ is a convex and closed subset in $H(\text{div}, \Omega)$, problem (4.1) has a unique solution (see theorem 1.5).

Algorithm for Solution of the Problem (4.1). Let $\{w^1, w^2, \ldots, w^N\}$ be a basis of the space $[V_h]^2$. Then, the equivalence

$$q \in \mathcal{U}_{0h} \Longleftrightarrow \left\{ q = \sum_{j=1}^{N} y_j w^j, \; By \geq 0 \right\}$$

evidently holds, where B is a $(p \times N)$-matrix with a rank $p, p < N$. The condition $By \geq 0$ results from the boundary condition $q \cdot \nu \geq 0$ on Γ. Notice that, by virtue of the fact that $q \cdot \nu$ is linear on each side of the polygonal boundary, it is sufficient to fulfill this condition solely at the nodes of the triangulation $\mathcal{T}_h$.

Thus, problem (4.1) can be written in the form

$$\mathcal{F}(y) = \frac{1}{2} y^T A y - y^T b = \min$$

$$\text{in the set } Y = \{y \in \mathbf{R}^N, \;\; By \geq 0\}. \tag{4.2}$$

where A is the Gramm matrix and b a fixed vector.

Problem (4.2) is a problem of *quadratic programming*. It can be solved, for example, in the following way:

1^0 At the nodes of the triangulation on Γ, we apply a local transformation of the form

$$\begin{aligned} z_j &= y_j \\ z_{j+1} &= y_j \nu_1 + y_{j+1} \nu_2 \end{aligned} \qquad \text{for } \; \nu_2 \neq 0,$$

$$\begin{aligned} z_j &= y_j \nu_1 + y_{j+1} \nu_2 \\ z_{j+1} &= y_{j+1} \end{aligned} \qquad \text{for } \; \nu_1 \neq 0,$$

setting

$$z_j = y_j \nu_1^{(1)} + y_{j+1} \nu_2^{(1)}$$

$$z_{j+1} = y_j \nu_1^{(2)} + y_{j+1} \nu_2^{(2)}$$

if the node coincides with a vertex of Γ with normals $\nu^{(j)}, j = 1, 2$.

At the nodes inside Ω we set $z_k = y_k$.

Altogether, we have a substitution

$$\mathbf{R}y = z,$$

with a regular matrix $\mathbf{R}$, where

$$By \geq 0 \Longleftrightarrow B\mathbf{R}^{-1}z \geq 0,$$

and $B\mathbf{R}^{-1}$ has only one nonzero element in each row.

Instead of problem (4.2) we now have the problem

$$G(z) = \mathcal{F}(\mathbf{R}^{-1}z) = \min$$

$$\text{in the set} \quad Z = \{z \in \mathbf{R}^N, \quad B\mathbf{R}^{-1} \geq 0\},$$

which we are able to effectively solve, for example, by the superrelaxation method with an additional projection (see section 1.1.3).

1.1.411. A Priori Error Bounds. In order to make an a priori estimate of the error $q^0 - q^h$, we will use the method of so called one-sided approximations (see Mosco and Strang (1974)).

Lemma 4.1. *Let $\mathcal{Y}(v)$ be a real-valued functional defined on a convex closed subset M of a Banach reflexive space X. Let us assume that $\mathcal{Y}$ has both the first and the second differential (in a Gateaux sense) and that there exist positive constants α_0, c such that*

$$\alpha_0\|z\|^2 \leq D^2\mathcal{Y}(u; z, z) \leq c\|z\|^2 \quad \forall u \in M, \ \forall z \in X. \tag{4.3}$$

Let $M_h \subset M$ be a convex closed subset. Denote by u and u_h the elements minimizing $\mathcal{Y}$ in M and M_h, respectively.

Let us assume that there exists $w_h \in M_h$ such that $2u - w_h \in M$. Then

$$\|u - u_h\| \leq \left(\frac{c}{\alpha_0}\right)^{1/2} \|u - w_h\|. \tag{4.4}$$

Proof. The Taylor theorem implies that there exists $\theta \in (0,1)$ such that

$$\mathcal{Y}(u_h) = \mathcal{Y}(u) + D\mathcal{Y}(u, u_h - u) + \frac{1}{2}D^2\mathcal{Y}(u + \theta(u_h - u); u_h - u, u_h - u)$$

$$\geq \mathcal{Y}(u) + \frac{1}{2}\alpha_0\|u_h - u\|^2, \tag{4.5}$$

as

$$D\mathcal{Y}(u, u_h - u) \geq 0.$$

For any $v \in M_h$ we may write

$$\begin{aligned} \mathcal{Y}(v) &= \mathcal{Y}(u) + D\mathcal{Y}(u, v-u) + \frac{1}{2}D^2\mathcal{Y}(u+\theta_1(v-u); v-u, v-u) \\ &\geq \mathcal{Y}(u_h). \end{aligned}$$

Substituting $v = w_h$ and $v = 2u - w_h$ into the condition $D\mathcal{Y}(u, v-u) \geq 0$, we obtain

$$D\mathcal{Y}(u, w_h - u) = 0,$$

hence

$$\begin{aligned} \mathcal{Y}(u_h) &\leq \mathcal{Y}(w_h) \\ &= \mathcal{Y}(u) + \frac{1}{2}D^2\mathcal{Y}(u+\theta_2(w_h-u); w_h-u, w_h-u) \\ &\leq \mathcal{Y}(u) + \frac{1}{2}c\|w_h - u\|^2. \end{aligned}$$

Combining (4.5) with this inequality we obtain (4.4). □

Let us apply lemma 1.2 with $\mathcal{Y} \equiv I$, $M = \mathcal{U}_0$, $M_h = \mathcal{U}_{0h}$, $X = H(\text{div}, \Omega)$, $\alpha_0 = c = 1$. If we find a vector $t^h \in \mathcal{U}_{0h}$ such that $2q - t^h \in \mathcal{U}_0$, then in virtue of (4.4) we shall have

$$\|q^0 - q^h\|_{H(\text{div},\Omega)} \leq \|q^0 - t^h\|_{H(\text{div},\Omega)}. \tag{4.6}$$

An answer to this question is given in the following theorem.

Theorem 4.1. *Let us assume that $q^0 \in [H^2(\Omega)]^2$ and $q^0 \cdot \nu \in H^2(\Gamma_m)$, $m = 1, 2, \ldots, \bar{m}$, where Γ_m denotes any one of the sides of the polygonal boundary Γ. Then there is $t^h \in \mathcal{U}_{0h}$ such that*

$$0 \leq t^h \cdot \nu \leq q^0 \cdot \nu \quad \text{a.e. on } \Gamma \tag{4.7}$$

and, if the system of triangulations $\{\mathcal{T}_h\}, 0 < h \leq h_0$, is (α, β)-regular[5] then

$$\|q^0 - t^h\|_{H(\text{div},\Omega)} \leq Ch\left(\sum_{j=1}^{2}\|q_j^0\|_2 + \sum_{m=1}^{\bar{m}}\|q^0 \cdot \nu\|_{2,\Gamma_m}\right). \tag{4.8}$$

We shall need two lemmas to prove the theorem.

[5] A system of triangulations $\{\mathcal{T}_h\}, 0 < h \leq h_0$ will be called (α, β)-regular, if positive constants α, β exist independent of h and such that (i) no inner angle among all the triangles is less than α (i.e., regularity), and (ii), the ratio of any two sides in $\mathcal{T}_h$ is less than β.

Lemma 4.2 (One-Sided Approximation of the Flow on the Boundary). *Let q^0 satisfy the assumptions of Theorem 4.1. Then there exist piecewise linear functions $\psi_h^m \in C(\bar{\Gamma}_m)$ with vertices determined by the triangulations $\mathcal{T}_h$, such that*

$$0 \le \psi_h^m \le q^0 \cdot \nu \quad \text{on } \bar{\Gamma}_m \quad \forall m, \tag{4.9}$$

$$\|q_I^0 \cdot \nu - \psi_h\|_{\Gamma,\infty}^2 \le h^3 \sum_{m=1}^{\bar{m}} |q^0 \cdot \nu|_{2,\Gamma_m}^2, \tag{4.10}$$

where q_I^0 denotes the piecewise linear interpolation of q^0 on $\mathcal{T}_h$,

$$\|\varphi\|_{\Gamma,\infty} = \max_{1 \le m \le \bar{m}} \sup_{s \in \Gamma_m} |\varphi(s)|.$$

$|\cdot|_{2,\Gamma_m}$ is the seminorm involving the second order derivatives on Γ_m.

Proof. Denote $q^0 \cdot \nu = t$ and let t_I be the linear interpolation of t on $\bar{\Gamma}_m$, with the nodes determined by the triangulation $\mathcal{T}_h$. Then, $t_I = q_I^0 \cdot \nu$. Let the coordinates corresponding to the nodes be $0 = s_1 < s_2 < \cdots < s_n$ on the segment Γ_m. Denote by φ_j the piecewise linear base functions ($\varphi_j(s_i) = \delta_{ji}$) on $\bar{\Gamma}_m$ and define the set

$$S_h = \left\{ a \in \mathbf{R}^n \mid 0 \le \sum_{j=1}^{n} a_j \varphi_j(s) \le t(s) \quad \forall s \in \bar{\Gamma}_m \right\}.$$

We will say that $a^0 \in S_h$ is a maximal element of S_h, if

$$\int_{\Gamma_m} \sum_{j=1}^{n} a_j^0 \varphi_j \, ds \ge \int_{\Gamma_m} \sum_{j=1}^{n} a_j \varphi_j \, ds \quad \forall a \in S_h.$$

A maximal element does exist. Indeed, S_h is bounded and closed in $\mathbf{R}^n$, since

$$0 \le a_j \le t(s_j) \le \|t\|_{C(\bar{\Gamma}_m)},$$

and the space $H^2(\Gamma_m)$ is continuously embedded into $C(\bar{\Gamma}_m)$ (see Nečas (1967)). The integral is a continuous function of the vector a; hence, it assumes its maximum in S_h.

Let us denote the maximal element of S_h by ψ_h^m. Then for every $j = 1, 2, \ldots, n$, at least one of the following conditions holds:

$$\psi_h^m(s_j) = t(s_j), \tag{C1}$$

$$\exists \sigma_j \in \langle s_{j-1}, s_j) \cup (s_j, s_{j+1}\rangle \quad \text{for } 1 < j < n, \tag{C2}$$

$$\sigma_1 \in (s_1, s_2\rangle, \quad \sigma_n \in \langle s_{n-1}, s_n),$$
$$\psi_h^m(\sigma_j) = t(\sigma_j), \quad \frac{d\psi_h^m}{ds}(\sigma_j) = \frac{dt}{ds}(\sigma_j).$$

(Notice that $dt/ds \in C(\bar{\Gamma}_m)$ provided $t \in H^2(\Gamma_m)$.)

Indeed, let neither (C1) nor (C2) hold for j_0. Then there evidently exists $\epsilon > 0$ such that

$$\psi^\epsilon \equiv \psi_h^m + \epsilon\varphi_{j_0} \le t \quad \forall s \in \bar{\Gamma}_m,$$

which leads to a contradiction with the definition of the maximal element.

If (C2) is fulfilled, then we can write

$$t(s_j) - \psi_h^m(s_j) = \int_{\sigma_j}^{s_j} \frac{d^2t}{ds^2}(z)(s_j - z)dz,$$

$$|t(s_j) - \psi_h^m(s_j)|^2 \le h^3 \| \frac{d^2t}{ds^2} \|_{0,\Gamma_m}^2 = h^3 |t|_{2,\Gamma_m}^2.$$

This yields the bound (4.10). □

Lemma 4.3. *Let $\varphi_m \in C(\bar{\Gamma}_m), 1 \le m \le m$, be piecewise linear functions with vertices determined by the triangulation $\mathcal{T}_h$. Then there exists a function $w^h \in [V_h]^2$ such that*

$$w^h \cdot \nu = \varphi_m \quad \text{on } \Gamma_m \quad \forall m, \tag{4.11}$$

$$\left(\sum_{j=1}^{2} \|w_j^h\|_1^2 \right)^{1/2} \le Ch^{-1/2} \|\varphi\|_{\Gamma,\infty},$$

where φ is a function whose restrictions to Γ_m coincide with φ_m for all m.

Proof. Let us consider a boundary strip Ω_h, which consists of all triangles $T \in \mathcal{T}_h$ such that $T \cap \Gamma \ne \emptyset$ (we regard T as a closed set). Put $w_j^h(b_i) = 0, j = 1, 2$, at all vertices $b_i \in \bar{\Omega} - \Gamma$. Then supp $w^h \subset \Omega_h$ and it is sufficient to determine and estimate the values $w^h(a_i)$ at the vertices $a_i \in \Gamma$.

1^0 Let a_i be a vertex of the polygonal boundary. Let us denote by φ^+, φ^- the limits of the function φ at the vertex a_i from the right and left, respectively, and by ν^+, ν^- the corresponding unit outer normals. The values w_j^h at the point a_i are given by

$$w^h \cdot \nu^- = \varphi^-, \quad w^h \cdot \nu^+ = \varphi^+$$

and

$$|w_j^h(a_i)| \le (|\varphi^+| + |\varphi^-|)\,|\sin\alpha_i|^{-1}, \quad j = 1, 2, \tag{4.12}$$

where α_i is the inner angle of Γ at the vertex a_i.

2^0 Let $a_i \in \Gamma_m$ be a vertex of $\mathcal{T}_h$ but not one of Γ. Let

$$|\nu_k| = \max\{|\nu_1|, |\nu_2|\}.$$

As $2\nu_k^2 \geq 1$, we have $|\nu_k| \geq 1/\sqrt{2}$. Let us choose

$$w_k^h(a_i) = \nu_k^{-1}\varphi(a_i),$$

while the remaining component $w_p^h(a_i)$ vanishes: $w_p^h(a_i) = 0$, $(p \neq k)$. Then, evidently

$$|w_j^h(a_i)| \leq \sqrt{2}|\varphi(a_i)|, \quad j = 1,2. \tag{4.13}$$

As a consequence of (4.12), (4.13) we obtain the estimate

$$\max_{j=1,2} \|w_j^h\|_{C(\bar{\Omega})} \leq C\|\varphi\|_{\Gamma,\infty}. \tag{4.14}$$

For an (α,β)-regular system of triangulations, the so-called inverse inequality holds (e.g., Ciarlet (1978)):

$$\|w_j^h\|_{1,\Omega} \leq Ch^{-1}\|w_j^h\|_{0,\Omega}. \tag{4.15}$$

On the other hand, we obtain from (4.14)

$$\|w_j^h\|_{0,\Omega} = \|w_j^h\|_{0,\Omega_h} \leq (\operatorname{mes}\Omega_h)^{1/2}\|w_j^h\|_{C(\Omega)} \leq Ch^{1/2}\|\varphi\|_{\Gamma,\infty} \tag{4.16}$$

as $\operatorname{mes}\Omega_h < Ch$. Combining (4.15) and (4.16), we arrive at the bound (4.11). □

Proof of Theorem 4.1. Let ψ_h be the function defined in lemma 4.2. Let us set

$$\varphi = q_I^0 \cdot \nu - \psi_h,$$

and construct the function $w^h \in [V_h]^2$ according to lemma 4.3. Then, the function

$$t^h = q_I^0 - w^h \in [V_h]^2$$

satisfies the conditions (4.7), (4.8). Indeed, on each side Γ_m we have

$$t^h \cdot \nu = q_I^0 \cdot \nu - \varphi_m = \psi_h^m,$$

and (4.9) yields (4.7). Besides,

$$\|(q^0 - q_I^0)_j\|_1 \leq Ch|q_j^0|_2 \quad (j = 1,2), \tag{4.17}$$

$$\|q\|_{H(\operatorname{div},\Omega)} \leq C\left(\sum_{j=1}^{2}\|q_j\|_1^2\right)^{1/2} \qquad \forall q \in H(\operatorname{div},\Omega). \tag{4.18}$$

From lemmas 4.2 and 4.3 we conclude

$$\|w^h\|_1 \le Ch^{-1/2}\|q_I^0 \cdot \nu - \psi_h\|_{\Gamma,\infty} \le Ch \left(\sum_{m=1}^{m} |q^0 \cdot \nu|_{2,\Gamma_m}^2\right)^{1/2}. \tag{4.19}$$

Finally, the inequality (in the $H(\operatorname{div}, \Omega)$-norm)

$$\|q^0 - t^h\| \le \|q^0 - q_I^0\| + \|w^h\|,$$

together with (4.17), (4.18) and (4.19) yields the bound (4.8). □

Corollary to Theorem 4.1. *Let the assumptions of Theorem 4.1 be fulfilled. Then, the following estimate holds for the approximations q^h of the dual problem* (1.18):

$$\|q^0 - q^h\|_{H(\operatorname{div},\Omega)} = 0(h).$$

Proof. Follows from theorem 4.1 and lemma 4.1. Indeed, first of all we have

$$(2q^0 - t^h) \cdot \nu \ge q^0 \cdot \nu - t^h \cdot \nu \ge 0 \quad \text{on } \Gamma,$$

by virtue of (4.7). Hence, $2q^0 - t^h \in \mathcal{U}_0, t^h \in \mathcal{U}_{0h}$, and we can apply (4.6) and (4.8). □

Remark 4.1. If q^h is an approximation of the dual problem (1.18), then

$$\lambda^h = \{q_1^h, q_2^h, f + \operatorname{div} q^h\} \in \mathcal{U}_1$$

is an approximation of the original dual problem (1.15). According to theorem 1.6 and the corollary to theorem 4.1, we have

$$\|q_i^h - \frac{\partial u}{\partial x_i}\|_0 = 0(h), \quad i = 1, 2,$$

$$\|\operatorname{div} q^h + f - u\|_0 = \|\operatorname{div}(q^h - q^0)\|_0 = 0(h)$$

for $h \to 0$.

1.1.412. A Posteriori Error Bounds and the Two-Sided Energy Bound. Let us assume that we have evaluated both the approximation $\tilde{u}_h \in K_{1h}$ of the primal problem, and the approximation $\tilde{q}^H \in \mathcal{U}_{0H}$ of the dual problem. ($\tilde{u}_h$ and $\tilde{q}^H$ may generally differ from the solutions u_h and q^H of the approximate problems (3.1) and (4.1), respectively.) Then it is possible to evaluate an error bound for both the primal and the dual approximation.

Theorem 4.2. *Let $\tilde{u}_h \in K_{1h}, \tilde{q}^H \in \mathcal{U}_{0H}$. Then,*

$$\|\tilde{u}_h - u\|_1^2 \leq \sum_{i=1}^{2} \|\tilde{q}_i^H - \frac{\partial \tilde{u}_h}{\partial x_i}\|_0^2 + \|f + \operatorname{div} \tilde{q}^H - \tilde{u}_h\|_0^2$$

$$+2\int_\Gamma \tilde{q}^H \cdot \nu\, \tilde{u}_h ds \equiv E(\tilde{q}^H, \tilde{u}_h), \tag{4.20}$$

$$\sum_{i=1}^{2} \|\tilde{q}_i^H - \frac{\partial u}{\partial x_i}\|_0^2 + \|\operatorname{div} \tilde{q}^H + f - u\|_0^2 \leq E(\tilde{q}^H, \tilde{u}_h). \tag{4.21}$$

Proof. From the variational inequality (1.8), we find

$$\begin{aligned} 2\mathcal{L}_1(v) - 2\mathcal{L}_1(u) &= \|v\|_1^2 - \|u\|_1^2 - 2(f, v-u)_0 \\ &\geq \|v\|_1^2 - \|u\|_1^2 - 2(u, v-u)_1 = \|v-u\|_1^2 \quad \forall v \in K_1. \end{aligned}$$

However, (1.20) yields

$$-\mathcal{L}_1(u) = \mathcal{S}_1(q^0) \leq \mathcal{S}_1(q) \quad \forall q \in \mathcal{U}_1.$$

Hence, for $\tilde{u}_h \in K_{1h} \subset K_1$ and $q = [\tilde{q}_1^H, \tilde{q}_2^H, f + \operatorname{div} \tilde{q}^H]$, we obtain

$$\|\tilde{u}_h - u\|_1^2 \leq 2\mathcal{L}_1(\tilde{u}_h) + 2\mathcal{S}_1(q) = \|\tilde{u}\|_1^2 - 2(f, \tilde{u}_h)_0 + \sum_{i=1}^{3} \|q_i\|_0^2$$

$$= \sum_{i=1}^{3} (\|q_i(\tilde{u}_h) - q_i\|_0^2 + 2(q_i, q_i(\tilde{u}_h))_0) - 2(f, \tilde{u}_h)_0, \tag{4.22}$$

where $q(\tilde{u}_h) = [\nabla \tilde{u}_h, \tilde{u}_h]$.

However, we have $q_3 = f + \operatorname{div} \tilde{q}^H$, hence,

$$\sum_{i=1}^{3} (q_i, q_i(\tilde{u}_h))_0 - (f, \tilde{u}_h)_0 = \int_\Gamma (q^h \cdot \nabla \tilde{u}_h + \tilde{u}_h \operatorname{div} \tilde{q}^H) dx$$

$$= \int_\Gamma \tilde{q}^H \cdot \nu \tilde{u}_h ds. \tag{4.23}$$

Substituting (4.23) into (4.22), we obtain (4.20).

In order to establish the bound (4.21), we will use the variational inequality which corresponds to the problem (1.15), that is (see theorem 1.1),

$$\sum_{i=1}^{3} (q_i^0, q_i - q_i^0)_0 \geq 0 \quad \forall q \in \mathcal{U}_1.$$

If we write for brevity $(\cdot,\cdot)$ for the scalar products in $[L^2(\Omega)]^3$, then we have for all $q \in \mathcal{U}_1$:

$$\begin{aligned} 2\mathcal{S}_1(q) - 2\mathcal{S}_1(q^0) &= \|q\|^2 - \|q^0\|^2 \geq \|q\|^2 - (q^0, q) \\ &= (q, q - q^0) - (q^0, q - q^0) + (q^0, q - q^0) \geq \|q - q^0\|^2. \end{aligned}$$

On the other hand, (1.20) implies

$$-\mathcal{S}_1(q^0) = \mathcal{L}_1(u) \leq \mathcal{L}_1(\tilde{u}_h),$$

hence,

$$\|q - q^0\|^2 \leq 2\mathcal{S}_1(q) + 2\mathcal{L}_1(\tilde{u}_h) \equiv E(\tilde{q}^H, \tilde{u}_h)$$

by virtue of (4.22) and (4.23). Now it is sufficient to substitute for the vector q^0 from theorem 1.6 and $q = [\tilde{q}_1^H, \tilde{q}_2^H, f + \operatorname{div} \tilde{q}^H]$. □

Remark 4.2. All summands in the expression for $E(\tilde{q}^H, \tilde{u}_h)$ are nonnegative, as follows from the definition of the set $\mathcal{U}_{0h}$. Further,

$$E(\tilde{q}^H, \tilde{u}_h) \to 0,$$

provided $\|\tilde{u}_h - u\|_1 \to 0$ for $h \to 0$, and $\|\tilde{q}^H - q^0\|_{H(\operatorname{div},\Omega)} \to 0$ for $h \to 0$. Indeed, in that case

$$\frac{1}{2}E(\tilde{q}^H, \tilde{u}_h) = \mathcal{L}_1(\tilde{u}_h) + \mathcal{S}_1([\tilde{q}_1^H, \tilde{q}_2^H, f + \operatorname{div} \tilde{q}^H]) \to \mathcal{L}_1(u) + \mathcal{S}_1(q^0) = 0.$$

Let us now find an interval to which the energy of the exact solution (i.e., $\|u\|_1^2$) or the work of the external forces $(u, f)_0$ belongs.

Theorem 4.3. *Let $\tilde{u}_h \in K_{1h}, \tilde{q}^H \in \mathcal{U}_{0h}$. Then,*

$$-2\mathcal{L}_1(\tilde{u}_h) \leq \|u\|_1^2 = (f, u)_0 \leq \sum_{i=1}^{2} \|\tilde{q}_i^H\|_0^2 + \|f + \operatorname{div} \tilde{q}^H\|_0^2.$$

Proof. The left inequality follows from (1.11), since

$$2\mathcal{L}_1(u) = -\|u\|_1^2 \leq 2\mathcal{L}_1(\tilde{u}_h).$$

The right inequality is obtained from (1.20), since

$$\|u\|_1^2 = -2\mathcal{L}_1(u) = 2\mathcal{S}_1(q^0) \leq 2\mathcal{S}_1([\tilde{q}_1^H, \tilde{q}_2^H, f + \operatorname{div} \tilde{q}^H]). \quad □$$

1.1.42. Problems Without Absolute Terms. Let us now consider problem $\mathcal{P}_2$ in a domain $\Omega \subset \mathbf{R}^2$ and its dual variational formulation (1.16).

We shall assume that the boundary Γ consists of a finite number of closed polygons $\partial\Omega_j$, i.e.,

$$\Gamma = \bigcup_{j=1}^{\bar{j}} \partial\Omega_j, \quad \partial\Omega_j \cap \partial\Omega_k = \emptyset \quad \text{for } j \neq k,$$

and

$$\text{mes}(\partial\Omega_j \cap \Gamma_u) > 0, \quad j = 1, \ldots, \bar{j}. \tag{4.24}$$

The definition of the set $\mathcal{U}_2$ involves the equation div $q + f = 0$. Similarly, as in the case of classical boundary value problems (see Haslinger and Hlaváček (1976)), it is useful first to find a particular solution λ of this equation (e.g.,

$$\lambda_1 = -\int_0^{x_1} f(t, x_2)dt, \lambda_2 = 0).$$

Then it is evident that

$$q \in \mathcal{U}_2 \Longleftrightarrow q - \lambda \equiv p \in \mathcal{U}_{20},$$

$$\mathcal{U}_{20} = \{p \in H\ (\text{div}, \Omega) \mid \text{div } p = 0,\ (p + \lambda) \cdot \nu \geq 0 \ \text{ on } \Gamma_a\}.$$

We will now work with vector solenoidal functions (that is, vectors with zero divergence). To this end, we shall use linear finite elements on triangles, which were introduced by Veubeke and Hogge (1972).

Let us recall the construction of these spaces. On each triangle T from the triangulation $\mathcal{T}_h$, we define a vector function

$$\mathcal{M}(T) = \{q \mid q \in [P_1(T)]^2,\ \text{div } q = 0\}.$$

Further, we introduce the space of solenoidal finite elements

$$\mathcal{N}_h = \{q | q|_T \in \mathcal{M}(T) \quad \forall T \in \mathcal{T}_h,$$

$$(q \cdot \nu)_T + (q \cdot \nu)_{T'} = 0 \quad \forall x \in T \cap T'\}.$$

The last condition means that the "flow" $q \cdot \nu$ is continuous when passing through the common side of any two adjacent triangles.

We easily verify that $\mathcal{N}_h$ is a linear finite-dimensional set and $\mathcal{N}_h \subset H\ (\text{div}, \Omega)$, since for each $q_h \in \mathcal{N}_h$ we have div $q_h = 0$ in the sense of distributions.

Let us define a linear continuous mapping

$$\Pi_T \in \mathcal{L}([H^1(T)]^2, [P_1(T)]^2)$$

by the conditions

$$\int_{S_i} [(q \cdot \nu)_{S_i} - (\Pi_T q \cdot \nu)_{S_i}] v \, ds = 0 \quad \forall v \in P_1(S_i)$$

on each side S_i of the triangle T.

Further, let

$$\mathcal{R}(\Omega) = \{q \in [H^1(\Omega)]^2 \mid \operatorname{div} q = 0\}.$$

If we define a mapping r_h on $\mathcal{R}(\Omega)$ so that

$$r_h q|_T = \Pi_T q \quad \forall T \in \mathcal{T}_h,$$

then it can be proved (see Haslinger and Hlaváček (1976)) that

$$r_h \in \mathcal{L}(\mathcal{R}(\Omega), \mathcal{N}_h), \tag{4.25}$$

$$\|q - r_h q\|_{0,\Omega} \leq C h^2 |q|_{2,\Omega} \quad \forall q \in [H^2(\Omega)]^2. \tag{4.26}$$

In addition, let us assume that there is such a function G that

$$G \in \mathcal{R}(\Omega), \quad G \cdot \nu = -\lambda \cdot \nu \quad \text{on } \Gamma_a. \tag{4.27}$$

Denote $-\lambda \cdot \nu = g$ and construct a function $g_h \in L^2(\Gamma_a)$ such that on each side $S \subset \Gamma_a, S \in \mathcal{T}_h$, the restriction $g_h|_S$ coincides with the $L^2(S)$-projection of the function g onto the subspace $P_1(S)$. Thus, g_h is piecewise linear and can have jumps at the nodes of the triangulation.

Define an approximation of the set $\mathcal{U}_{20}$,

$$\mathcal{U}_{20}^h = \{p \in \mathcal{N}_h \mid p \cdot \nu \geq g_h \text{ on } \Gamma_a\}.$$

(Since $g_h \geq g$ does not generally hold on Γ_a, the set $\mathcal{U}_{20}^h$ need not be a subset of $\mathcal{U}_{20}$.)

By substituting the particular solution we can transform the dual problem (1.16) to an *equivalent problem:*

$$\text{find } p^0 \in \mathcal{U}_{20} \text{ such that}$$

$$J(p^0) \leq J(p) \quad \forall p \in \mathcal{U}_{20}, \tag{4.28}$$

where

$$J(p) = \frac{1}{2}\|p\|_0^2 + (\lambda, p)_0.$$

A vector function $p^h \in \mathcal{U}_{20}^h$ will be called an approximation of the dual problem (4.28), if

$$J(p^h) \leq J(p) \quad \forall p \in \mathcal{U}_{20}^h. \tag{4.29}$$

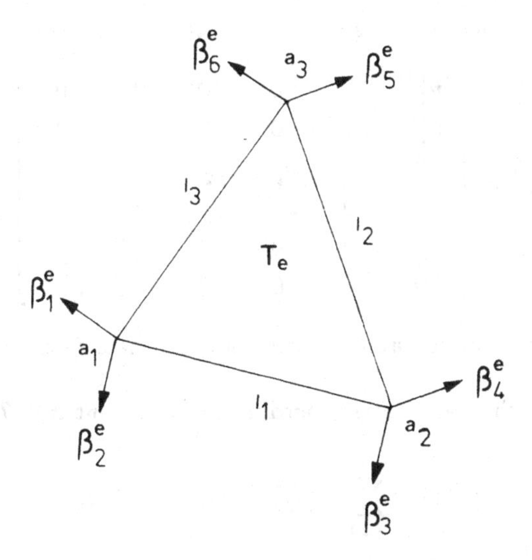

Figure 1

As $\mathcal{U}_{20}^h$ is convex and closed in $[L^2(\Omega)]^2$, problem (4.29) has a unique solution.

Algorithm for the Solution of Problem (4.29). For the nodal parameters in $\mathcal{M}(T_e)$ let us take the limit values of the "flow" $p \cdot \nu$ at the vertices, and let us denote them (see Figure 1) by

$$\beta^e = [\beta_1^e, \dots, \beta_6^e]^T, \quad T_e \in \mathcal{T}_h.$$

The following identity holds in each triangle $T_e \in \mathcal{T}_h$:

$$l_1(\beta_2^e + \beta_3^e) + l_2(\beta_4^e + \beta_5^e) + l_3(\beta_1^e + \beta_6^e) = 0. \tag{4.30}$$

In the triangle T_e let us denote $p|_{T_e} = p^e$ and

$$w^e = [p_1^e(a_1), p_2^e(a_1), p_1^e(a_2), p_2^e(a_2), p_1^e(a_3), p_2^e(a_3)]^T.$$

Then,

$$w^e = C_e \beta^e,$$

where the (6×6)-matrix C_e is regular, because its inverse matrix is

$$C_e^{-1} = \begin{bmatrix} \nu_1^{(3)} & \nu_2^{(3)} & 0 & 0 & 0 & 0 \\ \nu_1^{(1)} & \nu_2^{(1)} & 0 & 0 & 0 & 0 \\ 0 & 0 & \nu_1^{(1)} & \nu_2^{(1)} & 0 & 0 \\ 0 & 0 & \nu_1^{(2)} & \nu_2^{(2)} & 0 & 0 \\ 0 & 0 & 0 & 0 & \nu_1^{(2)} & \nu_2^{(2)} \\ 0 & 0 & 0 & 0 & \nu_1^{(3)} & \nu_2^{(3)} \end{bmatrix},$$

where $\nu^{(k)}$ stands for the unit outer normal to the side $S_k = a_k a_{k+1} (k = 1, 2, 3, a_4 \equiv a_1)$.

If $\varphi_j(x)$ are the barycentric coordinates of a point $x \in T_e, j = 1, 2, 3$, then

$$p_i^e(x) = \sum_{j=1}^{3} p_i^e(a_j)\varphi_j(x), \quad i = 1, 2.$$

Introducing the vector functions

$$\Phi_1 = [\varphi_1, 0, \varphi_2, 0, \varphi_3, 0]^T,$$

$$\Phi_2 = [0, \varphi_1, 0, \varphi_2, 0, \varphi_3]^T,$$

we easily derive that $p_i = (w^e)^T \Phi_i$ for $x \in T_e$, hence,

$$\begin{aligned} \sum_{i=1}^{2} \|p_i\|_0^2 &= \sum_{i=1}^{2} \sum_{T_e \in T_h} \int_{T_e} (w^e)^T \Phi_i \Phi_i^T w^e dx \\ &= \sum_{T_e} (w^e)^T \int_{T_e} (\Phi_1 \Phi_1^T + \Phi_2 \Phi_2^T) dx \; w^e \\ &= \sum_{T_e} (\beta^e)^T A_e \beta^e = \beta^T A \beta, \end{aligned}$$

where A is a symmetric positive definite $(N \times N)$-matrix. Indeed, all submatrices

$$A_e = C_e^T \int_{T_e} (\Phi_1 \Phi_1^T + \Phi_2 \Phi_2^T) dx \; C_e$$

are symmetric and positive definite, since

$$(\beta^e)^T A_e \beta^e = \|p^e\|_{0,T_e}^2 = 0 \Rightarrow w^e = 0 \Rightarrow \beta^e = 0.$$

However, the components of the vector $\beta \in \mathbf{R}^N$ are, as well as (4.30), subjected to the conditions of the form

$$\beta_i + \beta_k = 0, \tag{4.31}$$

which express the continuity of flows $p \cdot \nu$ on the interelement boundaries.

Further, it can be shown that

$$(\lambda, p)_0 = -b^T \beta,$$

where $b \in \mathbf{R}^N$ is a fixed vector. Finally, we express the conditions $p \cdot \nu \geq g_h$ on Γ_a in the form

$$\beta_j \geq g_h(a_j) \tag{4.32}$$

at all vertices of all sides $S \subset \bar{\Gamma}_a$.

Thus, we have

$$p \in \mathcal{U}_{20}^h \Longleftrightarrow \beta \in \mathcal{B} = \{\beta \in \mathbf{R}^N \mid \beta$$

which fulfills all conditions of the forms (4.30), (4.31), and (4.32)}, and the problem (4.29) is transformed to an equivalent problem:

$$\frac{1}{2}\beta^T A \beta - b^T \beta = \min \quad \text{in } \mathcal{B}. \tag{4.33}$$

In this way, we have obtained a quadratic programming problem, which can be solved, for example, by Uzawa's method (see Céa (1971), Chapter 4, Section 5.1), the method of feasible directions (Zoutendijk (1960), (1966)), etc.

Another algorithm arises from already choosing the base functions in $\mathcal{N}_h$ in such a way that they a priori fulfill the conditions of the forms (4.30) and (4.31).

1.1.421. A Priori Error Bound. In order to obtain a bound for the error $p^0 - p^h$, we again apply the method of one-sided approximations.

Let us first define the sets

$$C = \{q \in H(\text{div}, \Omega) \mid \text{div } q = 0, \; q \cdot \nu \geq 0 \quad \text{on } \Gamma_a\},$$

$$C_h = C \cap \mathcal{N}_h = \{q \in \mathcal{N}_h \mid q \cdot \nu \geq 0 \quad \text{on } \Gamma_a\}.$$

Under the assumption (4.27), we have

$$p^0 - G \equiv U \in C. \tag{4.34}$$

According to (4.25), we have $r_h G \in \mathcal{N}_h$ and, moreover,

$$(r_h G) \cdot \nu = g_h \quad \text{on } \Gamma_a,$$

because on each side $S \subset \Gamma_a, S \in \mathcal{T}_h, (r_h G) \cdot \nu$ is, by the definition of the mapping r_h equal to the $L^2(S)$-projection onto $P_1(S)$ of the function $G \cdot \nu = g$. Thus, we obtain

$$p^h - r_h G \equiv U_h \in C_h, \tag{4.35}$$

and this implies that

$$p \in \mathcal{U}_{20}^h \Longleftrightarrow p - r_h G \equiv V_h \in C_h.$$

Lemma 4.4. *Let there exist $W_h \in C_h$ such that $2U - W_h \in C$. Then,*

$$\|p^0 - p^h\|_0 \leq \|U - W_h\|_0 + \|G - r_h G\|_0. \tag{4.36}$$

Proof. Set $p = G + W_H$. Then, $p \in \mathcal{U}_{20}$ and

$$2p^0 - p = 2(G + U) - (G + W_h) = G + (2U - W_h) \in \mathcal{U}_{20}.$$

The solution p^0 satisfies the inequalities

$$DJ(p^0, p - p^0) \geq 0,$$

$$DJ(p^0, 2p^0 - p - p^0) = DJ(p^0, p^0 - p) \geq 0,$$

where

$$DJ(p, q) = (p, q)_0 + (\lambda, q)_0.$$

Thus, we have

$$0 = DJ(p^0, p - p^0) = (p^0, W_h - U)_0 + (\lambda, W_h - U)_0. \tag{4.37}$$

Second, set $p = G + U_h \in \mathcal{U}_{20}$. Since $p - p^0 = U_h - U$, we obtain

$$0 \leq DJ(p^0, p - p^0) = (p^0, U_h - U)_0 + (\lambda, U_h - U)_0. \tag{4.38}$$

Finally, setting $p = r_h G + W_h \in \mathcal{U}_{20}^h$, we can write

$$\begin{aligned} 0 \leq DJ(p^h, p - p^h) &= DJ(p^h, W_h - U_h) \\ &= (p^h, W_h - U_h)_0 + (\lambda, W_h - U_h)_0. \end{aligned} \tag{4.39}$$

By virtue of (4.37), (4.38), and (4.39), we derive

$$\begin{aligned} (p^0 - p^h, U_h - W_h)_0 &= (p^0, U - W_h - U + U_h)_0 + (p^h, W_h - U_h)_0 \\ &\geq (\lambda, W_h - U)_0 + (\lambda, U - U_h)_0 + (\lambda, U_h - W_h)_0 = 0. \end{aligned} \tag{4.40}$$

Since

$$p^0 - p^h = G + U - (r_h G + U_h) = G - r_h G + U - U_h,$$

using (4.40) we can write

$$\|p^0 - p^h\|_0^2 \leq (p^0 - p^h, G - r_h G)_0 + (p^0 - p^h, U - U_h + U_h - W_h)_0$$

$= (p^0 - p^h, G - r_h G + U - W_h)_0 \leq \|p^0 - p^h\|_0 \{\|G - r_h G\|_0 + \|U - W_h\|_0\}$. □

In the next step, we will verify the existence of a suitable element $W_h \in C_h$.

Theorem 4.4. *Let $U \equiv p^0 - G \in [H^2(\Omega)]^2$ and $U \cdot \nu \in H^2(\Gamma_a \cap \Gamma_m)$ on each side Γ_m of the polygonal boundary. Then there exists $W_h \in C_h$ such that $2U - W_h \in C$ and for any (α, β)-regular system $\{\mathcal{T}_h\}$, the following estimate holds:*

$$\|U - W_h\|_{0,\Omega} \leq C\{h^2 |U|_{2,\Omega} + h^{3/2} \sum_{m=1}^{\bar{m}} |U \cdot \nu|_{2,\Gamma_m \cap \Gamma_a}\}.$$

The proof is based on the two following lemmas.

Lemma 4.5 (One-Sided Approximation of the Flow on the Boundary). *Let the assumptions of Theorem 4.4 be fulfilled. Then there exists a piecewise linear function ψ_h on Γ, with nodes determined by the vertices of the triangulation $\mathcal{T}_h$ (the function being generally discontinuous at its nodes), such that*

$$\int_{\partial\Omega_j} \psi_h \, ds = \int_{\partial\Omega_j} (r_h U) \cdot \nu \, ds, \quad j = 1, \ldots, \bar{j}, \tag{4.41}$$

$$0 \leq \psi_h \leq U \cdot \nu \quad \text{on } \Gamma_a,$$

$$\|(r_h U) \cdot \nu - \psi_h\|_{0,\Gamma} \leq Ch^2 \sum_{m=1}^{\bar{m}} |U \cdot \nu|_{2,\Gamma_m \cap \Gamma_a}. \tag{4.42}$$

Proof. Denote $U \cdot \nu = t$ and $(r_h U) \cdot \nu = t_h$. Let s_i be the coordinates of the nodes of $\mathcal{T}_h$ on Γ. Let us consider an interval (a side) $S_i = (s_i, s_{i+1}) \subset \Gamma_a$. Let t_I be a linear function such that $t_I = t$ at the endpoints s_i and s_{i+1}. First we construct the function ψ_h on S_i.

1^0 If $t \geq t_I$ for all $s \in (s_i, s_{i+1})$, we put $\psi_h^i = t_I$. Then, obviously $\psi_h^i \geq 0$ on S_i, and

$$\|\psi_h^i - t_h\|_{0,S_i} \leq \|t_I - t\|_{0,S_i} + \|t - t_h\|_{0,S_i} \leq Ch^2 \|\frac{d^2 t}{ds^2}\|_{0,S_i}. \tag{4.43}$$

Indeed, t_h is the $L_2(S_i)$-projection of the function t onto $P_1(S_i)$, and we can apply the Bramble–Hilbert lemma to $t - t_h$ (see Ciarlet (1978)). The same lemma can also be applied to obtain an estimate of $t - t_I$.

2^0 Let there exist points $s_i \in S_i$ with $t < t_I$. Since $H^2(\Gamma_m \cap \Gamma_a) \subset C^1(\Gamma_m \cap \Gamma_a)$, we can find a point $\sigma \in S_i$ such that the tangent of the graph of the function t at the point σ lies completely under the graph of t

and, when denoting by ψ_h^i the function whose graph is the just mentioned tangent, then $\psi_h^i \geq 0$ on $\bar{S}_i$. Hence, we have

$$\|\psi_h^i - t_h\|_{0,S_i} \leq \|\psi_h^i - t\|_{0,S_i} + \|t - t_h\|_{0,S_i}. \tag{4.44}$$

In the same way as in the proof of lemma 4.2, we derive

$$\|t - \psi_h^i\|_{C(\bar{S}_i)} \leq h^{3/2} \|\frac{d^2t}{ds^2}\|_{0,S_i},$$

$$\|t - \psi_h^i\|_{0,S_i} \leq h^2 \|\frac{d^2t}{ds^2}\|_{0,S_i}.$$

This, together with (4.44) and with the estimate of $t - t_h$ as in 1^0, implies

$$\|t_h - \psi_h^i\|_{0,S_i} \leq Ch^2 \|\frac{d^2t}{ds^2}\|_{0,S_i}. \tag{4.45}$$

3^0 In this way, we can construct the function ψ_h on the whole part Γ_a by setting

$$\psi_{h|S_i} = \psi_h^i \quad \forall S_i \subset \Gamma_a.$$

The estimates (4.43) and (4.45) yield

$$\|\psi_h - t_h\|_{0,\Gamma_a}^2 \leq Ch^4 \sum_{m=1}^{\bar{m}} \|\frac{d^2t}{ds^2}\|_{0,\Gamma_m \cap \Gamma_a}^2. \tag{4.46}$$

On the part of the boundary $\partial\Omega_j \dot{-} \Gamma_a, j = 1, \ldots, \bar{j}$, we define

$$\psi_h = t_h + \Delta_j,$$

$$\Delta_j = [\text{mes}(\partial\Omega_j \dot{-} \Gamma_a)]^{-1} \int_{\partial\Omega_j \cap \Gamma_a} (t_h - \psi_h) ds,$$

$$\Delta_j = 0 \quad \text{provided} \quad \partial\Omega_j \cap \Gamma_a = \emptyset.$$

Recall that $\text{mes}(\partial\Omega_j \dot{-} \Gamma_a) = \text{mes}(\partial\Omega_j \cap \Gamma_u) > 0$ by (4.24).

We easily verify that (4.41) holds. Moreover,

$$\|\psi_h - t_h\|_{0,\partial\Omega_j \dot{-} \Gamma_a}^2 = \Delta_j^2 \,\text{mes}(\partial\Omega_j \dot{-} \Gamma_a) \leq C \|t_h - \psi_h\|_{0,\partial\Omega_j \cap \Gamma_a}^2$$

$$\leq C \|\psi_h - t_h\|_{0,\Gamma_a}^2. \tag{4.47}$$

Now (4.42) is derived by combining (4.46) and (4.47). □

Lemma 4.6. *Let a piecewise linear function φ on Γ be given, whose nodes coincide with the vertices of an (α, β)-regular system of triangulations $\{\mathcal{T}_h\}$ (φ is generally discontinuous at its nodes), such that*

$$\int_{\partial\Omega_j} \varphi \, ds = 0, \quad j = 1, \ldots, \bar{j}. \tag{4.48}$$

Then there exists a vector function $w^h \in \mathcal{N}_h$ such that

$$w^h \cdot \nu = \varphi \quad on\ \Gamma,$$

$$\|w^h\|_{0,\Omega} \leq Ch^{-1/2}\|\varphi\|_{0,\Gamma}. \tag{4.49}$$

Proof. Let us again consider a boundary layer $\Omega_h \subset \Omega$, which is formed by all (closed) triangles $T \in \mathcal{T}_h$ such that $T \cap \Gamma \neq \emptyset$. Evidently

$$\Omega_h = \bigcup_{j=1}^{\bar{j}} \Omega_h^j,$$

where Ω_h^j is adjacent to the polygon $\partial\Omega_j$.

We shall construct $w^h \in \mathcal{N}_h$ by means of suitably chosen parameters of the flow β (see algorithm for the solution of the problem (4.29)) in such a way that

$$\text{supp}\ w^h \subset \Omega_h.$$

Consider a layer Ω_h^j. On $\partial\Omega_j$ we choose the parameters of the flow equal to the corresponding values of the function φ, as we let them vanish on $\partial\Omega_h^j \dot{-} \partial\Omega_j$. On the sides which connect vertices of $\partial\Omega_j$ with those of $\partial\Omega_h^j \dot{-} \partial\Omega_j$, we set $\beta_k = 0$ at the inner vertices on $\partial\Omega_h^j \dot{-} \partial\Omega_j$, while the parameters β_i at the external vertices on $\partial\Omega_j$ remain free to be suitably chosen later. Each of the sides l_i, $i = 1, \ldots, n$ is associated with one unknown parameter β_i (see Figure 2).

1[0] Let us first assume that each $T_i \in \Omega_h^j$ has at most one side lying on $\partial\Omega_j$. The conditions of the form (4.30), (4.31) generate a system of n equations

$$\mathcal{A}\beta = b, \tag{4.50}$$

where

$$\begin{aligned} \mathcal{A}_{ii} &= -l_i, & i &= 1, 2, \ldots, n, \\ \mathcal{A}_{i,i+1} &= l_{i+1}, & i &= 1, \ldots, n-1, \\ \mathcal{A}_{n1} &= l_1 \end{aligned}$$

while all the other elements of the matrix $\mathcal{A}$ are zeros. Further,

$$b_i = -l_m(\varphi_m + \varphi_m^+)$$

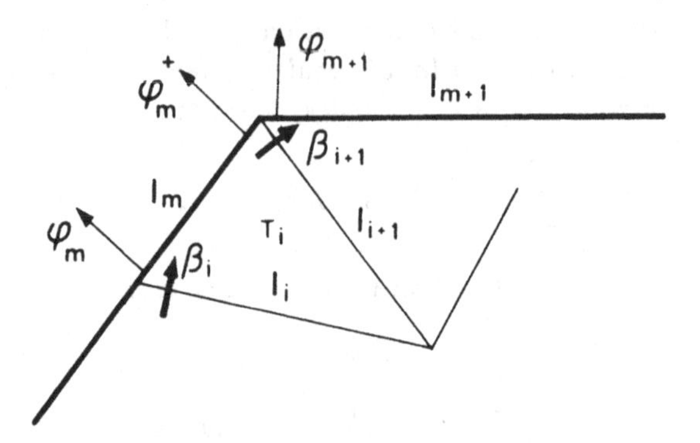

Figure 2

or $b_i = 0$ provided $T_i \cap \partial\Omega_j$ reduces to a single vertex.

The assumption (4.48) implies

$$\sum_{i=1}^{n} b_i = 0,$$

hence, we can omit the last equation of the system (4.50). If we put $\beta_1 = 0$, then the system has the solution

$$\beta_i = l_i^{-1} \sum_{p=1}^{i-1} b_p, \quad i = 2, 3, \ldots, n. \tag{4.51}$$

2^0 Let a triangle $T_k \in \partial\Omega_h^j$ have two sides l_q and l_{q+1} on $\partial\Omega_j$ (Figure 3). Then, we obtain the following equation for T_k:

$$(\beta_k^+ + \beta_k)l_k = -l_q(\varphi_q + \varphi_q^+) - l_{q+1}(\varphi_{q+1} + \varphi_{q+1}^+).$$

Then we can set $\beta_k^+ = 0$, for instance, and calculate β_k. The remaining system of equations again has the form (4.50), being of the shape of the system corresponding to the "truncated" triangulation $\Omega_h^j \dot{-} T_k$, with β_k and β_k^+ playing the roles of given external parameters of the flow.

3^0 Equation (4.51), together with the (α, β)-regularity of the system, $\mathcal{T}_h$ implies

$$|\beta_i| = 2l_i^{-1} \left| \int_0^{s_i} \varphi \, ds \right| \leq 2h_{\min}^{-1} \int_{\partial\Omega_j} |\varphi| \, ds \leq Ch^{-1} \|\varphi\|_{0,\Gamma}, \tag{4.52}$$

$$i = 1, \ldots, n; \quad j = 1, \ldots, \bar{j}.$$

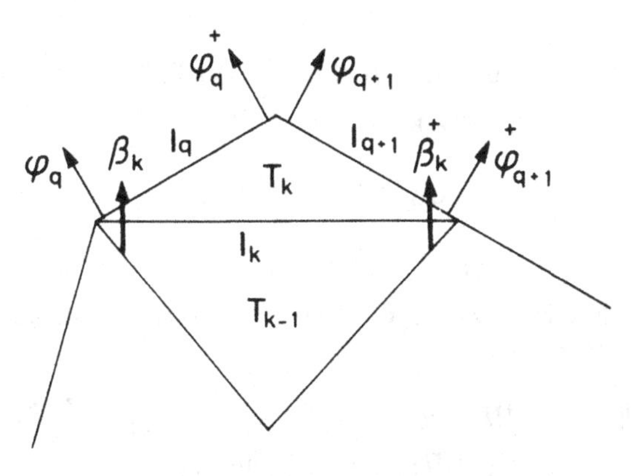

Figure 3

The same estimate is valid for β_k and the other parameters in 2^0.

The upper bound (4.52) is valid for the boundary parameters φ_m, φ_m^+ as well. Indeed, we have

$$\int_0^{l_m} \varphi^2 ds = \frac{1}{6} l_m [\varphi_m^2 + (\varphi_m^+)^2 + (\varphi_m + \varphi_m^+)^2] \geq \frac{1}{6} l_m (|\varphi_m|^2 + |\varphi_m^+|^2),$$

hence,

$$|\varphi_m|^2 + |\varphi_m^+|^2 \leq 6\, l_m^{-1} \|\varphi\|_{0,\Gamma}^2 \leq Ch^{-1} \|\varphi\|_{0,\Gamma}^2 \leq Ch^{-2} \|\varphi\|_{0,\Gamma}^2.$$

For any triangle $T \in \Omega_h^j$ and any vertex $Q \in T$, we have the inequality

$$|w_k^h(Q)| \leq (\sin \alpha)^{-1} C h^{-1} \|\varphi\|_{0,\Gamma}, \quad k = 1, 2;$$

consequently, the same bound is also valid at all points $x \in \Omega_h$. Hence,

$$\|w^h\|_0^2 = \sum_{k=1}^{2} \int_{\Omega_h} (w_k^h)^2 ds \leq Ch^{-2} \|\varphi\|_{0,\Gamma}^2 \text{ mes } \Omega_h \leq Ch^{-1} \|\varphi\|_{0,\Gamma}^2,$$

since mes $\Omega_h < Ch$. $\square$

Proof of Theorem 4.4. Let ψ_h be a one-sided approximation of the flow from lemma 4.5. Put

$$\varphi = (r_h U) \cdot \nu - \psi_h$$

and consider the extension $w^h \in \mathcal{N}_h$ of the function φ from lemma 4.6. Then, the function $W_h = r_h U - w^h$ satisfies the assumptions of the theorem. Indeed, $W_h \in \mathcal{N}_h$ and

$$W_h \cdot \nu = (r_h U) \cdot \nu - \varphi = \psi_h \quad \text{on } \Gamma.$$

Hence,

$$0 \le W_h \cdot \nu \le U \cdot \nu \quad \text{on } \Gamma_a.$$

Consequently, $W_h \in C_h$. Further,

$$\operatorname{div}(2U - W_h) = 0,$$

$$(2U - W_h) \cdot \nu \ge (U - W_h) \cdot \nu \ge 0 \quad \text{on } \Gamma_a,$$

which implies that $2U - W_h \in C$. Using estimate (4.26), and lemmas 4.5 and 4.6, we conclude that

$$\begin{aligned} \|U - W_h\|_0 &= \|U - r_h U + w^h\|_0 \le \|U - r_h\|_0 + \|w^h\|_0 \\ &\le C(h^2|U|_{2,\Omega} + h^{-1/2}\|\varphi\|_{0,\Gamma}) \\ &\le C\left\{h^2|U|_{2,\Omega} + h^{3/2}\sum_{m=1}^{\bar{m}} |U \cdot \nu|_{2,\Gamma_m \cap \Gamma_a}\right\}. \qquad \square \end{aligned}$$

Corollary to Theorem 4.4. *Let the assumptions of theorem 4.4 and inequality (4.24) hold, and let G from (4.27) belong to $[H^2(\Omega)]^2$. Then, the following estimate holds for any (α, β)-regular system of triangulations $\{\mathcal{T}_h\}$:*

$$\|p^0 - p^h\|_{0,\Omega} = 0(h^{3/2}).$$

Proof. Follows from lemma 4.4, theorem 4.4, and from estimate (4.26) as applied to the function G. $\square$

Remark 4.3. Let β^{hm} be an approximate solution of the quadratic programming problem (4.33). If we define the corresponding $p^{hm} = p(\beta^{hm})$ and the corresponding approximate solution of the problem (1.16) by $q^{hm} = \lambda + p^{hm}$, then our corollary to theorem 4.4 implies

$$\|q^0 - q^{hm}\|_0 = \|p^0 - p^{hm}\|_0 \le Ch^{3/2} + \|p^h - p^{hm}\|_0.$$

1.1.422. A Posteriori Error Bounds and the Two-Sided Energy Estimate. Let us assume that we have evaluated the approximation $\tilde{u}_h \in K_{2h}$ of the primal problem, as well as the approximation $q^{Hm} = \lambda + p^{Hm} \in \mathcal{U}_2$ of the dual problem. Then, we can establish error bounds for both the primal and dual approximations.

Theorem 4.5. *Let $\tilde{u}_h \in K_{2h}, q^{Hm} \in \mathcal{U}_2$. Then,*

$$|\tilde{u}_h - u|_1^2 \le \|q^{Hm} - \nabla \tilde{u}_h\|_0^2 + 2\int_{\Gamma_a} q^{Hm} \cdot \nu \tilde{u}_h ds \equiv E(q^{Hm}, \tilde{u}_h), \tag{4.53}$$

$$\|q^{Hm} - \nabla u\|_0 \le E(q^{Hm}, \tilde{u}_h). \tag{4.54}$$

Proof. Analogous to that of theorem 4.2. □

Remark 4.4. Both terms in the expression $E(q^{Hm}, \tilde{u}_h)$ are nonnegative, as follows from the definition of $\mathcal{U}_2$ and K_{2h}. However, let us notice that generally $\lambda + p^H \notin \mathcal{U}_2$, since $\mathcal{U}_{20}^H \subset \mathcal{U}_{20}$. In order to comply with this condition, we can proceed as follows. Let us look for a function $\lambda^f \in H(\mathrm{div}, \Omega)$ such that

$$\mathrm{div}\, \lambda^f = 0, \quad \lambda^f \cdot \nu = 0 \quad \text{on } \Gamma_a.$$

Replacing λ by the function λ^f, we obtain $g = -\lambda^f \cdot \nu = 0 = g_H$ on Γ_a and $q^H = \lambda^f + p^H \in \mathcal{U}_2$, since $\mathcal{U}_{20}^H = C_H \subset \mathcal{U}_{20} = C$.

The function λ^f can be constructed, for example, as the sum

$$\lambda^f = \lambda + \lambda^\omega, \quad \lambda^\omega = \left\{ -\frac{\partial \omega}{\partial x_2}, \frac{\partial \omega}{\partial x_1} \right\},$$

where $\omega \in H^2(\Omega)$ fulfills the boundary condition

$$\omega(s) = -\int_{s_0}^{s} (\lambda \cdot \nu)(t)dt \quad \forall s \in \Gamma_a.$$

Indeed, then

$$\frac{d\omega}{ds} = -\frac{\partial \omega}{\partial x_2}\nu_1 + \frac{\partial \omega}{\partial x_1}\nu_2 = \lambda^\omega \cdot \nu = -\lambda \cdot \nu \quad \text{on } \Gamma_a.$$

Then, it suffices that $p^{Hm} \in \mathcal{U}_{20}^H$. However, it is necessary to realize that some methods of quadratic programming (e.g., Uzawa's method) do not satisfy this condition

Theorem 4.6. *Let $\tilde{u}_h \in K_{2h}, q^{Hm} \in \mathcal{U}_2$. Then*

$$-2\mathcal{L}_2(\tilde{u}_h) \leq |u|_1^2 = (f, u)_0 \leq \|q^{Hm}\|_0^2.$$

Proof. Analogous to that of theorem 4.3. □

1.1.5 Solution of Mixed Problems by the Finite Element Method and Error Bounds

In this section we will study approximations of variational inequalities, starting with the mixed formulation of the given problem. Before moving to the general formulation of the problem, we will describe a method of proceeding when approximating problem (2.5) (that is, the mixed formulation of problem (2.4)).

Let $\Omega \subset \mathbf{R}^2$ be a bounded *polygonal* domain, $\{\mathcal{T}_h\}$, $h \to 0+$, a *regular* system of triangulations, whose nodes on Γ will be denoted by $a_1, a_2, \ldots, a_{m(h)}$. Each $\mathcal{T}_h$ is associated with a finite-dimensional space of functions V_h, which consists of piecewise linear functions on the triangles of the given triangulation (the same construction as in section 1.1.31). Further, let us define

$$L_h = \{\mu_h \in L^2(\Gamma) \mid \mu_h^i \equiv \mu_h|_{a_i a_{i+1}} \in P_0(a_i a_{i+1}), \quad i = 1, \ldots, m(h)\},$$

$$\Lambda_h = \{\mu_h \in L_h \mid |\mu_h| \le 1 \quad \text{on } \Gamma\}.$$

That is, L_h is the linear space of constant piecewise functions on Γ, the partition Γ being determined by the boundary nodes of the triangulation $\mathcal{T}_h$, while Λ_h is its bounded convex subset.

By an *approximation of problem* (2.5) we mean the problem of finding a saddle point (u_h, λ_h) of the Lagrangian $\tilde{\mathcal{H}}$ on $V_h \times \Lambda_h$. We are interested in the relationship between the approximate solutions (u_h, λ_h) and the exact solution (u, λ).

It is clear from our previous considerations that the approximation of the mixed variational formulation has one more advantage: this approximation takes into account the *simultaneous* approximation not only of the solution u itself, but of the Lagrangian multiplicator λ as well. This is especially important in those problems where the knowledge of λ—which usually has a good physical meaning—is desirable. In our particular case, λ_h are approximations of $\lambda = -\partial u/\partial \nu$. Even if it is correct that the approximate values of $\partial u/\partial \nu$ can be obtained by differentiating the approximate solution u_h, numerical experiments have shown that the results reached by this method are less satisfactory than those obtained by direct approximation of the mixed formulation.

Remark 5.1. For reasons that will be made apparent later, it is possible to consider even more general constructions of L_h and Λ_h. Let $b_1, b_2, \ldots, b_{M(H)}$ be different pairwise points from Γ. The partition of Γ that they determine is denoted by $\mathcal{T}_h$, where $H = \max |b_i b_{i+1}|$. This partition is independent of the triangulation $\mathcal{T}_h$. To each $\mathcal{T}_h$, we assign a finite-dimensional space

$$L_H = \{\mu_H \in L^2(\Gamma) \mid \mu_H^i \equiv \mu_H|_{b_i b_{i+1}} \in P_0(b_i b_{i+1}), \quad i = 1, \ldots, M(H)\},$$

and a convex bounded subset

$$\Lambda_H = \{\mu_H \in L_H \mid |\mu_H| \le 1 \quad \text{on } \Gamma\}.$$

By the approximation of (2.5) we will then understand the problem of finding a saddle point (u_h, λ_H) of the Lagrangian $\tilde{\mathcal{H}}$ on $V_h \times \Lambda_H$.

Remark 5.2. We proceed analogously when approximating problems $\mathcal{P}_i$, $i = 1, 2$. We define finite-dimensional sets

$$V_h = \{v_h \in C(\bar{\Omega}) \mid v_h|_T \in P_1(T) \ \ \forall T \in \mathcal{T}_h, \ \ v_h = 0 \ \text{ on } \Gamma_u\}$$

(in the case $\mathcal{P}_1, \Gamma_u = \emptyset$),

$$L_H = \{\mu_H \in L^2(\Gamma) \mid \mu_H^i \in P_0(b_i b_{i+1}), \ \ i = 1, \dots, M(H)\}$$

and

$$\Lambda_H = \{\mu_H \in L_H \mid \mu_H \geq 0 \quad \text{on } \Gamma_a\}.$$

By the approximation of problems $\mathcal{P}_i$ we mean the problem of finding a saddle point $\{u_h, \lambda_H\}$ corresponding to the Lagrangian on $V_h \times \Lambda_H$.

1.1.51. Mixed Variational Formulations of Elliptic Inequalities. We see from the results of section 1.1.2 that the mixed formulation leads to the problem of finding a saddle point of a certain Lagrangian in a certain convex closed subset. Let us now generally formulate this problem.

Let V, L be the two real Hilbert spaces with norms $\|\cdot\|$, $|\cdot|$, and let V', L' be the corresponding spaces of continuous linear functionals. If $f \in V', v \in V$, or $g \in L', \mu \in L$, then the values of the functionals at the corresponding points will be denoted by $\langle f, v\rangle$ or $[g, \mu]$, respectively. Let

$$\mathcal{Y}(v) = \frac{1}{2}a(v, v) - \langle f, v\rangle \tag{5.1}$$

be a quadratic functional determined by a continuous, V-elliptic and symmetric bilinear form a, and by $f \in V'$. Let $b : V \times L \to \mathbf{R}$ be another continuous bilinear form, that is,

$$\exists M_1 = \text{const} > 0 : |b(v, \mu)| \leq M_1 \|v\| |\mu| \quad \forall (v, \mu) \in V \times L, \tag{5.2}$$

and let us define a functional $\mathcal{H} : V \times L \mapsto \mathbf{R}$ by

$$\mathcal{H}(v, \mu) = \mathcal{Y}(v) + b(v, \mu) - [g, \mu], \tag{5.3}$$

where $g \in L'$ is fixed.

Finally, let $K \subseteq V, \Lambda \subseteq L$ be nonempty, convex, closed subsets. In the following, we will assume that Λ is either (CC) a convex cone with its vertex at θ_L [6] (the zero element of L) and $K = V$, or (BC), a bounded subset of L.

[6] This means that if $\mu \in \Lambda$, then $\rho\mu \in \Lambda$ for every $\rho \geq 0$. In what follows θ_X denotes the zero element of a linear space X.

Let (u, λ) be a saddle point of $\mathcal{H}$ in $K \times \Lambda$:

$$\mathcal{H}(u, \mu) \leq \mathcal{H}(u, \lambda) \leq \mathcal{H}(v, \lambda) \quad \forall (v, \mu) \in K \times \Lambda, \tag{$\tilde{\mathcal{P}}$}$$

or, equivalently (see Ekeland and Temam (1974))

$$(u, \lambda) \in K \times \Lambda \text{ such that}$$

$$a(u, v-u) + b(v-u, \lambda) \geq \langle f, v-u \rangle \quad \forall v \in K, \tag{$\tilde{\mathcal{P}}'$}$$

$$b(u, \mu - \lambda) \leq [g, \mu - \lambda] \quad \forall \mu \in \Lambda.$$

We easily verify that the first component of u is a solution of a certain minimization problem. Indeed, we have

Lemma 5.1. *Set* $j(v) = \sup_\Lambda \{b(v, \mu) - [g, \mu]\}$. *Then,*

$$\mathcal{Y}(u) + j(u) = \min_K \{\mathcal{Y}(v) + j(v)\}. \tag{$\mathcal{P}$}$$

Proof. The first inequality in $\tilde{\mathcal{P}}$ yields

$$j(u) = \sup_\Lambda \{b(u, \mu) - [g, \mu]\} \leq b(u, \lambda) - [g, \lambda].$$

On the other hand, the converse inequality is evident and hence,

$$j(u) = b(u, \lambda) - [g, \lambda].$$

This, together with the second inequality in $(\tilde{\mathcal{P}})$, immediately implies the assertion. □

Definition 5.1. Problem $\tilde{\mathcal{P}}$ will be called the *mixed* variational formulation of the minimization problem $\mathcal{P}$.

As concerns the very *existence*, or even the *uniqueness* of (u, λ) satisfying $\tilde{\mathcal{P}}$, it is possible to make use of the well-known results of convex analysis (e.g., Céa (1971), Ekeland and Temam (1974), etc.). We recall here those results, which will be referred to frequently in the following.

As a consequence of the V-ellipticity of the form a, the functional $\mathcal{Y}(v) + j(v)$ is *strictly convex* in V; hence, the first component of the saddle point (if it exists), is uniquely determined.

Let us now discuss the conditions which would guarantee the uniqueness of the second component of the saddle point. Let us additionally assume that K is a convex closed cone with its vertex at θ_V. Let us define

$$K^* = \{v \in K \mid -v \in K\}.$$

Lemma 5.2. *Let*

$$b(v,\mu) = 0 \quad \forall v \in K^* \Longrightarrow \mu = \theta_L. \tag{5.4}$$

The second component of the saddle point also is uniquely determined.

Proof. Let $(u,\lambda_1), (u,\lambda_2)$ be saddle points of $\mathcal{H}$ in $K \times \Lambda$. As K is a cone with its vertex at θ_V, we can first choose $v = \theta_V$, and then $v = 2u$, thus eventually obtaining

$$a(u,v) + b(v,\lambda_i) \geq \langle f, v\rangle \quad \forall v \in K, \quad i = 1,2.$$

Restricting ourselves to the trial functions $v \in K^*$, the previous inequalities reduce to equations as a consequence of the linearity of K^*:

$$a(u,v) + b(v,\lambda_i) = \langle f, v\rangle \quad \forall v \in K^*, \quad i = 1,2.$$

Subtracting one from the other, we find

$$b(v, \lambda_1 - \lambda_2) = 0 \quad \forall v \in K^*,$$

and (5.4) yields $\lambda_1 = \lambda_2$. □

Next we will introduce conditions guaranteeing the *existence* of a solution (u,λ).

Lemma 5.3. *Let (CC) hold and let there exists a constant $\beta > 0$ such that*

$$\sup_V \frac{b(v,\mu)}{\|v\|} \geq \beta|\mu| \quad \forall \mu \in L. \tag{5.5}$$

Then there exists exactly one solution (u,λ) of problem $\tilde{\mathcal{P}}$.

Proof. The uniqueness follows from the fact that (5.5) implies (5.4), from $K^* = V$, and finally, from the V-ellipticity of the form a. Its existence is proved in Brezzi, Hager, and Raviart (1979). □

If we assume that Λ is bounded, the situation is much simpler.

Lemma 5.4. *Let (BC) hold. Then there exists a solution (u,λ) for problem $\tilde{\mathcal{P}}$.*

Proof. The assertion is a consequence of a more general result (see proposition 2.2, Chapter 6, Ekeland and Temam (1974)). □

Remark 5.3. In concrete cases, the proof of existence of a solution usually proceeds in such a way that we first "guess" the solution, and only then verify that it really satisfies $\tilde{\mathcal{P}}$ or $\tilde{\mathcal{P}}'$.

Remark 5.4. Until now, we have assumed that the form a is symmetric. If a is a general, continuous, V-elliptic form (not necessarily symmetric), then we start from $\tilde{\mathcal{P}}'$. We are looking for $(u, \lambda) \in K \times \Lambda$ that satisfies the inequalities from $\tilde{\mathcal{P}}'$. Analogously to lemma 5.1, it is possible to verify that $u \in K$ satisfies the inequality

$$a(u, v-u) + j(v) - j(u) \geq \langle f, v-u \rangle \quad \forall v \in K. \qquad (\mathcal{P}')$$

In this case, $\tilde{\mathcal{P}}'$ will be called the *mixed formulation of* $\mathcal{P}'$. Thus, it is evident that $\tilde{\mathcal{P}}'$ is more general than $\tilde{\mathcal{P}}$, as it does not require the symmetry of a. It is possible to formulate conditions analogous to those mentioned before, which guarantee uniqueness and existence of the solution (u, λ) for problem $\tilde{\mathcal{P}}'$. However, throughout this book we will only encounter problems with a symmetric form a.

1.1.52. Approximation of the Mixed Variational Formulation and Error Bounds. Let $\{V_h\}$ and $\{L_H\}$, $h, H \in (0,1)$ be systems of finite-dimensional subspaces of V and L, respectively. Let K_h and Λ_H be nonempty convex closed subsets of V_h and L_H, respectively. It need not generally hold that $K_h \subset K$, $\Lambda_H \subset \Lambda$. Each pair $(h, H) \in (0,1)$ will be associated with the set $K_h \times \Lambda_H$. Further, we assume Λ_H either to be (CC_H) a cone with its vertex at θ_L, and $K_h = V_h \ \ \forall h, H \in (0,1)$, or (BC_H) a uniformly bounded subset of L, that is, there exists a positive constant $c > 0$ such that

$$|\mu_H| \leq c \quad \forall \mu_H \in \Lambda_H \quad \forall H \in (0,1).$$

By an approximation of problem $\tilde{\mathcal{P}}$ we mean to find a saddle point (u_h, λ_H) of the Lagrangian $\mathcal{H}$ on $K_h \times \Lambda_H$:

$$\mathcal{H}(u_h, \mu_H) \leq \mathcal{H}(u_h, \lambda_H) \leq \mathcal{H}(v_h, \lambda_H) \quad \forall (v_h, \mu_H) \in K_h \times \Lambda_H \qquad (\tilde{\mathcal{P}}_{hH})$$

or equivalently

$$\text{find } (u_h, \lambda_H) \in K_h \times \Lambda_H \ \text{ such that}$$

$$\begin{aligned} a(u_h, v_h - u_h) + b(v_h - u_h, \lambda_H) &\geq \langle f, v_h - u_h \rangle \quad \forall v_h \in K_h, \\ b(u_h, \mu_H - \lambda_H) &\leq [g, \mu_H - \lambda_H] \quad \forall \mu_H \in \Lambda_H. \end{aligned} \qquad (\tilde{\mathcal{P}}'_{hH})$$

Denote

$$j_H(v_h) = \sup_{\Lambda_H} \{ b(v_h, \mu_H) - [g, \mu_H] \}.$$

As in the continuous case, we can verify that the first component $u_h \in K_h$ is a solution of the minimization problem

$$\mathcal{Y}(u_h) + j_H(u_h) \leq \mathcal{Y}(v_h) + j_H(v_h) \quad \forall v_h \in K_h. \qquad (5.6)$$

The proof is left to the reader as an easy exercise.

The question of *existence* or *uniqueness* of the solution of problems $\tilde{\mathcal{P}}_{hH}$ in a finite dimension is simpler than it was with the continuous case, as is evident from the following lemma.

Lemma 5.5. *Let* (BC_H) *hold. Then there exists a solution of problem* $\tilde{\mathcal{P}}_{hH}$. *Moreover, its first component is uniquely determined.*

Proof. The existence of a solution is a consequence of a more general assertion (see the proof of lemma 5.4); the uniqueness of the first component is a consequence of the V-ellipticity of a. □

Lemma 5.6. *Let* (CC_H) *hold and let*

$$K_{hH} = \{v_h \in V_h \mid j_H(v_h) < +\infty\}.$$

If $K^0_{hH} \neq \emptyset$ *(that is,* K_{hH} *is a set with nonempty interior), then there exists a solution of* $\tilde{\mathcal{P}}_{hH}$ *with a uniquely determined first component.*

Proof. We easily see that

$$K_{hH} = \{v_h \in V_h \mid b(v_h, \mu_H) \leq [g, \mu_H] \quad \forall \mu_H \in \Lambda_H\}.$$

Indeed, if $v_h \in K_{hH}$, then $j_H(v_h) = 0$; otherwise, it would be $j_H(v_h) = +\infty$ (see section 1.1.2). By virtue of (5.6), we have

$$u_h \in K_{hH}, \quad \mathcal{Y}(u_h) \leq \mathcal{Y}(v_h) \quad \forall v_h \in K_{hH}.$$

Now the existence of a solution (u_h, λ_H) is a consequence of the fact that K^0_{hH} is nonempty (e.g., Céa (1971)). The uniqueness of the first component follows from the V-ellipticity of the form a. □

Remark 5.5. Let

$$K^*_h = \{v_h \in K_h \mid -v_h \in K_h\}.$$

If

$$b(v_h, \mu_H) = 0 \quad \forall v_h \in K^*_h \Longrightarrow \mu_H = \theta_L,$$

then even the second component of the saddle point is uniquely determined. The proof completely coincides with that for the continuous case.

Now we will study the mutual relation between (u_h, λ_H) and (u, λ). To this end, we first establish a bound for the error $\|u - u_h\|$.

Lemma 5.7. *Let* $\{u, \lambda\}$ *and* $\{u_h, \lambda_H\}$ *be solutions of* $\tilde{\mathcal{P}}$ *and* $\tilde{\mathcal{P}}_{hH}$, *respectively. Then*

$$\|u - u_h\|^2 \leq c[\|u - v_h\|^2 + |\lambda - \mu_H|^2 + A_1(v_h) + A_2(v)$$

$$+\{b(u,\lambda_H-\mu)-[g,\lambda_H-\mu]\}+\{b(u,\lambda-\mu_H) \qquad (5.7)$$
$$-[g,\lambda-\mu_H]\}+|\lambda-\lambda_H|^2]$$

holds for any $v_h\in K_h, v\in K, \mu_H\in\Lambda_H, \mu\in\Lambda$. *Here,*

$$A_1(v_h)=a(u,v_h-u)+b(v_h-u,\lambda)+\langle f,u-v_h\rangle,$$

$$A_2(v)=a(u,v-u_h)+b(v-u_h,\lambda)+\langle f,u_h-v\rangle,$$

and c *is a positive constant independent of* $h,H\in(0,1)$.

Proof. See Haslinger (1981). □

Estimate (5.7) makes it possible to derive other useful relations between u_h and u or λ_H and λ. For their proofs, the reader is referred again to Haslinger (1981).

Theorem 5.1′. *Let* $(CC),(CC_H)$ *be fulfilled, and moreover, let there exist a positive constant* $\tilde\beta$ *independent of* h,H *such that*

$$\sup_{V_h}\frac{b(v_h,\mu_H)}{\|v_h\|}\ge\tilde\beta|\mu_H|\quad\forall\mu_H\in L_H. \qquad (5.8)$$

Then there exists a positive constant c *independent of* $h,H\in(0,1)$ *and such that*

$$\|u-u_h\|^2\le c[\|u-v_h\|^2+|\lambda-\mu_H|^2+\{b(u,\lambda_H-\mu)$$

$$-[g,\lambda_H-\mu]\}+\{b(u,\lambda-\mu_H)-[g,\lambda-\mu_H]\}], \qquad (5.9)$$

$$|\lambda-\lambda_H|\le c\{\|u-u_h\|+|\lambda-\mu_H|\}, \qquad (5.10)$$

hold for arbitrary $v_h\in V_h,\mu\in\Lambda,\mu_H\in\Lambda_H$.

Remark 5.6. If $\Lambda_H\subset\Lambda\ \ \forall H\in(0,1)$, then we can set $\mu=\lambda_H$ in (5.9), obtaining a simpler formula:

$$\|u-u_h\|^2\le c[\|u-v_h\|^2+|\lambda-\mu_H|^2+\{b(u,\lambda-\mu_H)$$

$$-[g,\lambda-\mu_H]\}]\quad\forall\mu_H\in\Lambda_H. \qquad (5.9')$$

Theorem 5.2. *Let* $(BC),(BC_H)$ *be fulfilled. Then,*

if $K=V, K_h=V_h$ *and the condition* (5.8) *is*

fulfilled, (5.9), (5.10) *hold,* (5.11)

or,

$$\|u-u_h\|^2\le c[\|u-v_h\|^2+|\lambda-\mu_H|^2+A_1(v_h)+A_2(v)$$

$$+\|u-v_h\|+\{b(u,\lambda_H-\mu)-[g,\lambda_H-\mu]\}+\{b(u,\lambda-\mu_H)-[g,\lambda-\mu_H]\}\} \quad (5.12)$$

$$\forall v_h \in K_h, v \in K, \mu \in \Lambda, \mu_H \in \Lambda_H.$$

Remark 5.7. If $K_h \subset K, \Lambda_H \subset \Lambda \quad \forall h, H \in (0,1)$, then we can set $v = u_h, \mu = \lambda_H$ in the preceding formula.

This choice of v, μ yields

$$A_2(v) = 0, \quad b(u,\lambda_H - \mu) - [g, \lambda_H - \mu] = 0.$$

In the following, we will consider only such pairs (h, H) that satisfy

$$h \to 0_+ \Longleftrightarrow H \to 0_+.$$

The relations (5.9), (5.10), (5.12) can be used to derive the rate of convergence of u_h to u and λ_H to λ, provided u, λ are sufficiently smooth. The following two theorems on the convergence of the approximate solutions u_h, λ_H to the exact ones u, λ are also consequences of these relations as well.

Theorem 5.3. *Let $(BC), (BC_H)$ be fulfilled and moreover, let*

$$\forall v \in K \ \ \exists v_h \in K_h : v_h \to v, h \to 0_+ \ \ in \ \ V; \quad (5.13)$$

$$\forall \mu \in \Lambda \ \ \exists \mu_H \in \Lambda_H : \mu_H \to \mu, H \to 0_+ \ \ in \ \ L; \quad (5.14)$$

$$v_h \in K_h, v_h \rightharpoonup v, h \to 0_+ \ \ (weakly) \ in \ \ V \Rightarrow v \in K; \quad (5.15)$$

$$\mu_H \in \Lambda_H, \mu_H \rightharpoonup \mu, H \to 0_+ \ \ (weakly) \ in \ \ L \Rightarrow \mu \in \Lambda. \quad (5.16)$$

Let the solution $(u, \lambda) \in K \times \Lambda$ of problem $\tilde{\mathcal{P}}$ be unique. Then,

$$u_h \to u, h \to 0_+ \ \ in \ \ V,$$

$$\lambda_H \rightharpoonup \lambda, H \to 0_+ \ \ (weakly) \ in \ \ L. \quad (5.17)$$

Proof. First of all, $\{u_h\}, \{\lambda_H\}$ are bounded. This follows for $\{\lambda_H\}$ from (BC_H), for $\{u_h\}$ from (5.13) and $(\tilde{\mathcal{P}}'_{hH})_2$. Consequently, there exist subsequences $\{u_{\underline{h}}\} \subset \{u_h\}, \{\lambda_{\underline{H}}\} \subset \{\lambda_H\}$ and pairs $(u^*, \lambda^*) \in V \times L$ such that

$$u_{\underline{h}} \rightharpoonup u^*, \underline{h} \to 0_+; \lambda_{\underline{H}} \rightharpoonup \lambda^*, \underline{H} \to 0_+. \quad (5.18)$$

(5.15), (5.16) now imply that $(u^*, \lambda^*) \in K \times \Lambda$. We will show that (u^*, λ^*) is a solution of $\tilde{\mathcal{P}}$.

Let $(v, \mu) \in K \times \Lambda$ be an arbitrary but fixed element. Then (5.13), (5.14) imply the existence of sequences $\{v_h\}, \{\mu_H\}$, $v_h \in K_h, \mu_H \in \Lambda_H$ such that

$$v_h \to v, h \to 0_+; \ \ \mu_H \to \mu, H \to 0_+. \quad (5.19)$$

The pair $(u_{\underline{h}}, \lambda_{\underline{H}})$ is a solution of $\tilde{\mathcal{P}}'_{\underline{hH}}$, that is,

$$a(u_{\underline{h}}, u_{\underline{h}} - v_{\underline{h}}) + b(u_{\underline{h}} - v_{\underline{h}}, \lambda_{\underline{H}}) \leq \langle f, u_{\underline{h}} - v_{\underline{h}} \rangle, \tag{5.20}$$

$$b(u_{\underline{h}}, \mu_{\underline{H}} - \lambda_{\underline{H}}) \leq [g, \mu_{\underline{H}} - \lambda_{\underline{H}}]. \tag{5.21}$$

Passing to the limit for $\underline{h}, \underline{H} \to 0_+$, in (5.20) and using (5.18), we obtain

$$a(u^*, u^* - v) + \varliminf_{\underline{h}, \underline{H} \to 0_+} b(u_{\underline{h}}, \lambda_{\underline{H}}) - b(v, \lambda^*) \leq \langle f, u^* - v \rangle. \tag{5.22}$$

Similarly, by passing to the limit in (5.21), we find

$$b(u^*, \mu) - [g, \mu - \lambda^*] \leq \varliminf_{\underline{h}, \underline{H} \to 0_+} b(u_{\underline{h}}, \lambda_{\underline{H}}). \tag{5.23}$$

In particular, for $\mu = \lambda^*$ this yields

$$b(u^*, \lambda^*) \leq \varliminf_{\underline{h}, \underline{H} \to 0_+} b(u_{\underline{h}}, \lambda_{\underline{H}}). \tag{5.24}$$

Finally, substituting (5.24) into (5.22), we arrive at the inequality

$$a(u^*, u^* - v) + b(u^* - v, \lambda^*) \leq \langle f, u^* - v \rangle. \tag{5.25}$$

Setting $v = u^*$ in (5.22), we find

$$\varliminf_{\underline{h}, \underline{H} \to 0_+} b(u_{\underline{h}}, \lambda_{\underline{H}}) \leq b(u^*, \lambda^*),$$

which together with (5.23) yields

$$b(u^*, \mu - \lambda^*) \leq [g, \mu - \lambda^*].$$

Since $(v, \mu) \in K \times \Lambda$ was arbitrarily chosen, this together with (5.25) implies that (u^*, λ^*) is a solution of $\tilde{\mathcal{P}}'$. As a consequence of the uniqueness of this problem we have $(u^*, \lambda^*) = (u, \lambda)$. Moreover, not only the subsequences, but the entire sequences $\{u_h\}, \{\lambda_H\}$ weakly converge to u and λ, respectively, in the corresponding spaces.

Let us show that u_h converge to u strongly in V. Let $\{\bar{v}_h\}, \{\bar{\mu}_H\}, \bar{v}_h \in K_h, \bar{\mu}_H \in \Lambda_H$ be such sequences that

$$\bar{v}_h \to u, h \to 0_+; \ \bar{\mu}_H \to \lambda, H \to 0_+.$$

The existence of such sequences is guaranteed by (5.13). Setting $v = u, \mu = \lambda, v_h = \bar{v}_h, \mu_H = \bar{\mu}_H$ in (5.12) and using the weak convergence of (u_h, λ_H) to (u, λ) proven above, we conclude that $\|u_h - u\| \to 0, h \to 0_+$. □

Remark 5.8. If $K_h \subset K \quad \forall h \in (0,1)$ or $\Lambda_H \subset \Lambda \quad \forall H \in (0,1)$, then the condition (5.15) or (5.16), respectively, is automatically fulfilled (see remark 3.7).

Theorem 5.4. *Let* $(CC), (CC_H)$ *and* (5.8) *be fulfilled and moveover, let*

$$\forall v \in V \ \exists v_h \in V_h : v_h \to v, h \to 0_+ \quad in\ V; \tag{5.26}$$

$$\forall \mu \in \Lambda \ \exists \mu_H \in \Lambda_H : \mu_H \to \mu, H \to 0_+ \ \ in\ \ L; \tag{5.27}$$

$$\mu_H \in \Lambda_H, \mu_H \rightharpoonup \mu, H \to 0_+ \ \ (weakly)\ in\ \ L \Rightarrow \mu \in \Lambda; \tag{5.28}$$

there is a real number d, *a positive constant* c *and a bounded sequence* $\{\bar{v}_h\}, \bar{v}_h \in V$, *such that*

$$j_H(v_h) \geq d \quad \forall v_h \in V_h, j_H(\bar{v}_h) \leq c \quad \forall h, H \in (0,1). \tag{5.29}$$

Let the solution (u, λ) *of problem* $\tilde{\mathcal{P}}$ *be unique. Then*

$$u_h \to u, h \to 0_+ \ \ in\ V,$$

$$\lambda_H \to \lambda, H \to 0_+ \ \ in\ \ L.$$

Proof. We will show that the sequences $\{u_h\}, \{\lambda_H\}$ are bounded. The rest of the proof then coincides with that of the preceding theorem.

As a consequence of (5.6), the first component of u_h satisfies the variational inequality

$$a(u_h, v_h - u_h) + j_H(v_h) - j_H(u_h) \geq \langle f, v_h - u_h \rangle \quad \forall v_h \in V_h.$$

In particular, putting $v_h = \bar{v}_h$ here and using (5.29) we obtain

$$\alpha \|u_h\|^2 + d \leq a(u_h, u_h) + j_H(u_h) \leq a(u_h, \bar{v}_h) + j_H(\bar{v}_h) - \langle f, \bar{v}_h - u_h \rangle,$$

which immediately implies the boundedness of $\{u_h\}$. Hence, from (5.10) we at once obtain the boundedness of $\{\lambda_H\}$. □

Remark 5.9. If $\Lambda_H \subset \Lambda \ \forall H \in (0,1)$, then (5.28) is automatically fulfilled.

1.1.53. Numerical Realization of Mixed Variational Formulations. In section 1.1.42 we mentioned Uzawa's method of solving a certain quadratic programming problem. Since this method is one of the most effective means of solution, let us discuss it in more detail. The reader will find a full account of this method in the book by Ekeland and Temam (1974), where the motivation for the algorithm given below can be found, as well as the proof of convergence.

Let V, L be two Hilbert space, $\mathcal{A} \subseteq V$, $\mathcal{B} \subseteq L$ nonempty convex closed subsets. Consider the Lagrangian $\mathcal{H} : V \times L \to \mathbf{R}$ defined by formula (5.3). Let the form a fulfill all the assumptions of section 1.1.51. Let (u, λ) be a saddle point of $\mathcal{H}$ in $\mathcal{A} \times \mathcal{B}$. Uzawa's method consists of constructing two sequences of elements $\{u^n\}, \{\lambda^n\}$, $u^n \in \mathcal{A}$, $\lambda^n \in \mathcal{B}$, according to the following rule: choose $\lambda^0 \in \mathcal{B}$ arbitrarily; use it to calculate $u^0 \in \mathcal{A}$, then λ^1, u^1, etc. If we know $\lambda^n \in \mathcal{B}$, then we look for $u_n \in \mathcal{A}$ such that

$$\mathcal{H}(u^n, \lambda^n) \leq \mathcal{H}(v, \lambda^n) \quad \forall v \in \mathcal{A}, \tag{5.30}$$

then replacing λ^n by λ^{n+1}:

$$\lambda^{n+1} = \Pi_{\mathcal{B}}(\lambda^n + \rho_n \Phi(u^n)), \quad n = 1, 2, \ldots. \tag{5.31}$$

Here, $\Pi_{\mathcal{B}}$ stands for the operator of projection of L to a convex closed subset $\mathcal{B}$, $\rho_n > 0$ is a given parameter, and $\Phi : V \mapsto L$ is the mapping defined by

$$(\mu, \Phi(v)) = b(v, \mu) - [g, \mu] \quad \forall (v, \mu) \in V \times L,$$

where $(\ ,\ )$ denotes the scalar product in L. Since $b : V \times L \to \mathbf{R}$ is continuous, we have

$$|\Phi(v) - \Phi(w)| \leq M_1 \|v - w\| \quad \forall v, w \in V.$$

Thus, all the assumptions of section 1, Chap. 7 of Ekeland and Temam (1974) are fulfilled, and hence we have:

Theorem 5.5. *There exist numbers ρ_1, ρ_2, $0 < \rho_1 < \rho_2$, such that for $\rho_n \in (\rho_1, \rho_2)$ the algorithm defined by the formulas* (5.30), (5.31) *is convergent in the following sense:*

$$u^n \to u, \quad n \to \infty \ \text{ in } V.$$

Remark 5.10. This theorem guarantees only the convergence to the first component u. The convergence of λ^n to λ will be dealt with in Chapter 2, in connection with the Signorini problem with friction.

In order to acquire a better comprehension of this and preceding sections, let us go back to the approximation of the problem (2.5). Let V_h, L_H, Λ_H be defined in the same way as in remark 5.1. Let (u_h, λ_H) be a saddle point of $\tilde{\mathcal{H}}$ in $V_h \times \Lambda_H$. In this case $K = V$, $K_h = V_h$, $\Lambda_H \subset \Lambda\ \forall H \in (0, 1)$ and it is again possible to verify the validity of each of the conditions (5.13)–(5.17). Hence, $u_h \to u$, $h \to 0_+$ in $H^1(\Omega)$, $\lambda_H \to \lambda$, $H \to 0_+$ in $L^2(\Gamma)$. Moreover, u_h is uniquely determined. As we know, a sufficient condition for the uniqueness of λ_H is (see remark 5.5 and the definition of $\tilde{\mathcal{H}}$):

$$\int_\Gamma v_h \mu_H \, ds = 0 \quad \forall v_h \in K_h^* = V_h \Rightarrow \mu_H = \Theta_L. \tag{5.32}$$

It is evident that this condition need *not* be generally fulfilled. Let $h = H$ (that is the partition $\mathcal{T}_H$ of the boundary Γ is generated by the boundary nodes of the triangulation $\mathcal{T}_h$ of the domain $\bar{\Omega}$), and let us assume that the nodes $a_1, \ldots, a_{m(h)}$ form an equidistant partition of Γ. Then, (5.32) is equivalent to the system of linear algebraic equations

$$\mu_1 + \mu_2 = 0$$

$$\mu_2 + \mu_3 = 0$$

$$\ldots$$

$$\mu_1 + \mu_m = 0, \quad \mu_i = \mu_H|_{a_i a_{i+1}}.$$

If m is an even number, then this system has a nontrivial solution. On the other hand, we easily find out when (5.32) is fulfilled. Let $M(H)$ be the number of segments of the partition $\mathcal{T}_H$, and $m(h)$ the number of boundary nodes of $\mathcal{T}_h$. Then, (5.32) represents a homogeneous system of $m(h)$ equations for $M(H)$ unknowns. Hence, it is sufficient that this system be "overdetermined" and therefore have solely the trivial solution.

To get an approximation of the saddle point (u_h, λ_H) of the Lagrangian $\tilde{\mathcal{H}}$ in $V_h \times \Lambda_H$, we use Uzawa's algorithm. In this particular case we set $V = V_h$, $L = L_H$, $\mathcal{A} = V_h$, $\mathcal{B} = \Lambda_H$, $\mathcal{H} = \tilde{\mathcal{H}}$. Now, we can write (5.30) and (5.31) in the following explicit form: choose $\lambda_H^0 \in \Lambda_H$ arbitrarily; use it to calculate $u_h^0 \in V_h$, then λ_H^1, u_h^1, etc. If we know $\lambda_H^n \in \Lambda_H$, then we find $u_h^n \in V_h$ such that

$$(\nabla u_h^n, \nabla v_h)_0 = (f, v_h)_0 - \int_\Gamma \lambda_H^n v_h ds \quad \forall v_h \in V_h; \tag{5.30'}$$

Then,

$$\lambda_H^{n+1} = \prod_{\Lambda_H} (\lambda^n - \rho_n u_h^n), \quad \rho_n > 0, \quad n = 1, 2, \ldots,$$

where

$$\prod_{\Lambda_H} (q)|_{b_i b_{i+1}} = \begin{matrix} 1 & \text{if } \pi_i(q) > 1 \\ \pi_i(q) & \text{if } \pi_i(q) \in [-1, 1]. \\ -1 & \text{if } \pi_i(q) < -1 \end{matrix}$$

Here $\pi_i(q)$ denotes the mean value of the function q in $b_i b_{i+1}$; that is,

$$\pi_i(q) = \frac{1}{|b_i b_{i+1}|} \int_{b_i b_{i+1}} q \, ds.$$

Thus, (5.30′) represents the problem of finding a solution for a system of linear equations in which only some components of the right-hand side change, while the matrix remains unchanged in the course of the iteration process. (5.31′) is the projection of the corresponding function into Λ_H. Effective solution methods for problems of this type will be discussed in Chapter 2. Theorem 5.5 implies the existence of ρ_1, ρ_2, $0 < \rho_1 < \rho_2$, such that $u_h^n \to u_h$, $n \to \infty$, for $\rho_n \in (\rho_1, \rho_2)$.

Remark 5.11. In order to determine the rate of convergence of (u_h, λ_H) to (u, λ), we need to verify condition (5.8). Let us assume that the problem

$$-\Delta v + v = 0 \quad \text{in } \Omega$$

$$\partial v / \partial n = \mu \quad \text{on } \Gamma$$

is regular in the following sense: for every $\mu \in H^{-1/2+\epsilon}(\Gamma)$, $\epsilon > 0$, its solution v fulfills $v \in H^{1+\epsilon}(\Omega)$ and

$$\|v\|_{1+\epsilon,\Omega} \leq c(\epsilon) \|\mu\|_{-1/2+\epsilon,\Gamma}.^7$$

Under these assumptions we can show (Haslinger and Lovišek (1980)) that there is $\tilde{\beta} > 0$ independent of $h, H > 0$ such that

$$\sup_{V_h} \frac{\int_\Gamma v_h \mu_H \, ds}{\|v_h\|_1} \geq \tilde{\beta} \|\mu_H\|_{-1/2,\Gamma} \quad \forall \mu_H \in L_H,$$

provided that h/H is "suitably" small. In the above quoted paper, the rate of convergence of (u_h, λ_H) to (u, λ) is analyzed in detail for problem $\mathcal{P}_2$. We will come back to these problems once more in the next chapter when studying contact problems with friction.

1.1.6 Semicoercive Problems

In this section, we will study a problem analogous to $\mathcal{P}_2$ except for the additional assumption $\Gamma_u = \emptyset$. Thus, let us consider the following one-sided boundary value problem:

$$-\Delta u = f \quad \text{in } \Omega,$$

[7] For the definition of the Sobolev spaces with a fractional derivative we refer the reader to Nečas (1967) and Aubin (1972).

$$u \geq 0, \quad \partial u/\partial \nu \geq 0, \quad u\partial/u\partial\nu = 0 \quad \text{on } \Gamma. \tag{6.1}$$

Set

$$K = \{v \in H^1(\Omega) \,|\, v \geq 0 \quad \text{a.e. on } \Gamma\},$$

$$\mathcal{Y}(v) = \frac{1}{2}|v|_1^2 - (f, v)_0, \quad f \in L^2(\Omega).$$

A function $u \in K$ will be called a *variational solution* of problem (6.1) if

$$\mathcal{Y}(u) \leq \mathcal{Y}(v) \quad \forall v \in K. \tag{6.2}$$

If a variational solution is sufficiently smooth, then it fulfills the point relations (6.1). The proof proceeds similarly to that of problem $\mathcal{P}_1$ in section 1.1.1.

Let us now deal with the problem of existence and uniqueness of (6.2). Since $\Gamma_u = \emptyset$, Friedrich's inequality does not hold, and hence the functional $\mathcal{Y}$ is *not* coercive in $H^1(\Omega)$. On the other hand, as follows from theorem 1.5, section 1.1.11, the coercivity of $\mathcal{Y}$ only on K is already sufficient for the existence of a solution. We shall now formulate conditions which guarantee this property.

First of all, let us show that a solution of the problem (6.2) cannot exist for arbitrary right-hand sides $f \in L^2(\Omega)$. Indeed, we have

Lemma 6.1. *If there exists a solution of problem* (6.2), *then*

$$(f, 1)_0 \leq 0. \tag{6.3}$$

Proof. (6.2) is equivalent to

$$(\nabla u, \nabla v - \nabla u)_0 \geq (f, v - u)_0 \quad \forall v \in K.$$

Substituting here the function $u + 1$ for v, where obviously $u + 1 \in K$, we immediately conclude (6.3).

Thus, a necessary condition for the existence of a solution is that the mean value of f in Ω be nonpositive. If the mean value of f is even negative, then we have

Theorem 6.1. *Let*

$$(f, 1)_0 < 0. \tag{6.4}$$

Then there is one and only one solution of (6.2).

Proof. Uniqueness. Let u_1, u_2 be solutions of (6.2), that is,

$$(\nabla u_i, \nabla v - \nabla u_i) \geq (f, v - u_i)_0 \quad \forall v \in K.$$

Substituting first the function u_2, and then u_1 for v, after subtracting the resulting inequalities we obtain

$$|u_2 - u_1|_1^2 \leq 0,$$

or $u_2 = u_1 + c$, $c \in R$. Let us assume $c \neq 0$. Then,

$$\mathcal{Y}(u_1 + c) = \mathcal{Y}(u_1) \Longrightarrow (f, u_1 + c)_0 = (f, u_1)_0 \Longrightarrow (f, 1)_0 = 0,$$

which contradicts (6.4). Hence, $c = 0$, that is, $u_1 = u_2$. □

Existence. Let $\Gamma_0 \subset \Gamma$ be an arbitrary open subset of the boundary Γ. Let us define

$$\bar{v} = (\text{mes}\ \Gamma_0)^{-1} \int_{\Gamma_0} \gamma\, v ds \quad \forall v \in H^1(\Omega),$$

and set

$$\tilde{v} = v - \bar{v}.$$

Then, evidently

$$\int_{\Gamma_0} \gamma\, \tilde{v} ds = 0,$$

and moreover, there exists a positive constant c such that (see Nečas (1967))

$$|\tilde{v}|_1 \geq c\|\tilde{v}\|_1. \tag{6.5}$$

Now,

$$\begin{aligned} \mathcal{Y}(v) = \frac{1}{2}|\tilde{v}|_1^2 - (f, \tilde{v})_0 &- \bar{v}(f, 1)_0 \geq \frac{1}{2}c^2\|\tilde{v}\|_1^2 - c_1\|\tilde{v}\|_1 \\ &- \bar{v}(f, 1)_0. \end{aligned}$$

Let $v \in K$, $\|v\|_1 \to \infty$. Then $\bar{v} \geq 0$ and at least one of the norms $\|\tilde{v}\|_1$, $\|\bar{v}\|_1 = \bar{v}(\text{mes}\,\Omega)^{1/2}$ increase to infinity. Hence, it follows from (6.5) that $\mathcal{Y}$ is coercive on K. The existence of a minimizing element is now a consequence of theorem 1.5.

Remark 6.1. While in problems $\mathcal{P}_1$, $\mathcal{P}_2$ the coercivity of $\mathcal{Y}$ in K was a consequence of the coercivity of $\mathcal{Y}$ in $H^1(\Omega)$ (and this again is a consequence of the $H^1(\Omega)$-ellipticity of the form a), the present situation is more complicated. The coercivity of $\mathcal{Y}$ in K is a consequence of the proper sign of the right-hand side f (in the sense of (6.4)), and the form a itself *is not* $H^1(\Omega)$-elliptic. Generally, if $|\cdot|$ stands for a seminorm in V, and if there is a positive constant α such that

$$a(v, v) \geq \alpha|v|^2 \quad \forall v \in V,$$

then we will use the attribute *semicoercive* when formulating problem $\mathcal{P}$, in contradistinction to *coercive* problems in which the form a is V-elliptic (i.e., (3.5) holds).
Let us turn to the remaining case, namely $(f,1)_0 = 0$.

Theorem 6.2. *Let*

$$(f,1)_0 = 0. \tag{6.6}$$

Let $w \in H^1(\Omega)$ be a weak solution of the Neumann problem

$$-\Delta w = f \quad in\ \Omega, \tag{6.7}$$

$$\frac{\partial w}{\partial \nu} = 0, \ on\ \Gamma, \quad \int_{\Gamma_0} \gamma\, w ds = 0,$$

where $\Gamma_0 \subset \Gamma$ is an arbitrary nonempty open subset of the boundary. Then (6.2) has a solution if and only if the trace of γw is bounded from below on Γ, and all the solutions of (6.2) have the form $u = w + c$, where $c \in R$ is such that $\gamma w + c \geq 0$ on Γ.

Proof. Let u be a solution of (6.2). Then, the Green formula (1.10) and (6.6) imply

$$\langle \frac{\partial u}{\partial \nu}, 1\rangle = (f,1)_0 = 0.$$

On the other hand, $\partial u/\partial \nu \geq 0$, and consequently, $\partial u/\partial \nu = 0$. The function $w = u - c$, where

$$c = (\operatorname{mes} \Gamma_0)^{-1} \int_{\Gamma_0} \gamma\, u ds,$$

is thus a solution of (6.7). The converse assertion follows analogously. □

Remark 6.2. If all the assumptions of theorem 6.2 are fulfilled, then there exist infinitely many solutions of (6.2), and all of them can be determined from a single solution of problem (6.7).

In the same way as in section 1.1.11, it is possible to derive the dual formulation of problem (6.2). Let us denote

$$\mathcal{U} = \{q \mid q \in H(\operatorname{div} \Omega), \ \operatorname{div} q + f = 0, \ q \cdot \nu \geq 0 \ \text{ on } \Gamma\},$$

$$\mathcal{S}(q) = \frac{1}{2} \sum_{i=1}^{n} \|q_i\|_0^2.$$

The problem

$$\text{find } q^0 \in \mathcal{U} \ : \ \mathcal{S}(q^0) \leq \mathcal{S}(q) \quad \forall q \in \mathcal{U} \tag{6.8}$$

will be called *dual* to (6.2). The relation between (6.2) and (6.8) is expressed by:

Theorem 6.3. *Let* (6.3) *hold and let problem* (6.2) *have a solution* u (*or* $u+c$ *for the case* (6.6)). *Then there is a unique solution* q^0 *of* (6.8), *and:*

$$q^0 = \nabla u,$$

$$\mathcal{Y}(u) + \mathcal{S}(q^0) = 0.$$

Proof. Follows the same lines as that of theorem 1.7. □

1.1.61. Solution of the Primal Problem by the Finite Element Method and the Error Bounds. In what follows we shall assume that $\Omega \subset \mathbf{R}^2$ is a *bounded polygonal* domain, $\{\mathcal{T}_h\}$, $h \to 0_+$ is a regular system of triangulations. Each $\mathcal{T}_h$ is associated with a finite-dimensional subspace

$$V_h = \{v \in C(\bar{\Omega}) \; |v_h|_T \in P_1(T) \quad \forall T \in \mathcal{T}_h\},$$

and a convex closed set K_h, given by the relation

$$K_h = \{v_h \in V_h \,|\, v_h(a_i) \geq 0 \quad \forall i = 1, \dots, m(h)\},$$

where $a_1, \dots, a_{m(h)}$ are the nodes of $\mathcal{T}_h$ which belong to Γ. Analogously to section 1.1.33, we mean by an approximation of (6.2) to find $u_h \in K_h$ such that

$$\mathcal{Y}(u_h) \leq \mathcal{Y}(v_h) \quad \forall v_h \in K_h. \tag{6.9}$$

Theorem 6.4. *Let*

$$(f, 1)_0 < 0.$$

Then for every $h > 0$ *there is exactly one solution* u_h *of problem* (6.9). *Moreover, there is a constant* $c > 0$ *such that*

$$\|u_h\|_1 \leq c \quad \forall h \in (0,1). \tag{6.10}$$

Proof. It follows from the definition of K_h that $K_h \subset K$ $\forall h \in (0,1)$, and hence the functional $\mathcal{Y}$ is coercive in each K_h. Now, the existence of a solution is a direct consequence of theorem 1.5. Uniqueness is proved in exactly the same way as in the continuous case. Let us prove (6.10). Similarly to theorem 3.3, we prove that the system $\{K_h\}$, $h > 0$ is complete in K, that is,

$$\forall v \in K \; \exists v_h \in K_h \;:\; \|v_h - v\|_1 \to 0, \; h \to 0_+.$$

Let $u \in K$ be a solution of (6.2). The existence of $\bar{v}_h \in K_h$ such that

$$\bar{v}_h \to u, \; h \to 0_+ \quad \text{in } H^1(\Omega) \tag{6.11}$$

follows from the above considerations. From (6.9) and the continuity of $\mathcal{Y}$ we obtain

$$\mathcal{Y}(u_h) \leq \mathcal{Y}(\bar{v}_h) \rightarrow \mathcal{Y}(u), \ h \rightarrow 0_+.$$

Consequently, the sequence $\{\mathcal{Y}(u_h)\}$, $h \in (0,1)$ is bounded from above. This, together with the coercivity of $\mathcal{Y}$ in K implies (6.10). □

Taking into account remark 3.9 and proceeding in the same way as in the proof of theorem 3.2, we obtain

Theorem 6.5. *Let*

$$(f,1)_0 < 0$$

and let a solution u of problem (6.2) *satisfy all the assumptions of theorem* 3.2. *Then,*

$$|u - u_h|_1 \leq c(u)h, \ \ h \rightarrow 0_+. \tag{6.12}$$

Remark 6.3. It follows from (6.10) and (6.12) that u_h converge to u, not only in the seminorm of $H^1(\Omega)$, but even in the norm of $H^1(\Omega)$. Indeed, as $\{u_h\}$ is bounded in $H^1(\Omega)$, there exists a subsequence $\{u_{h'}\} \subset \{u_h\}$ and an element $u^* \in H^1(\Omega)$, such that

$$u_{h'} \rightharpoonup u^*, \ h' \rightarrow 0_+ \quad \text{in } H^1(\Omega).$$

Since $K_h \subset K \ \forall h \in (0,1)$, we also have $u^* \in K$. Moreover, the functional $\mathcal{Y}$ is weakly lower semicontinuous in $H^1(\Omega)$, and consequently

$$\mathcal{Y}(u) \leq \mathcal{Y}(u^*) \leq \lim_{h \rightarrow 0_+} \inf \mathcal{Y}(u_h) \leq \lim_{h \rightarrow 0_+} \mathcal{Y}(\bar{v}_{h'}) = \mathcal{Y}(u),$$

where $\bar{v}_h \in K_h$ are elements with the property (6.11). As the solution of (6.2) is unique (in virtue of (6.4)), we conclude that $u^* = u$, and not only the subsequence but even the original sequence weakly converges to u in the norm of $H^1(\Omega)$. The identical mapping of $H^1(\Omega)$ into $L^2(\Omega)$ is completely continuous (see Nečas (1967)), hence,

$$u_h \rightarrow u, \ h \rightarrow 0_+ \quad \text{in } L^2(\Omega).$$

This, together with (6.12), implies the convergence of the approximate solutions to the exact one in the norm of $H^1(\Omega)$. However, here we lose the information concerning the rate of this convergence.

Remark 6.4. If the solution u is not regular (that is, we merely know $u \in K$), then we can prove the convergence of u_h to u in the norm of $H^1(\Omega)$ by virtue of the result contained in remark 3.9, setting there $V = H^1(\Omega)$, $H = L^2(\Omega)$.

A question which is much more interesting is that of the numerical realization of (6.9). It again leads to a problem of quadratic programming, which is at first sight analogous to the type of problem studied in section 1.1.31. Nevertheless, there is an essential difference, consisting of the fact that the stiffness matrix A is only *positive semidefinite*, and therefore the immediate application of the supperrelaxation method with additional projection does not necessarily lead to satisfactory results. Before suggesting how to proceed in this case, let us prove one important property of u_h.

Lemma 6.2. *Let* (6.4) *hold. Then,* $c_h \equiv \inf_\Gamma u_h = 0$.

Proof. Since $u_h \in K_h$, we have $c_h \geq 0$. Let us assume $c_h > 0$. Then we can find $\bar{c}_h > 0$, such that $u_h - \bar{c}_h > 0$ on Γ, and hence $u_h - \bar{c}_h \in K_h$. By virtue of (6.4) we have

$$\mathcal{Y}(u_h - \bar{c}_h) = \mathcal{Y}(u_h) + \bar{c}_h(f, 1)_0 < \mathcal{Y}(u_h),$$

which contradicts the assumption that $u_h \in K_h$ is a minimizing element for $\mathcal{Y}$ in K_h. □

This lemma implies that there exists at least one node $a_i \in \Gamma$ (we do not know it explicitly) such that $u_h(a_i) = 0$. This fact is the basis for a method of numerical solution. Let us arbitrarily choose a tangent point $\tilde{a}_i \in \Gamma$ and define

$$\tilde{V}_h = \{v_h \in V_h \mid v_h(\tilde{a}_i) = 0\},$$

$$K_h = \{v_h \in \tilde{V}_h \mid v_h \geq 0 \ \text{ on } \Gamma\}.$$

Now we replace problem (6.9) by

$$\text{find } \tilde{u}_h \in \tilde{K}_h \ : \ \mathcal{Y}(\tilde{u}_h) \leq \mathcal{Y}(\tilde{v}_h) \quad \forall \tilde{v}_h \in \tilde{K}_h. \tag{6.9'}$$

We easily verify that the stiffness matrix $\tilde{A}$ formed from the base elements of $\tilde{V}_h$ already is *positive definite* and results from A through deletion of the corresponding row and column of A. Thus, to solve problem (6.9′) we can again, for example, use the superrelaxation method with additional projection. Nevertheless, it is then necessary to verify whether $\tilde{a}_i$ was correctly chosen; that is, whether $u_h(a_i) = 0$ really holds. A critierion for this verification may be, for instance, the calculation of $\partial\tilde{u}_h/\partial\nu$ in a neighborhood of $\tilde{a}_i$. If $\partial\tilde{u}_h/\partial\nu < 0$, then it is necessary to choose another point, at which we fix the solution. However, the method of *conjugate gradients* (see Pšeničnii and Danilin (1975)) is much serviceable. This method makes it possible to find the minimum of the quadratic form given by a *positive semidefinite* matrix A with constraints of the form

$$(\beta_i, \alpha)_{\mathbf{R}^n} = b_i, \quad b_i, c_i \in \mathbf{R}_i$$

$$(\gamma_i, \alpha)_{\mathbf{R}^n} \leq c_i, \quad \gamma_i, \beta_i \in \mathbf{R}^n. \tag{6.13}$$

The method has several advantages. First, A is admitted to be *semidefinite*. Further, when constructing the matrix A we need not respect the boundary condition $\tilde{u}_h(\tilde{a}_i) = 0$ but we may include it directly among the equality constraints in (6.13). This is especially convenient if we do not succeed in choosing $\tilde{a}_i$ correctly on the first try. Moreover, when applying the conjugate gradients method, some dual quantities are obtained as extra results, and their signs enable us to find out easily whether the choice of $\tilde{a}_i$ was correct. We shall discuss this method in detail in the next chapter.

Remark 6.5. If $(f, 1)_0 = 0$, then the approximate solutions of (6.2) are obtained from the approximate solutions of the Neumann problem (6.7) (see theorem 6.2).

1.1.62. Solution of the Dual Problem by the Finite Element Method and Error Bounds. We replace the dual problem (6.8) by an equivalent one. Let $\lambda \in H(\text{div}, \Omega)$ be a particular solution of the equation

$$\text{div}\,\lambda + f = 0 \quad \text{in } \Omega.$$

Denote

$$g_0 = (f, 1)_0/\text{mes}\,\Gamma.$$

There exists a vector function $\lambda^f \in H\,(\text{div}, \Omega)$, such that

$$\text{div}\,\lambda^f + f = 0, \quad \lambda^f \cdot \nu = -g_0 \quad \text{on } \Gamma \tag{6.14}$$

(hence $\lambda^f \in \mathcal{U}$, for we assume the validity of the condition $(f, 1)_0 \le 0$, which is necessary for the existence of a solution). Indeed, we can write

$$\lambda^f = \lambda + z^0, \quad z^0 = \nabla w,$$

$$\Delta w = 0, \quad \frac{\partial w}{\partial \nu} = -\lambda \cdot \nu - g_0 \quad \text{on } \Gamma,$$

and $w \in H^1(\Omega)$ exists, for it is a solution of the Neumann problem and

$$\langle \lambda \cdot \nu + g_0, 1\rangle = \langle \lambda \cdot \nu, 1\rangle + (f, 1)_0 = (f + \text{div}\,\lambda, 1)_0 = 0.$$

If $\lambda \cdot \nu$ is piecewise linear on Γ, then we can find the vector z^0 even without solving the Neumann problem, namely, in the space $\mathcal{N}_h$ with a suitable triangulation $\mathcal{T}_h$ of the domain Ω; the following identity must be fulfilled:

$$z^0 \cdot \nu = -\lambda \cdot \nu - g_0 \quad \text{on } \Gamma.$$

If $\lambda \cdot \nu$ is not piecewise linear, then we can construct z^0 in the following way. Let a function $\omega \in H^2(\Omega)$ satisfy the boundary condition

$$\omega(s) = -\int_{s_0}^{s} (\lambda \cdot \nu + g_0)dt \quad \text{on } \Gamma.$$

Then, the vector

$$z^0 = \left\{ -\frac{\partial \omega}{\partial x_2}, \frac{\partial \omega}{\partial x_1} \right\}$$

satisfies the boundary condition

$$\frac{\partial \omega}{\partial s} = -\frac{\partial \omega}{\partial x_2}\nu_1 + \frac{\partial \omega}{\partial x_1}\nu_2 = z^0 \cdot \nu = -\lambda \cdot \nu - g_0.$$

In some cases the function ω can be constructed, for example, by the finite element method with polynomials of higher degrees on a suitable triangulation, the nodal parameters inside Ω being set equal to zero.

If we introduce the set

$$\mathcal{U}_0 = \{p \in H(\mathrm{div}, \Omega) \,|\, \mathrm{div}\, p = 0, \; (p + \lambda^f) \cdot \nu \geq 0 \;\text{ on } \Gamma\},$$

and the functional

$$J(p) = \frac{1}{2}\|p\|_0^2 + (\lambda^f, p)_0,$$

then the problem to find $p^0 \in \mathcal{U}_0$ such that

$$J(p^0) \leq J(p) \quad \forall p \in \mathcal{U}_0 \tag{6.15}$$

is equivalent to the original dual problem (6.8). The relationship between their solutions is expressed by the identity $q^0 = p^0 + \lambda^f$.

As in section 1.1.42, we will assume that the boundary Γ consists of a finite number of closed polygons:

$$\Gamma = \bigcup_{j=1}^{J} \partial\Omega_j, \quad \partial\Omega_j \cap \partial\Omega_k = \emptyset \quad \text{for } j \neq k.$$

Let us again consider a triangulation $\mathcal{T}_h$ of the domain Ω and an approximation of the set $\mathcal{U}_0$

$$\mathcal{U}_0^h = \mathcal{U}_0 \cap \mathcal{N}_h = \{p \in \mathcal{N}_h \,|\, p \cdot \nu \geq g_0 \quad \text{on } \Gamma\}.$$

We say that p^h is an approximation of problem (6.15) if

$$p^h \in \mathcal{U}_0^h, \quad J(p^h) \leq J(p) \quad \forall p \in \mathcal{U}_0^h. \tag{6.16}$$

If (6.4) holds, then problem (6.16) has a unique solution. Indeed, $\mathcal{U}_0^h$ is nonempty ($0 \in \mathcal{U}_0^h$), convex, and closed in $[L^2(\Omega)]^2$. $J(p)$ is strictly convex and continuous.

Lemma 6.3. *Let there exist $W^h \in \mathcal{U}_0^h$ such that $2p^0 - W^h \in \mathcal{U}_0$. Then,*

$$\|p^0 - p^h\|_0 \leq \|p^0 - W^h\|_0. \tag{6.17}$$

Proof. See lemma 4.1, in which we set $X = [L^2(\Omega)]^2$, $\mathcal{Y} = J$, $M = \mathcal{U}_0$, $M_h = \mathcal{U}_0^h$, $\alpha_0 = c = 1$. □

Lemma 6.4. *Let us assume that $(f,1)_0 < 0$, $p^0 \in [H^2(\Omega)]^2$, $p^0 \cdot \nu \in H^2(\Gamma_m)$ on each side Γ_m of the polygonal boundary, $m = 1, \ldots, \bar{m}$.*

Then, for sufficiently small h there exists a piecewise linear function ψ_h with nodes determined by the triangulation $\mathcal{T}_h$ (generally discontinuous at the nodes), such that

$$\int_{\partial\Omega_j} \psi_h ds = \int_{\partial\Omega_j} p^0 \cdot \nu \, ds, \quad j = 1, 2, \ldots, \bar{\jmath}, \tag{6.18}$$

$$g_0 \leq \psi_h \leq 2p^0 \cdot \nu - g_0 \quad \text{on } \Gamma, \tag{6.19}$$

$$\|\psi_h - (r_h p^0) \cdot \nu\|_{0,\Gamma} \leq Ch^2 \sum_{m=1}^{\bar{m}} |p^0 \cdot \nu|_{2,\Gamma_m}. \tag{6.20}$$

Remark 6.6. In contradistinction to the coercive problem $\mathcal{P}_2$, which was considered in section 1.1.42, we cannot use a one-sided approximation of the flow $p^0 \cdot \nu$ here. Indeed, if we set

$$g_0 \leq \psi_h \leq p^0 \cdot \nu \quad \text{on } \Gamma,$$

then (6.18) yields

$$0 \leq \int_\Gamma (p^0 \cdot \nu - \psi_h) ds = 0 \Longrightarrow \psi_h = p^0 \cdot \nu \quad \text{on } \Gamma,$$

which is generally impossible, as $p^0 \cdot \nu$ need not be piecewise linear.

Proof of Lemma 6.4. For brevity, let us denote $p^0 \cdot \nu = t$. By the definition of the mapping r_h, the linear function $(r_h p^0) \cdot \nu$ is defined as the $L^2(S)$-projection of t to the subspace $P_1(S)$ on each side $S \subset \Gamma$, $S \in \mathcal{T}_h$. Further, let us denote $(r_h p^0) \cdot \nu = t_h$.

The condition $(f,1)_0 < 0$ implies the following property of the solution of the primal problem: there exists a set $E \subset \Gamma$, mes $E > 0$, such that $\partial u/\partial \nu > 0$ on E, $\partial u/\partial \nu = 0$ on $\Gamma \dot{-} E$. (Notice also that $\partial u/\partial \nu = \lambda^0 \cdot \nu = (\lambda^f + p^0) \cdot \nu = -g_0 + p^0 \cdot \nu \in H^2(\Gamma_m)$ on each Γ_m.) Consequently,

$$t = p^0 \cdot \nu = \lambda^0 \cdot \nu + g_0 > g_0 \quad \text{on } E,$$

$$t = g_0 \quad \text{on } \Gamma \dot{-} E.$$

The assumption $t \in H^2(\Gamma_m)$ implies that $t \in C^1(\bar{\Gamma}_m)$ for all m, hence,

$$\text{supp}(t - g_0) = \bigcup_{m=1}^{\bar{m}} \bigcup_j I_j^{(m)},$$

where $I_j^{(m)} \subset \Gamma_m$ are closed intervals of positive lengths.

Let us consider an arbitrary interval $I_j^{(m)} \equiv \langle \sigma, \bar{\sigma} \rangle$ and let $s_0 \leq \sigma < s_1 < \bar{\sigma}$, where $\langle s_{k-1}, s_k \rangle$ corresponds to the side $S_k \in \mathcal{T}_h$, $k = 1, 2, \dots$. Set $\psi_h = g_0$ on $\langle s_0, s_2 \rangle$. (If $\lim_{j\to\infty}(\text{mes}\, I_j^{(m)}) = 0$, $I_j^{(m)} = \langle \sigma_j, \bar{\sigma}_j \rangle$, $\sigma_j \to \sigma$, $\bar{\sigma}_j \to \sigma$, $\lim_{s\to\sigma+}(t - g_0)(s) = 0$, then we also set $\psi_h = g_0$ on a suitable interval $\langle s_0, s_k \rangle$, where $t(s_k) > g_0$.)

Let $t - g_0 > 0$ at all vertices $Q_k \in \mathcal{T}_h$, with parameters $s_1 < s_2 < \dots < s_{n-1} < \bar{\sigma}$ and let $\bar{\sigma} \leq s_n$. Set $\psi_h = g_0$ on $\langle s_{n-2}, s_n \rangle$ and $\psi_h = t_h + a_j$ in the intervals $\langle s_{k-1}, s_k \rangle$, $k = 3, 4, \dots, n-2$, where

$$a_j = (s_{n-2} - s_2)^{-1} \left\{ \int_\sigma^{s_2} (t - g_0)ds + \int_{s_{n-2}}^{\bar{\sigma}} (t - g_0)ds \right\}, \tag{6.21}$$

(provided $s_{n-2} > s_2$).

There exists a point $\theta \in \langle \sigma, s_2 \rangle$ such that

$$\int_\sigma^{s_2} (t - g_0)ds = (t - g_0)(\theta)(s_2 - \sigma),$$

and

$$(t - g_0)(\xi) = \int_\sigma^\xi \frac{d^2t}{ds^2}(s)(\xi - s)ds \leq (2h)^{3/2}\|t''\|_{0,\Gamma_m} \quad \forall \xi \in \langle \sigma, s_2 \rangle, \tag{6.22}$$

(where $t'' = d^2t/ds^2$). This yields an upper bound for the first of the integrals in (6.21). The other integral can be similarly estimated. Hence, we have

$$a_j \leq 2^{5/2}(s_{n-2} - s_2)^{-1}h^{5/2}\|t''\|_{0,\Gamma_m}.$$

Denoting $l_j = \bar{\sigma} - \sigma$, we have

$$(s_{n-2} - s_2)^{-1} \leq (l_j - 4h)^{-1} \leq 2/l_j$$

for sufficiently small h. Without loss of generality we may consider only a finite number of intervals $I_j^{(m)}$, hence

$$l_j \geq \min\, l_j = c > 0.$$

(If $l_j \to 0$ for $j \to \infty$, then we replace the interval $I_j^{(m)}$ by a suitable union $\cup_{j=k}^\infty I_j^{(m)}$.) Thus, we obtain

$$a_j \leq 2^{7/2}c^{-1}h^{5/2}\|t''\|_{0,\Gamma_m}, \tag{6.23}$$

where c is independent of h.

Let us consider the interval $\langle s_0, s_2\rangle = S_1 \cup S_2$. We have

$$\|\psi_h - t_h\|_{0,S_i} \le \|g_0 - t\|_{0,S_i} + \|t - t_h\|_{0,S_i}, \quad i = 1, 2.$$

By virtue of (6.22) we obtain

$$\|t - g_0\|_{0,S_i} \le Ch^2\|t''\|_{0,\Gamma_m},$$

and a similar estimate is valid for $\|t - t_h\|_{0,S_i}$ as well (see the proof of lemma 4.5). Hence, we have

$$\|\psi_h - t_h\|_{0,(s_0,s_2)} \le Ch^2\|t''\|_{0,\Gamma_m}, \tag{6.24}$$

as well as a similar estimate for the interval (s_{n-2}, s_n). Combining (6.23) and (6.24) we conclude

$$\|\psi_h - t_h\|^2_{0,(s_0,s_n)} = \|\psi_h - t_h\|^2_{0,(s_0,s_2)} + \|\psi_h - t_h\|^2_{0,(s_{n-2},s_n)}$$
$$+ \int_{s_2}^{s_{n-2}} a_j^2\, ds \le (2Ch^4 + C_1 l_j h^5)\|t''\|^2_{0,\Gamma_m} \le C_2 h^4 |t|^2_{2,\Gamma_m}.$$

Further, set $\psi_h = g_0$ on $\Gamma_m \dot{-} \cup_j I_j^{(m)}$. Due to the finite number of the intervals involved, we obtain the same estimate for $\|\psi_h - t_h\|^2_{0,\Gamma_m}$ as well, and this again yields (6.20).

The relation (6.21) together with

$$\int_{S_k} (t_h - t)ds = 0 \quad \forall S_k \subset \Gamma$$

implies

$$\int_{\sigma}^{\bar\sigma} (\psi_h - t)ds = \int_{\sigma}^{s_2} (g_0 - t)ds + \int_{s_{n-2}}^{\bar\sigma} (g_0 - t)ds + a_j(s_{n-2} - s_2) = 0,$$

hence (6.18). The inequalities (6.19) are also fulfilled, provided h is sufficiently small. □

Theorem 6.6. *Let $(f,1)_0 < 0$, and let the assumptions of lemma 6.4 be fulfilled. Denote $q^h = \lambda^f + p^h$, $q^0 = \lambda^f + p^0$, where λ^f satisfies (6.14), and p^0 and p^h are solutions of problems (6.15) and (6.16), respectively. Then, for an (α,β)-regular system of triangulations we have the bound*

$$\|q^h - q^0\|_{0,\Omega} \le Ch^{3/2}\{|p^0|_{2,\Omega} + \sum_{m=1}^{\bar m} |p^0 \cdot \nu|_{2,\Gamma_m}\}.$$

Proof. Let ψ_h be an approximation of the flow from lemma 6.4. Set

$$\varphi = (r_h p^0) \cdot \nu - \psi_h \equiv t_h - \psi_h.$$

By lemma 4.6 there exists $w^h \in \mathcal{N}_h$ such that $w^h \cdot \nu = \varphi$ on Γ,

$$\|w^h\|_{0,\Omega} \leq C h^{-1/2} \|\varphi\|_{0,\Gamma}, \tag{6.25}$$

since

$$\int_{\partial\Omega_j} (\psi - t_h) ds = \int_{\partial\Omega_j} (\psi_h - t) ds = 0 \quad \forall j$$

by virtue of (6.18). The function $W^h = r_h p^0 - w^h$ fulfills the conditions of lemma 6.3. Indeed, $W^h \in \mathcal{N}_h$, $W^h \cdot \nu = t_h - \varphi = \psi_h \geq g_0$ on Γ, hence, $W^h \in \mathcal{U}_0^h$. Now (6.19) implies

$$W^h \cdot \nu \leq 2p^0 \cdot \nu - g_0 \quad \text{on } \Gamma \Longrightarrow (2p^0 - W^h) \cdot \nu - g_0 \geq 0,$$

hence $2p^0 - W^h \in \mathcal{U}_0$. By (4.26), (6.25), and (6.20) we obtain

$$\|p^0 - W^h\|_0 \leq \|p^0 - r_h p^0\|_0 + \|r_h p^0 - W^h\|_0 \leq \|p^0 - r_h p^0\|_0$$

$$+ \|w^h\|_0 \leq C h^2 \|p^0\|_{2,\Omega} + C_1 h^{3/2} \sum_{m=1}^{\bar{m}} |p^0 \cdot \nu|_{2,\Gamma_m}.$$

Now the assertion of the theorem follows from (6.17). □

Remark 6.7. As concerns *the algorithm of solution* of problem (6.16), we can essentially say the same as when dealing with the solution of the coercive dual problem (4.29) (see section 1.1.42).

A Posteriori Error Bounds and Two-Sided Energy Bounds

Theorem 6.7. *Let $\tilde{u}_h \in K_h$ be an approximation of the primal problem, and $q^{Hm} = \lambda^f + p^{Hm}$, $p^{Hm} \in \mathcal{U}_0^H$, be an approximation of the dual problem. Then,*

$$|\tilde{u}_h - u|_1^2 \leq \|q^{Hm} - \nabla \tilde{u}_h\|_0^2 + 2 \int_\Gamma q^{Hm} \cdot \nu \, \tilde{u}_h ds \equiv E(q^{Hm}, \tilde{u}_h),$$

$$\|q^{Hm} - \nabla u\|_0^2 \leq E(q^{Hm}, \tilde{u}_h).$$

Proof. Similar to that of theorem 4.2. □

Remark 6.8. The condition $p^{Hm} \in \mathcal{U}_0^H$ is not fulfilled for some quadratic programming methods (for example, Uzawa's method).

Theorem 6.8. *Let $\tilde{u} \in K_h$, $q^{Hm} = \lambda^f + p^{Hm}$, $p^{Hm} \in \mathcal{U}_0^H$. Then,*

$$-2\mathcal{L}(\tilde{u}_h) \leq |u|_1^2 = (f, u)_0 \leq 2\mathcal{S}(q^{Hm}).$$

Proof. Similar to that of theorem 4.3. □

1.1.63. Convergence of the Dual Finite Element Method. In the previous section we established an a priori error bound, which obviously implies the convergence of approximations in $[L^2(\Omega)]^2$. However, it was necessary to assume that the solution, (or as the case may be, its part p^0) is sufficiently regular. Generally, an appropriate degree of regularity cannot be expected. Thus, a question arises as to whether the dual method converges at all, provided the solution fails to be sufficiently regular.

In this section we will answer the above question—at least for a certain class of convex domains Ω. To this end, we will use the fundamental theorem 3.1 on the convergence of Ritz–Galerkin approximations, setting

$$V = Q_0(\Omega) = \{p \in H(\mathrm{div}, \Omega) \mid \mathrm{div}\, p = 0\},$$

$$\|p\| = \|p\|_{0,\Omega}, \quad K = \mathcal{U}_0, \quad K_h = \mathcal{U}_0^h, \quad \mathcal{Y} = J,$$

$$u = p^0, \quad u_h = p^h.$$

The main problem is to prove condition (3.8) or (3.8′). We will first prove the existence of a smooth function

$$\tilde{p} \in \mathcal{U}_0 \cap [C^\infty(\bar{\Omega})]^2,$$

which is arbitrarily close to p^0 in the space $Q_0(\Omega)$.

Theorem 6.9. *Let Ω be a convex polygonal domain such that the sum of any two adjacent angles is not less than π. Let $\lambda^f \in [H^1(\Omega)]^2$. Then, for every $\eta > 0$ there exists $\tilde{p} \in \mathcal{U}_0 \cap [C^\infty(\bar{\Omega})]^2$ such that*

$$\|p^0 - \tilde{p}\|_{0,\Omega} < \eta. \tag{6.26}$$

Remark 6.9. If Ω is a convex polygonal domain, then the solution u of the primal problem (6.2) belongs to $H^2(\Omega)$ (see Grisvard and Ioos (1976)). Then, $p^0 = \nabla u - \lambda^f \in H^1(\Omega)$.

The proof of theorem 6.9 is based on several lemmas.

Lemma 6.5. *Let the assumptions of Theorem 6.9 be fulfilled. Let Ω^* be a bounded polygonal domain such that $\Omega^* \supset \bar{\Omega}$, the sides of $\partial\Omega^*$ being parallel to those of $\partial\Omega$, and let us denote $G = \Omega^* \dot{-} \bar{\Omega}$.*

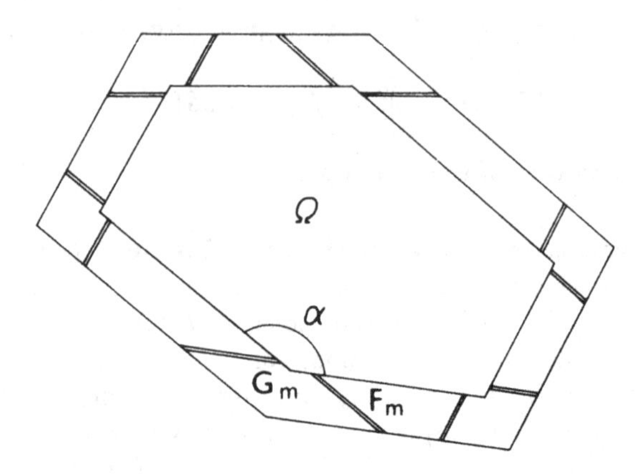

Figure 4

Let $k_0 > 1$ and $\mathcal{H}_0 > 0$ be real parameters; let the origin of the coordinate system (x_1, x_2) belong to Ω. Then there exists an extension $p \in Q_0(G)$ of the function p^0, such that

$$p \cdot \nu = p^0 \cdot \nu \quad on\ \Gamma, \tag{6.27}$$

$$p(z) \cdot \nu(x) \geq g_0 \tag{6.28}$$

holds for almost all $x \in \Gamma$ and $z \in G$ satisfying $|z - kx| < k\mathcal{H}_0$, $k \in [1, k_0)$.

Proof. Let us define neighborhoods G_m of the vertices of the boundary Γ such that the entire boundary strip G is divided by line segments, parallel to the sides of Γ, into subdomains G_m and trapezoids F_m, $m = 1, \ldots, M$ (see Figure 4).

1^0 Consider any domain G_m and introduce oblique coordinates by the affine transformation

$$y = \mathcal{F}(x) \equiv \begin{array}{l} y_1 = x_1 \sin \alpha - x_2 \cos \alpha, \\ y_2 = x_2. \end{array}$$

The system (y_1, y_2) corresponds to the new basis $\{e^1 \sin^{-1} \alpha, e^2 \sin^{-1} \alpha\}$ (see Figure 5), where e^1, e^2 are unit tangent vectors. Then the domain Ω is mapped into the first quadrant $\{y_1 > 0, y_2 > 0\}$, and the rays Γ_1 and Γ_2 into the positive halfaxes y_1 and y_2, respectively.

Every vector p admits the representation

$$p = \sum_{j=1}^{2} p^{(j)} e^j \sin^{-1} \alpha.$$

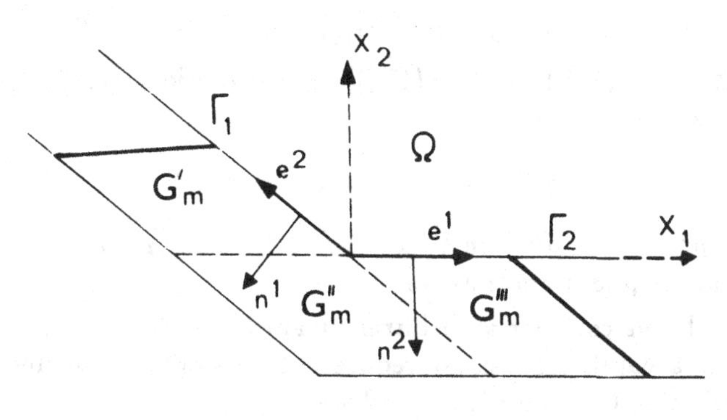

Figure 5

Let n^j be the unit outer normals on Γ_j. Since $e^j \cdot n^j = -\sin\alpha$, we have

$$p^j = -p \cdot n^j, \quad j = 1, 2.$$

(These are the coordinates of the contravariant vector.)

By definition, $p \in Q_0(G_m)$ provided

$$\int_{G_m} p \cdot \nabla v dx = 0 \quad \forall v \in C_0^\infty(G_m).$$

By means of the mapping $\mathcal{F}^{-1}$, which is inverse to $\mathcal{F}$, we obtain

$$\int_{G_m} p \cdot \nabla v dx = \sin^{-1}\alpha \int_{\mathcal{F}(G_m)} \sum_{j=1}^{2} p^{(j)} \frac{\partial \tilde{v}}{\partial y_j} dy, \tag{6.29}$$

where $\tilde{v}(y) = v(\mathcal{F}^{-1}(y))$, $\tilde{v} \in C_0^\infty(\mathcal{F}(G_m))$. This implies that if the integral on the right-hand side of (6.29) vanishes for all $\tilde{v} \in C_0^\infty(\mathcal{F}(G_m))$, then the corresponding vector $p = (p_1, p_2) \in Q_0(G_m)$.

By remark 6.9, $p^0 = q^0 - \lambda^f = \nabla u - \lambda^f \in [H^1(\Omega)]^2$, and hence the traces $\gamma p_i^0 \in H^{1/2}(\Gamma)$, $i = 1, 2$. Then,

$$\gamma p^{0(j)} = -\gamma p^0 \cdot n^j \in H^{1/2}(\Gamma_j), \quad j = 1, 2,$$

and we can set

$$\left.\begin{aligned} p^{(1)}(y_1, y_2) &= \gamma p^{0(1)}(y_2) \\ p^{(2)}(y_1, y_2) &= -g_0 \end{aligned}\right. \quad \text{in } \mathcal{F}(G'_m) \text{ (for } y_1 < 0, y_2 > 0), \tag{6.30}$$

$$\left.\begin{aligned} p^{(1)} &= -g_0 \\ p^{(2)} &= \gamma p^{0(2)}(y_1) \end{aligned}\right. \quad \text{in } \mathcal{F}(G'''_m) \text{ (for } y_1 > 0, y_2 < 0), \tag{6.31}$$

$$p^{(1)} = p^{(2)} = -g_0 \quad \text{in } \mathcal{F}(G''_m) \text{ (for } y_1 < 0, y_2 < 0). \tag{6.32}$$

Evidently $(p^{(1)}, p^{(2)}) \in Q_0(\mathcal{F}(G_m))$; the conditions (6.27), (6.28) are fulfilled for $z \in G_m$, since

$$-\gamma p^{0(j)} \geq g_0 \quad \text{on } \Gamma_j, \quad j = 1, 2. \tag{6.33}$$

Indeed, $p^0 \cdot \nu = -\gamma p^{0(j)}$ on Γ_j and $(p^0 + \lambda^f) \cdot \nu = p^0 \cdot \nu - g_0 \geq 0$ holds a.e. on Γ, hence (6.33) holds as well.

2^0 Now let us consider an arbitrary trapezoidal domain F_m, $1 \leq m \leq M$. If F_m is a parallelogram, we reduce it to a single line segment $\bar{G}_m \cap \bar{G}_{m+1}$ by dilating G_m and G_{m+1}, and we set

$$p^{(1)} = 0 \quad \text{in } \mathcal{F}(G''_m \cup G'''_m),$$

$$\tilde{p}^{(2)} = 0 \quad \text{in } \mathcal{F}(G'_{m+1} \cup G''_{m+1})$$

in (6.30), (6.31), (6.32), where $\tilde{p}^{(j)}$ are components of the vector $\tilde{p}$ in the local basis of G_{m+1}. Thus, we obtain the continuity of the flow $p \cdot \nu$ on the segment $\bar{G}_m \cap \bar{G}_{m+1}$.

Thus, let us consider a general trapezoid F_m. Let us introduce a new coordinate system (y_1, y_2) by the mapping

$$x = Ty \equiv \begin{matrix} x_1 = y_1 y_2 \\ x_2 = y_2 \end{matrix} \tag{6.34}$$

where the origin of the local Cartesian system (x_1, x_2) is at the point of intersection of the lines $\bar{A}\bar{B}$ and $\bar{C}\bar{D}$ (see Figure 6). Then the trapezoid

$$F_m = \left\{ x \in \mathbf{R}^2 \,|\, a < \frac{x_1}{x_2} < b, \;\; 0 < c < x_2 < d \right\}$$

is the image $T\mathcal{R}_m$ of the rectangle

$$\mathcal{R}_m = \{ y \in \mathbf{R}^2 \,|\, a < y_1 < b, \;\; c < y_2 < d \}.$$

Then we have

$$\int_{F_m} p \cdot \nabla v \, dx = \int_{\mathcal{R}_m} \sum_{i=1}^{2} \tilde{p}^i \frac{\partial \tilde{v}}{\partial y_i} y_2 \, dy$$

$$= -\int_{\mathcal{R}_m} \tilde{v} \left(y_2 \frac{\partial \tilde{p}^1}{\partial y_1} + y_2 \frac{\partial \tilde{p}^2}{\partial y_2} + \tilde{p}^2 \right) dy,$$

where $v(T(y)) \equiv \tilde{v}(y) \in C_0^\infty(\mathcal{R}_m)$ provided $v \in C_0^\infty(F_m)$, and

$$\tilde{p}^i = \sum_{k=1}^{2} p_k \frac{\partial y_i}{\partial x_k}, \quad p_i = \sum_{k=1}^{2} \tilde{p}^k \frac{\partial x_i}{\partial y_k}.$$

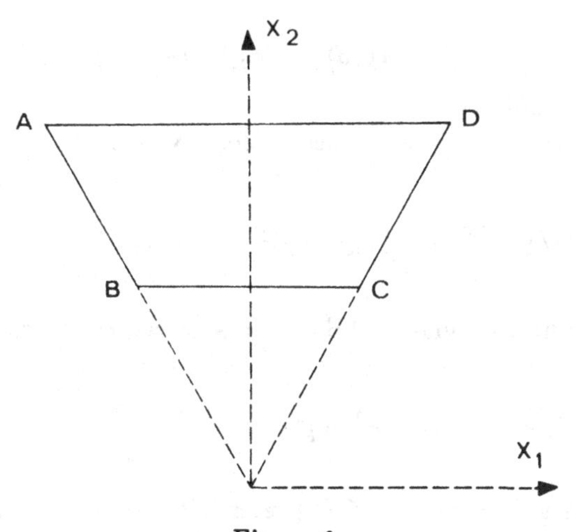

Figure 6

Evidently,

$$y_2\left(\frac{\partial\tilde{p}^1}{\partial y_1}+\frac{\partial\tilde{p}^2}{\partial y_2}\right)+\tilde{p}^2=0\quad\text{in } \mathcal{R}_m \Longrightarrow p\in Q_0(F_m). \tag{6.35}$$

Moreover, if

$$-p_2=p\cdot\nu|_{AD}=-\tilde{p}^2\geq g_0\quad\text{in } \mathcal{R}_m \tag{6.36}$$

and if p is defined by (6.30)–(6.32) in G_m, G_{m+1}, then (6.28) holds for $z\in G_m\cup F_m\cup G_{m+1}$.

Let us define

$$\tilde{p}^2=\gamma\tilde{p}^{02}(y_1)+\frac{\mathcal{A}}{b-a}\left(\frac{d}{y_2}-1\right), \tag{6.37}$$

$$\tilde{p}^1=-g_0\frac{(1+a^2)^{1/2}}{y_2}+\int_a^{y_1}\left[-\frac{\partial\tilde{p}^2}{\partial y_2}-\frac{1}{y_2}\tilde{p}^2(t,y_2)\right]dt, \tag{6.38}$$

where

$$\mathcal{A}=g_0(\sqrt{1+a^2}+\sqrt{1+b^2})+\int_a^b\gamma\tilde{p}^{02}(y_1)dy_1,$$

$$\gamma\tilde{p}^{02}(y_1)=\gamma p_2^0(x_1),\quad x_1=y_1d.$$

Then, it is easily verified that condition (6.35) holds. Further, as $\gamma\tilde{p}^{02}\leq -g_0$, we have

$$\mathcal{A}\leq g_0(\sqrt{1+a^2}+\sqrt{1+b^2}-(b-a))<0.$$

Hence,

$$\tilde{p}^2 \leq \gamma p_2^0(y_1 d) \leq -g_0, \quad y \in \mathcal{R}_m,$$

and condition (6.36) is fulfilled.

Finally, let us evaluate the values of the flow $p \cdot \nu$ on AB and CD. For $y_1 = a$ we have

$$\tilde{p}^{(1)} = -\frac{g_0}{y_2}\sqrt{1+a^2} \Rightarrow -p \cdot n^1 = p^{(1)} = \tilde{p}^{(1)} \frac{y_2}{\sqrt{1+a^2}} = -g_0, \tag{6.39}$$

where n^1 is the unit normal to AB (outer with respect to F_m). For $y_1 = b$ we obtain

$$\tilde{p}^{(1)} = \frac{g_0}{y_2}\sqrt{1+b^2} \Rightarrow -p \cdot \bar{n}^2 = \bar{p}^{(2)} = -\tilde{p}^{(1)} \frac{y_2}{\sqrt{1+b^2}} = -g_0, \tag{6.40}$$

where $\bar{n}^2$ is the unit normal to CD (outer with respect to F_m), and $\bar{p}^2$ is the component of $\bar{p}$ in the local basis of the domain G_{m+1}.

Comparing (6.31) with (6.39) and (6.30) with (6.40), we conclude that the flow $p \cdot \nu$ passes continuously across the lines AB and CD.

In this way we can construct a vector field p which satisfies all the conditions required by lemma 6.5. □

Let us now define an extension of p^0:

$$Ep^0 = \begin{matrix} p^0 & \text{in } \Omega \\ p & \text{in } G. \end{matrix} \tag{6.41}$$

It is easily seen that $Ep^0 \in Q_0(\Omega^*)$.

Further, put $k = 1 + \epsilon$, $0 \leq \epsilon < k_0 - 1$ and define

$$p^\epsilon(y) = Ep^0(ky) \quad \text{for } y \in k^{-1}\Omega^*. \tag{6.42}$$

By regularizing the field p^ϵ by means of the kernel $\omega_{\mathcal{H}}$ (whose support is in a ball with radius $\mathcal{H}$—see Nečas (1967)), we obtain the field

$$\tilde{p}(x) = \mathbf{R}_{\mathcal{H}} p^\epsilon(x) = \int_{|x-y|<\mathcal{H}} \omega_{\mathcal{H}}(x-y) p^\epsilon(y) dy. \tag{6.43}$$

Lemma 6.6. *Let all the assumptions of theorem 6.9 be fulfilled. Let us define $\tilde{p}$ by (6.41)–(6.43), where p is an extension of lemma 6.5. Then, $\tilde{p} \in \mathcal{U}_0 \cap [C^\infty(\bar{\Omega})]^2$ and the inequality (6.26) holds, provided both ϵ and $\mathcal{H}(\epsilon)$ are sufficiently small.*

Proof. 1^0 Denote

$$\Omega_e = \{x \mid \text{dist}(x, \bar{\Omega}) < e\}, \quad \bar{\Omega}_e \subset \Omega^*.$$

Then, $p^\epsilon \in Q_0(\Omega_e)$ for ϵ sufficiently small. Indeed, let $v \in H_0^1(\Omega_e)$ and let $\tilde{v}(y) = v(y/k)$. Since $k\bar{\Omega}_e \subset \Omega^*$ for ϵ sufficiently small and $\tilde{v} \in H_0^1(k\Omega_e)$, we can continue $\tilde{v}$ by zero, obtaining $P\tilde{v} \in H_0^1(\Omega^*)$. Then

$$\begin{aligned}\int_{\Omega_e} p^\epsilon \cdot \nabla v dx &= \int_{\Omega_e} Ep^0(kx) \cdot \nabla v(x) dx \\ &= \frac{1}{k}\int_{k\Omega_e} Ep_i^0(y)\frac{\partial \tilde{v}}{\partial y_i} dy = \frac{1}{k}\int_{\Omega} Ep^0 \cdot \nabla P\tilde{v} dy = 0,\end{aligned}$$

by virtue of $Ep^0 \in Q_0(\Omega^*)$. Hence, $p^\epsilon \in Q_0(\Omega_e)$. Further, we can evaluate

$$\frac{\partial \tilde{p}_i}{\partial x_i}(x) = -\int_{|x-y|<\varkappa} \frac{\partial \omega_\varkappa(x-y)}{\partial y_i} p_i^\epsilon(y) dy, \quad x \in \Omega.$$

If $\varkappa < e$, then $\omega_\varkappa(x - \cdot) \in C_0^\infty(\Omega_e)$, and thus div $\tilde{p}(x) = 0$.

2^0 By virtue of

$$g_0 = \int_{|x-y|<\varkappa} \omega_\varkappa(x-y) g_0 dy \quad \forall x \in \Gamma,$$

we can write

$$\nu_i(x)\tilde{p}_i(x) - g_0 = \int_{|x-y|<\varkappa} \omega_\varkappa(x-y)[\nu_i(x)p_i^\epsilon(y) - g_0] dy.$$

Evidently

$$\Gamma \subset k^{-1}G \quad \forall k = 1 + \epsilon > 1.$$

Consequently, if $\varkappa < \operatorname{dist}(\Gamma, k^{-1}\Gamma)$ and $\varkappa < \varkappa_0$, then

$$|y - x| < \varkappa \Longrightarrow ky \in G, \quad |ky - kx| < k\varkappa < k\varkappa_0,$$

and by lemma 6.5, (6.28), we obtain

$$\nu_i(x)p_i^\epsilon(y) = \nu_i(x)p_i(ky) \geq g_0.$$

Since $\omega_\varkappa \geq 0$, this implies $\nu \cdot \tilde{p} - g_0 \geq 0$ on the whole boundary Γ; hence $\tilde{p} \in \mathcal{U}_0 \cap [C^\infty(\bar{\Omega})]^2$.

3^0 We still have to prove (6.26). There exists a sequence $p^n \in [C_0^\infty(\Omega)]^2$ such that $\|p^n - p^0\|_{0,\Omega} \to 0$ for $n \to \infty$. Let $E_0 p_i^n$ be the extension of p_i^n $(i = 1, 2)$ by zero outside Ω; define $(p^n)_i^\epsilon(y) = E_0 p_i^n(ky)$. Then the following estimates are valid:

$$\|p_i^n - (p^n)_i^\epsilon\|_{0,\Omega} \leq \epsilon C_1(p_i^n),$$

$$\|p_i - (p^n)_i^\epsilon\|_{0,\Omega} \le \|p_i^n - p_i^0\|_{0,\Omega} + \|Ep_i^0\|_{0,k\Omega \dot{-} \Omega}.$$

(For the detailed proofs we refer the reader to the paper Hlaváček (1980b).) If we use these estimates and the identity $Ep^0 = p$ from lemma 6.5, we infer

$$\|p^\epsilon - p^0\|_0 \le \|p^\epsilon - (p^n)^\epsilon\|_0 + \|(p^n)^\epsilon - p^n\|_0 + \|p^n - p^0\|_0$$
$$\le 3\|p^n - p^0\|_0 + \epsilon C_2(p^n) + \|p\|_{k\Omega \dot{-} \Omega}. \tag{6.44}$$

For a given $\eta > 0$ we can find p^n and ϵ (depending on p^n) such that each of the three terms on the right-hand side of (6.44) will be less than $\frac{1}{4}\eta$. Finally, we choose $\mathcal{H}$ (depending on p^ϵ) sufficiently small and such that

$$\|\mathbf{R}_{\mathcal{H}} p^\epsilon - p^\epsilon\|_0 < \frac{1}{4}\eta. \tag{6.45}$$

By virtue of (6.43–6.45), we obtain the inequality (6.26). This completes the proofs of both lemma 6.6 and theorem 6.9. □

Theorem 6.10. *Let Ω fulfill the assumptions of theorem 6.9, let $\lambda^f \in [H^1(\Omega)]^2$ and $(f,1)_0 < 0$. Then, for any (α,β)-regular system of triangulations and $q^0 = \lambda^f + p^0$, $q^h = \lambda^f + p^h$ we have*

$$\|q^0 - q^h\|_{0,\Omega} \to 0, \quad h \to 0.$$

Proof. According to theorem 3.1 and remark 3.5 it suffices only to verify the condition (3.8′). Applying theorem 6.9, we obtain a field $\tilde{p} \in \mathcal{U}_0 \cap [C^\infty(\bar{\Omega}]^2$, which satisfies (6.26).

Now let us apply lemmas 6.4 and 4.6 to $\tilde{p}$ in order to construct a field $W^h \in \mathcal{U}_0^h$, such that

$$\|\tilde{p} - W^h\|_0 \le C(\tilde{p})h^{3/2}.$$

Then we can write

$$\|p^0 - W^h\|_0 \le \|p^0 - \tilde{p}\|_0 + \|\tilde{p} - W^h\|_0 \le \eta + C(\tilde{p})h^{3/2}.$$

The condition (3.8′) is thus fulfilled, provided we choose $v_h = W^h$, $\tilde{K} = \mathcal{U}_0 \cap [H^1(\Omega)]^2$. □

1.1.7 Problems with Nonhomogeneous Boundary Obstacle

In this section we generalize the results from sections 1.1.1, 1.1.3, and 1.1.4 to problems with nonhomogeneous obstacles on the boundary. Thus, we shall consider a model problem $\mathcal{P}_g$:

$$-\Delta u + u = f \quad \text{in } \Omega \subset \mathbf{R}^n,$$

with boundary conditions

$$u - g \geq 0, \quad \frac{\partial u}{\partial \nu} \geq 0, \quad (u-g)\frac{\partial u}{\partial \nu} = 0 \quad \text{on } \partial\Omega \equiv \Gamma,$$

where f and g are given functions. Let $f \in L^2(\Omega)$.

Let us assume that there exists a function $G \in H^2(\Omega)$ such that $G = g$ on the boundary $\partial\Omega$. (Some sufficient conditions for the existence of the function G can be found in the paper by Jakovlev (1961).)

Let us define the set

$$\mathcal{K}_g = \{v \in H^1(\Omega) \mid \gamma v \geq g \quad \text{on } \Gamma\},$$

and the functional

$$\mathcal{L}_1(v) = \frac{1}{2}\|v\|_1^2 - (f, v)_0.$$

Then the problem: find $u \in \mathcal{K}_g$ such that

$$\mathcal{L}_1(u) \leq \mathcal{L}_1(v) \quad \forall v \in \mathcal{K}_g \tag{7.1}$$

is the primal variational formulation of problem $\mathcal{P}_g$.

The relation between $\mathcal{P}_g$ and this variational formulation can be proved in the same way as for problem $\mathcal{P}_1$ in section 1.1.1.

In order to formulate the dual variational problem, let us recall the set of admissible vector fields (see 1.1.11),

$$\mathcal{U}_1 = \{q \mid q \in [L^2(\Omega)]^{n+1}, \quad q = [\bar{q}, q_{n+1}], \quad \bar{q} \in H(\text{div}, \Omega),$$
$$q_{n+1} = f + \text{div}\, \bar{q}, \quad \bar{q} \cdot \nu \geq 0 \quad \text{on } \Gamma\}$$

and introduce the functional—the complementary energy—

$$\mathcal{S}_g(q) = \frac{1}{2}\sum_{i=1}^{n+1} \|q_i\|_0^2 - \langle \bar{q} \cdot \nu, g \rangle.$$

The problem: find $q^0 \in \mathcal{U}_1$ such that

$$\mathcal{S}_g(q^0) \leq \mathcal{S}_g(q) \quad \forall q \in \mathcal{U}_1 \tag{7.2}$$

is called the *dual variational formulation* with respect to problem $\mathcal{P}_g$.

Problem (7.1) (as well as (7.2)) has a unique solution, which follows from theorem 1.5 (see the proof of theorem 1.4). The dual problem (7.2) is transformed by the substitution $q_{n+1} = f + \text{div}\, \bar{q}$ to an equivalent one: find such $\bar{q}^0 \in \mathcal{U}_0$ that

$$\mathcal{I}_g(\bar{q}^0) \leq \mathcal{I}_g(\bar{q}) \quad \forall \bar{q} \in \mathcal{U}_0, \tag{7.3}$$

where

$$\mathcal{U}_0 = \{\bar{q} \in H(\text{div}, \Omega) \mid \bar{q} \cdot \nu \geq 0 \quad \text{on } \Gamma\},$$

$$I_g(\bar{q}) = \frac{1}{2}\|\bar{q}\|^2_{H(\text{div},\Omega)} + (f, \text{div } \bar{q})_0 - \langle \bar{q} \cdot \nu, g \rangle.$$

Theorem 7.1. *If u is a solution of the primal problem (7.1) and q^0 a solution of the dual problem (7.2), then*

$$q_i^0 = \partial u / \partial x_i, \quad i = 1, \ldots, n, \tag{7.4}$$

$$q_{n+1}^0 = u,$$

$$S_g(q^0) + \mathcal{L}_1(u) = 0. \tag{7.5}$$

Proof. Proceeds analogously to that of theorem 1.6, where S_1 is replaced by the functional S_g, and K_1 by the set $\mathcal{K}_g$. Instead of lemma 1.1, we then have

$$\inf_{v \in \mathcal{K}_g} \mathcal{H}_2(v, q) = \begin{cases} \mathcal{H}_2(G, q) & \text{for } q \in \mathcal{U}_1, \\ -\infty & \text{for } q \notin \mathcal{U}_1. \end{cases}$$

Indeed, $v \in \mathcal{K}_g \Longleftrightarrow v = G + w$, $w \in K_1$. Therefore, we can write

$$\inf_{v \in \mathcal{K}_g} \mathcal{H}_2(v, q) = \mathcal{H}_2(G, q) + \inf_{w \in K_1} \mathcal{H}_2(w, q),$$

and apply lemma 1.1.

For $q \in \mathcal{U}_1$ we have $q_{n+1} - f = \text{div } \bar{q}$; hence,

$$\mathcal{H}_2(G, q) = \int_\Omega (\bar{q} \cdot \nabla G + (q_{n+1} - f)G)dx = \langle \bar{q} \cdot \nu, g \rangle.$$

Now we can write

$$S_0(q) = \inf_{K_h \times M} \mathcal{H}(v, \mathcal{N}; q) = -\frac{1}{2}\|q\|^2 + \langle \bar{q} \cdot \nu, g \rangle = -S_g(q) \text{ for } q \in \mathcal{U}_1,$$

$$S_0(q) = -\infty \quad \text{for } q \notin \mathcal{U}_1,$$

$$\mathcal{L}_1(u) \geq \sup_M S_0(q) = \sup_{\mathcal{U}_1}[-S_g(q)] = -\inf_{\mathcal{U}_1} S_g(q) = -S_g(q^0). \tag{7.6}$$

If we set $q = \hat{q} = [\nabla u, u]$, then again as in the proof of theorem 1.6, we find $\hat{q} \in \mathcal{U}_1$. Indeed, according to theorem 1.1, inequality (1.8) holds for all $v \in \mathcal{K}_g$. Substituting $v - u = \pm\varphi$, $\varphi \in C_0^\infty(\Omega)$, we find that $\Delta u = u - f$, hence $\text{div } \hat{q} = \hat{q}_{n+1} - f$. Insert $u = G + \omega$, $v = G + w$ in (1.8), where $\omega \in K_1$, $w \in K_1$. Then

$$(u, w - \omega)_1 \geq (f, w - \omega)_0 \quad \forall w \in K_1 \tag{7.7}$$

and setting $w = 0$, $w = 2\omega$ we obtain

$$(u, \omega)_1 = (f, \omega)_0, \tag{7.8}$$

$$(u, w)_1 \geq (f, w)_0 \quad \forall w \in K_1. \tag{7.9}$$

Hence for all $w \in K_1$,

$$\begin{aligned} \langle \hat{q} \cdot \nu, \gamma w \rangle &= (\nabla u, \nabla w)_0 + (w, \Delta u)_0 = (\nabla u, \nabla w)_0 + (w, u)_0 \\ &- (f, w)_0 = (u, w)_1 - (f, w)_0 \geq 0. \end{aligned}$$

Consequently, $\hat{q} \in \mathcal{U}_1$.

Further, using (7.8) we obtain

$$\begin{aligned} \langle \frac{\partial u}{\partial \nu}, \gamma u - g \rangle &= (\nabla u, \nabla \omega)_0 + (\omega, \Delta u)_0 \\ &= (\nabla u, \nabla \omega)_0 + (\omega, u - f)_0 = (u, \omega)_1 - (f, \omega)_0 = 0; \end{aligned}$$

hence

$$\begin{aligned} \langle \hat{q} \cdot \nu, g \rangle = \langle \frac{\partial u}{\partial \nu}, \gamma u \rangle &= (\nabla u, \nabla u)_0 + (u, \Delta u)_0 \\ = (\nabla u, \nabla u)_0 + (u, u - f)_0 &= \|u\|_1^2 - (f, u)_0. \end{aligned}$$

Substituting $q = \hat{q}$ into S_g, we obtain

$$-S_g(\hat{q}) = -\frac{1}{2}\|\hat{q}\|^2 + \langle \hat{q} \cdot \nu, g \rangle$$

$$= -\frac{1}{2}\|u\|_1^2 + \|u\|_1^2 - (f, u)_0 = \mathcal{L}_1(u). \tag{7.10}$$

By virtue of (7.6), this implies that the functional $(-S_g)$ assumes its maximum at the point $\hat{q}$. The uniqueness of problem (7.2) implies $\hat{q} = q^0$. Relation (7.5) follows from this and (7.10). □

1.1.71. Approximation of the Primal Problem. Let us now consider problem $\mathcal{P}_g$ in a planar polygonal domain $\Omega \subset \mathbf{R}^2$, whose vertices will be denoted by $A_1, \ldots, A_n$. Let $\{\mathcal{T}_h\}$, $h \to 0_+$ be a regular system of triangulations of $\bar{\Omega}$. Each $\mathcal{T}_h$ will again be associated with a finite-dimensional space

$$V_h = \{v_h \in C(\bar{\Omega}) \,|\, v_h|_T \in P_1(T) \quad \forall T \in \mathcal{T}_h\},$$

and a convex closed set

$$\mathcal{K}_{gh} = \{v_h \in V_h \,|\, v_h(a_i) \geq g(a_i), \quad i = 1, \ldots, m(h)\},$$

where a_i $i = 1, \ldots, m(h)$ are the nodes of the triangulation $\mathcal{T}_h$, which lie on Γ. The approximation of problem $\mathcal{P}_g$ is defined in the usual way: find $u_h \in \mathcal{K}_{gh}$ such that

$$\mathcal{L}_1(u_h) \leq \mathcal{L}_1(v_h) \quad \forall v_h \in \mathcal{K}_{gh}. \tag{7.11}$$

As $\mathcal{L}_1$ fulfills all the assumptions of theorem 1.5 on $\mathcal{K}_{gh}$, for every $h > 0$ there exists a unique solution u_h of problem (7.11).

Let us study the relation between u, u_h. The situation is now more complicated due to the fact that $\mathcal{K}_{gh}$ are generally not subsets of $\mathcal{K}_g$. First, we will deal with the rate of convergence of u_h to u, provided the solution u is sufficiently smooth.

Theorem 7.2. *Let the solution u fulfill $u \in H^2(\Omega) \cap \mathcal{K}_g$, $u \in W^{1,\infty}(\Gamma)$ and let the set of points from Γ at which $u = g$ changes to $u > g$ be finite. Moreover, let $g \in H^2(A_i, A_{i+1})$, $i = 1, \ldots, n$. Then*

$$\|u - u_h\|_1 \leq c(u, g)h, \quad h \to 0_+.$$

Proof. As $\mathcal{K}_{gh}$ generally is an external approximation, we will establish an error bound via formula (3.10), which for our particular case reads

$$\|u - u_h\|_1^2 \leq (f, u - v_h)_0 + (f, u_h - v)_0 + (u_h - u, v_h - u)_1$$
$$+ (u, v - u_h)_1 + (u, v_h - u)_1 \quad \forall v \in \mathcal{K}_g,\ v_h \in \mathcal{K}_{gh}. \tag{7.12}$$

Now, using the Green's formula (1.10) and taking into account the pointwise relations fulfilled by the solution u and formulated at the beginning of this section, we can write (7.12) in the form

$$\|u - u_h\|_1^2 \leq (u - u_h, u - v_h)_1$$
$$+ \int_\Gamma \frac{\partial u}{\partial \nu}(v_h - u)ds + \int_\Gamma \frac{\partial u}{\partial \nu}(v - u_h)ds \quad \forall v \in \mathcal{K}_g,\ v_h \in \mathcal{K}_{gh}. \tag{7.13}$$

The first and second terms on the right-hand side of (7.13) are estimated in the same way as in theorem 3.2. Let us estimate the third term. To this end let us define on the boundary Γ a function w by

$$w = \sup\{g, u_h\}.$$

It is possible to show that this function can be extended from Γ to Ω in the sense that there exists a function $v \in H^1(\Omega)$ such that $v = w$ on Γ. Since $w \geq g$ on Γ, we have $v \in \mathcal{K}_g$. Now we shall use this function to establish a bound for the third term. By the definition of w,

$$\int_\Gamma \frac{\partial u}{\partial \nu}(v - u_h)ds = \int_{\Gamma_-} \frac{\partial u}{\partial \nu}(g - u_h)ds,$$

where

$$\Gamma_- = \{x \in \Gamma \mid u_h(x) < g(x)\}.$$

Taking into account that $u_h(a_i) \geq g(a_i)\ \forall i = 1, \ldots, m(h)$, which follows from the definition of $\mathcal{K}_{gh}$, we infer that $u_h \geq R_h g$, where $R_h g$ is the piecewise linear Lagrangian interpolation of the function g on Γ. We simultaneously have $\partial u/\partial \nu \geq 0$ a.e. on Γ and consequently

$$\int_{\Gamma_-} \frac{\partial u}{\partial \nu}(g - u_h)ds \leq \int_{\Gamma_-} \frac{\partial u}{\partial \nu}(g - R_h g)ds \leq \left\| \frac{\partial u}{\partial \nu} \right\|_{L^2(\Gamma)} \|g - R_h g\|_{L^2(\Gamma)}$$

$$\leq ch^2 \left\| \frac{\partial u}{\partial \nu} \right\|_{L^2(\Gamma)} \sum_{i=1}^{n} |g|_{H^2(A_i, A_{i+1})}.$$

Thus, all three terms on the right-hand side of (7.13) are of the order $0(h^2)$. This together with (7.13) implies the assertion of the theorem. □

We can also deal with the problem of convergence (resigning on its rate) in the case of *nonregular solution*. In that case, we must verify not only (3.8), but also (3.9). Let us formulate only the result: a detailed proof can be found, for example, in Haslinger (1977).

Theorem 7.3. *Given an arbitrary regular system of triangulations, we have*

$$\|u - u_h\|_1 \to 0, \quad h \to 0_+.$$

The superrelaxation method with additional projection (see section 1.1.32) can be successfully applied to solve the problem (7.11) numerically. The individual variables, which have the meaning of an approximate solution at the given node, either are subjected to no constraint at all (for internal nodes of $\mathcal{T}_h$), or belong to the interval $[g(a_i), +\infty)$, $i = 1, \ldots, m(h)$, provided the index i corresponds to a boundary node of $\mathcal{T}_h$.

1.1.72. Solution of the Dual Problem by the Finite Element Method. We shall start with the equivalent dual problem (7.3). Let us consider an (α, β)-regular system of triangulations $\{\mathcal{T}_h\}$ of a given domain $\Omega \subset \mathbf{R}^2$ and the space V_h of piecewise linear functions continuous on $\bar{\Omega}$. Let us introduce the subset

$$\mathcal{U}_{0h} = \mathcal{U}_0 \cap [V_h]^2 = \{q \in [V_h]^2 \mid q \cdot \nu \geq 0 \ \text{ on } \Gamma\}$$

and define: by an approximation $q^h \in \mathcal{U}_{0h}$ of the dual problem we mean a solution of the problem

$$I_g(q^h) \leq I_g(\bar{q}) \quad \forall \bar{q} \in \mathcal{U}_{0h}. \tag{7.14}$$

This problem has a unique solution, since $\mathcal{U}_{0h}$ is convex and closed in $H(\text{div}, \Omega)$, I_g is quadratic and strictly convex. The last term of the functional I_g reduces to the integral

$$-\langle \bar{q} \cdot \nu, g \rangle = -\int_\Gamma \bar{q} \cdot \nu g ds \quad \forall \bar{q} \in [V_h]^2. \tag{7.15}$$

As concerns the error $\bar{q}^0 - q^h$, we can apply the method from section 1.1.41, only adding the term (7.15) to the functional I. Hence, we obtain

Theorem 7.4. *Let us assume $\bar{q}^0 \in [H^2(\Omega)]^2$ and $\bar{q}^0 \cdot \nu \in H^2(\Gamma_m)$ on each side Γ_m of the polygonal boundary.*

Then

$$\sum_{i=1}^{2} \|\bar{q}_i^0 - q_i^h\|_0^2 + \|\text{div}(\bar{q}^0 - q^h)\|_0^2)^{1/2} = 0(h)$$

holds for any (α, β)-regular system of triangulations.

Remark 7.1. If q^h is a solution of problem (7.14), then

$$\lambda^h = \{q_1^h, q_2^h, f + \text{div}\, q^h\} \in \mathcal{U}_1$$

is an approximation of the original dual problem (7.2). By virtue of theorems 7.1 and 7.4, we obtain, under the assumptions listed above,

$$\sum_{i=1}^{2} \left\| q_i^h - \frac{\partial u}{\partial x_i} \right\|_0 = 0(h), \quad \|\text{div}\, q^h + f - u\|_0 = 0(h).$$

1.1.73. A Posteriori Error Bounds and a Two-Sided Energy Estimate. Let us assume that we have evaluated approximations $\tilde{u}_h \in \mathcal{K}_g$ and $\tilde{q}^H \in \mathcal{U}_{0H}$ of the primal and the dual problem, respectively. (Notice that if $g_h \geq g$ is not valid on Γ, then $\mathcal{K}_{gh}$ is not a subset of $\mathcal{K}_g$, and hence the above defined approximation by the finite element cannot be used!) Then, error bounds for both primal and dual approximations can be established.

Theorem 7.5. *Let $\tilde{u}_h \in \mathcal{K}_g$ and $\tilde{q}^H \in \mathcal{U}_{0H}$. Then*

$$\|\tilde{u}_h - u\|_1^2 \leq \sum_{i=1}^{2} \left\| \tilde{q}_i^H - \frac{\partial \tilde{u}_h}{\partial x_i} \right\|_0^2 + \|f + \text{div}\, \tilde{q}^H - \tilde{u}_h\|_0^2$$

$$+2 \int_\Gamma \tilde{q}^H \cdot \nu(\tilde{u}_h - g) ds \equiv E(\tilde{q}^H, \tilde{u}_h),$$

$$\sum_{i=1}^{2} \left\| \tilde{q}_i^h - \frac{\partial u}{\partial x_i} \right\|_0^2 + \|f + \operatorname{div} \tilde{q}^H - u\|_0^2 \leq E(\tilde{q}^H, \tilde{u}_h).$$

Proof. Analogous to that of theorem 4.2 in section 1.1.412. It is based on the variational inequality (1.8) and on the relation of the extrema (7.5). □

Remark 7.2. If we know the function G which continues g inside the domain Ω, then we can put

$$\tilde{u}_h = u_h + G - G_I$$

in theorem 7.5, where $G_I \in V_h$ is the Lagrangian linear interpolation of the function G on the triangulation $\mathcal{T}_h$.

Theorem 7.6. *Under the assumptions of theorem 7.5 we have*

$$2\mathcal{L}_1(G) - 2\mathcal{L}_1(\tilde{u}_h) \leq \|u - G\|_1^2 = (f, u - G)_0 - (G, u - G)_1$$

$$\leq \sum_{i=1}^{2} \left\| \tilde{q}_i^H - \frac{\partial G}{\partial x_i} \right\|_0^2 + \|f - G + \operatorname{div} \tilde{q}^h\|_0^2.$$

Proof. Analogous to that of theorem 4.3 and makes use of the identities (7.9), (7.5). □

Remark 7.3. If $f = 0$, then theorem 7.6 easily yields a bound for the norm $\|u\|_1^2$, since

$$\|u\|_1^2 = \|u - G\|_1^2 + \|G\|_1^2 + 2(u - G, G)_1 = \|G\|_1^2 - \|u - G\|_1^2.$$

1.2 Problems with Inner Obstacles for Second-Order Operators

1.2.1 Primal and Dual Variational Formulations

We shall now leave the one-sided boundary value problems, where the conditions are given on the boundary in the form of inequalities, and we shall engage in another type of problem, in which the inequality conditions are prescribed in the whole domain Ω. For simplicity, let us consider the following variational problem:

$$\text{find } u \in K \text{ such that}$$

$$\mathcal{Y}(u) \leq \mathcal{Y}(v) \quad \forall v \in K, \tag{1.1}$$

where

$$\mathcal{Y}(v) = \frac{1}{2}|v|_1^2 - (f, v)_0, \quad f \in L^2(\Omega),$$

$$K = \{v \in H_0^1(\Omega) \mid v \geq \varphi \quad \text{a.e. in } \Omega\}.$$

Further, we assume $\varphi \in C(\bar{\Omega})$, $\varphi \leq 0$ on Γ. Thus, it is immediately seen that K is a nonempty convex closed subset of $H_0^1(\Omega)$. Theorem 1.5 implies that *there is exactly one solution u*. Moreover, this solution can equivalently be characterized as the element $u \in H_0^1(\Omega)$ satisfying

$$(\nabla u, \nabla v - \nabla u)_0 \geq (f, v - u)_0 \quad \forall v \in K. \tag{1.1'}$$

Assuming that the solution u is sufficiently smooth, taking into account the above inequality and applying the Green formula, we find that the function u almost everywhere fulfills the following pointwise relations:

$$-\Delta u \geq f, \quad (-\Delta u - f)(u - \varphi) = 0,$$

$$u \geq \varphi \quad \text{a.e. in } \Omega, \quad u = 0 \quad \Gamma. \tag{1.2}$$

The domain Ω can be divided into two subsets Ω_0, Ω_+, where

$$\Omega_0 = \{x \in \Omega \mid u(x) = \varphi(x)\},$$

$$\Omega_+ = \{x \in \Omega \mid u(x) > \varphi(x)\}.$$

If $x \in \Omega_+$, then (1.2) implies that $-\Delta u(x) = f(x)$. Notice that the partition of Ω into Ω_0, Ω_+ is not a priori known. Therefore, this partition is one of the unknowns of the problem considered.

Definition 1.1. The problem (1.1) will be called the *primal variational formulation* of the problem with an inner obstacle. Its classical (pointwise) formulation is given by (1.2).

For the same reasons as in section 1.1.11, we will formulate problem (1.2) in terms of the gradient components. In what follows *we assume* $\varphi \in H_0^1(\Omega)$. Let

$$Q_f^- = \{q \in [L^2(\Omega)]^2 \mid (q, \nabla v)_0 \geq (f, v)_0 \quad \forall v \in K_0\},$$

where

$$K_0 = \{v \in H_0^1(\Omega) \mid v \geq 0 \quad \text{a.e. on } \Omega\}.$$

Remark 1.1. We easily find that

$$q \in Q_f^- \Longleftrightarrow \operatorname{div} q + f \leq 0 \quad \text{in } \Omega$$

in the sense of distributions.

Let us define the functional of the complementary energy by

$$S(q) = \frac{1}{2}\|q\|_0^2 - (q, \nabla\varphi)_0,$$

and let us consider the problem

$$\text{find } q^* \in Q_f^- \text{ such that}$$

$$S(q^*) \leq S(q) \quad \forall q \in Q_f^-. \tag{1.3}$$

The mutual relation of solutions to problems (1.1) and (1.3) is described in

Theorem 1.1. *There exists exactly one solution q^* of the problem* (1.3). *Moreover,*

$$q^* = \nabla u, \tag{1.4}$$

where u is the solution of the primal variational formulation (1.1).

Proof. Q_f^- is a nonempty convex closed subset of $[L^2(\Omega)]^2$. S is strictly convex, coercive, and continuous in $[L^2(\Omega)]^2$. The existence and uniqueness of the minimizing element q^* follows from theorem 1.5. Let us prove (1.4).

First, we show that $\nabla u \in Q_f^-$. As $\varphi \in H_0^1(\Omega)$, we have $K = \varphi + K_0$, that is, every element $v \in K$ can be written in the form

$$v = \varphi + v^*, \quad v^* \in K_0. \tag{1.5}$$

In particular,

$$u = \varphi + u^*, \quad u^* \in K_0. \tag{1.6}$$

Substituting for v into (1.1′) successively $v = \varphi$, $v = \varphi + 2u^*$, which certainly are elements from K, we obtain

$$(\nabla u, \nabla u^*)_0 = (f, u^*)_0. \tag{1.7}$$

This together with (1.5), (1.6), and (1.1′) then implies

$$(\nabla u, \nabla v^*)_0 \geq (f, v^*)_0 \quad \forall v^* \in K_0,$$

that is, $\nabla u \in Q_f^-$.

Let us now prove

$$S(\nabla u) \leq S(q) \quad \forall q \in Q_f^-,$$

or equivalently

$$(\nabla u, q - \nabla u)_0 \geq (q - \nabla u, \nabla\varphi)_0 \quad \forall q \in Q_f^-.$$

By virtue of (1.7) we obtain

$$(\nabla u, q - \nabla u)_0 - (q - \nabla u, \nabla \varphi)_0 = (q, \nabla(u - \varphi))_0 - (\nabla u, \nabla(u - \varphi))_0$$
$$= (q, \nabla u^*)_0 - (\nabla u, \nabla u^*)_0 = (q, \nabla u^*)_0 - (f, u^*)_0 \geq 0,$$

since $u^* \in K_0$ and $q \in Q_f^-$. As q^* is uniquely determined, the identity $q^* = \nabla u$ obviously holds. □

Definition 1.2. Problem (1.3) will be called the *dual variational formulation* of the problem with an inner obstacle.

Remark 1.2. Let us introduce still another equivalent formulation of problem (1.3), which will be useful for an approximation of the original problem. Let $\bar{q} \in [L^2(\Omega)]^2$ be such that

$$\operatorname{div} \bar{q} = -f \quad \text{in } \Omega.$$

Then evidently

$$Q_f^- = \bar{q} + Q_0^-,$$

where

$$Q_0^- = \{q \in [L^2(\Omega)]^2 \mid (q, \nabla v)_0 \geq 0 \quad \forall v \in K_0\}$$

is the set of vector functions whose divergence (in the sense of distributions) is nonpositive. Denote

$$\bar{S}(q) = S(\bar{q} + q), \quad q \in Q_0^-.$$

Let $q^0 \in Q_0^-$ be the element for which $\bar{S}$ assumes its minimum in Q_0^- (such an element exists and is unique, as $\bar{S}$ again is a quadratic, strictly convex, and continuous functional in Q_0^-):

$$\bar{S}(q^0) \leq \bar{S}(q) \quad \forall q \in Q_0^-. \tag{1.8}$$

Then $\bar{q} + q^0$ minimizes S in Q_f^-, that is, $q^* = \bar{q} + q_0$.

Remark 1.3. Let a solution u of problem (1.2) be smooth enough to guarantee $\Delta u = \operatorname{div}(\nabla u) \in L^2(\Omega)$. Then instead of Q_f^- we can consider the set $Q_f^-(\operatorname{div}, \Omega)$ defined by

$$Q_f^-(\operatorname{div}, \Omega) = \{q \in H(\operatorname{div}, \Omega) \mid (\operatorname{div} q + f, v)_0 \leq 0 \quad \forall v \in K_0\}.$$

The definition of the operator of divergence implies the following equivalent version of the functional S:

$$\tilde{S}(q) \equiv S(q) = \frac{1}{2}\|q\|_0^2 + (\operatorname{div} q, \varphi)_0 \quad \forall q \in Q_f^-(\operatorname{div}, \Omega).$$

As an exercise, the reader can prove the following assertion: let the solution u of (1.1) satisfy $\Delta u \in L^2(\Omega)$.

Then $q^* = \nabla u \in H(\mathrm{div}, \Omega)$ minimizes $\tilde{S}$ in $Q_f^-(\mathrm{div}, \Omega)$, that is,

$$\tilde{S}(q^*) \leq \tilde{S}(q) \quad \forall q \in Q_f^-(\mathrm{div}, \Omega).$$

The proof is analogous to that of theorem 1.1.

1.2.2 The Mixed Variational Formulation

Problem (1.3) represents a minimization problem with a constraint. This constraint can be removed by introducing the Lagrange multiplier in a suitable way. Let us now describe this procedure.

Let us define the Lagrangian $\mathcal{H}$ in $[L^2(\Omega)]^2 \times H_0^1(\Omega)$ by

$$\mathcal{H}(q, v) = \frac{1}{2}\|q\|_0^2 - (q, \nabla\varphi)_0 + (f, v)_0 - (q, \nabla v)_0,$$

$$(q, v) \in [L^2(\Omega)]^2 \times H_0^1(\Omega). \tag{2.1}$$

Theorem 2.1. *There exists exactly one saddle point (q^*, v^*) of the Lagrangian $\mathcal{H}$ in $[L^2(\Omega)]^2 \times K_0$, and*

$$v^* = u - \varphi, \quad q^* = \nabla u, \tag{2.2}$$

where u is a solution of the problem (1.1).

Proof. (i) First we will show that every saddle point of $\mathcal{H}$ in $[L^2(\Omega)]^2 \times K_0$ satisfies (2.2). Let (q^*, v^*) be a saddle point of $\mathcal{H}$ in the set just mentioned. Then necessarily

$$(q^*, q)_0 - (\nabla\varphi, q)_0 - (\nabla v^*, q)_0 = 0 \quad \forall q \in [L^2(\Omega)]^2, \tag{2.3}$$

$$(f, v^0 - v^*)_0 - (q^*, \nabla(v - v^*))_0 \leq 0 \quad \forall v \in K_0. \tag{2.4}$$

The last inequality implies

$$q^* = \nabla(\varphi + v^*); \tag{2.5}$$

hence, by substituting in (2.4) we obtain

$$(\nabla(\varphi + v^*), \ \nabla(v - v^*))_0 \geq (f, v - v^*)_0 \quad \forall v \in K_0.$$

This inequality can be written in the form

$$(\nabla(\varphi+v^*), \nabla(v+\varphi)-\nabla(v^*+\varphi))_0 \geq (f, (v+\varphi-(v^*+\varphi))_0 \ \forall v \in K_0. \tag{2.6}$$

The function $v + \varphi \in K$ for every $v \in K_0$. Now (2.6) is equivalent to the fact that the function $v^* + \varphi$ is a solution of (1.1). The rest follows from (2.5).

(ii) Let $u \in K$ be a solution of (1.1). Then we can write

$$u = \varphi + u^*, \quad u^* \in K_0,$$

and consequently

$$\nabla u = \nabla \varphi + \nabla u^*.$$

Evaluating the scalar product of both sides of this identity with $q \in [L^2(\Omega)]^2$, we obtain (2.3) with $q^* = \nabla u$, $v^* = u - \varphi$. On the other hand, u as a solution of (1.1) satisfies (1.1'), which by virtue of (i) can be equivalently written in the form

$$(\nabla u, \nabla v - \nabla u^*)_0 \geq (f, v - u^*)_0 \quad \forall v \in K_0,$$

which is exactly (2.4) with $q^* = \nabla u$, $v^* = u^* = u - \varphi$. The pair $(\nabla u, u - \varphi) \in [L^2(\Omega)]^2 \times K_0$ is thus a saddle point of $\mathcal{H}$ on the above mentioned set.

□

Definition 2.1. The problem of finding a saddle point of $\mathcal{H}$ given by (2.1) on the set $[L^2(\Omega)]^2 \times K_0$ is called the *mixed variational formulation* of the problem with an inner obstacle.

Remark 2.1. If $\Delta u \in L^2(\Omega)$, then we can replace the Lagrangian $\mathcal{H}$ by $\tilde{\mathcal{H}}$, which is in the set $H(\mathrm{div}, \Omega) \times L^2_+(\Omega)$ defined by

$$\tilde{\mathcal{H}}(q, v) = \frac{1}{2}\|q\|_0^2 + (\mathrm{div}\ q, \varphi)_0 + (f, v)_0 + (\mathrm{div}\ q, v)_0. \tag{2.7}$$

Then we look for a saddle point (q^*, v^*) of the Lagrangian $\tilde{\mathcal{H}}$ in the set $H(\mathrm{div}, \Omega) \times L^2_+(\Omega)$. Again, (2.2) can be shown to be valid.

1.2.3 Solution of the Primal Problem by the Finite Element Method

In what follows we shall assume that $\Omega \subset \mathbf{R}^2$ is a *bounded polygonal domain*. Let $\{\mathcal{T}_h\}$, $h \to 0_+$ be a *regular* system of triangulation. We associate each $\mathcal{T}_h$ with a finite-dimensional space $V_h \subset H_0^1(\Omega)$, where

$$V_h = \{v_h \in C(\bar{\Omega})\ |v_h|_T \in P_1(T) \quad \forall T \in \mathcal{T}_h,\ v_h = 0 \text{ on } \Gamma\}.$$

We denote the inner nodes of the triangulation $\mathcal{T}_h$ successively by $a_1, a_2, \ldots, a_{m(h)}$, and define

$$K_h = \{v_h \in V_h \mid v_h(a_i) \geq \varphi(a_i) \quad \forall i = 1, 2, \ldots, m(h)\}.$$

K_h is a convex closed subset of V_h for every $h \in (0,1)$. Nevertheless, $K_h \subset K$ does not generally hold.

By an *approximation of the primal formulation* of the problem with an inner obstacle, we mean a function $u_h \in K_h$ which satisfies

$$\mathcal{Y}(u_h) \le \mathcal{Y}(v_h) \quad \forall v_h \in K_h.$$

Our aim is to establish the order of the error $\|u - u_h\|_1$. We have

Theorem 3.1. *Let $\varphi \in H^2(\Omega)$, $u \in H^2(\Omega) \cap K$. Then*

$$\|u - u_h\|_1 \le c(u, f, \varphi)h, \quad h \to 0_+.$$

Proof. We apply (3.10) of section 1.1.32 to our particular case:

$$|u - u_h|_1^2 \le (f, u - v_h)_0 + (f, u_h - v)_0 + (\nabla(u_h - u),\ \nabla(v_h - u))_0$$
$$+(\nabla u, \nabla(v - u_h))_0 + (\nabla u, \nabla(v_h - u))_0 \quad \forall v \in K,\ \forall v_h \in K_h. \tag{3.1}$$

The Green formula together with the inclusion K_h, $K \subset H_0^1(\Omega)$ $\forall h > 0$ implies

$$(\nabla u, \nabla(v - u_h))_0 = (-\Delta u, v - u_h)_0 \quad \forall v \in K,$$
$$(\nabla u, \nabla(v_h - u))_0 = (-\Delta u, v_h - u)_0 \quad \forall v_h \in K_h.$$

By substituting into (3.1) we obtain

$$|u - u_h|_1^2 \le (-\Delta u - f, v - u_h)_0 + (-\Delta u - f, v_h - u)_0$$
$$+ (\nabla(u_h - u), \nabla(v_h - u))_0 \quad \forall v \in K,\ \forall v_h \in K_h. \tag{3.1'}$$

In (3.1′) we set $v_h = r_h u$, where $r_h u$ is the piecewise linear Lagrangian interpolation of the function u. This choice is justified by $r_h u \in K_h$. The following inequalities now follow from the well-known interpolation properties of V_h:

$$|(-\Delta u - f, r_h u - u)_0| \le \|-\Delta u - f\|_0 \|r_h u - u\|_0 \le c(u, f)h^2, \tag{3.2}$$

$$|(\nabla(u_h - u), \nabla(r_h u - u))_0| \le \frac{1}{2}|u_h - u|_1^2 + \frac{1}{2}|r_h u - u|_1^2$$
$$\le \frac{1}{2}|u_h - u|^2 + c(u)h^2. \tag{3.3}$$

The estimate of the first term on the right-hand side of (3.1′) is a little more complicated. The function used to establish a bound for this term is defined by $v = \sup\{\varphi, u_h\}$. It can be shown that $v \in K$, and therefore this function can be used to estimate (3.1′). Let

$$\Omega^- = \{x \in \Omega \mid u_h(x) \le \varphi(x)\},$$

$$\Omega^+ = \{x \in \Omega \mid u_h(x) > \varphi(x)\}.$$

The definition of K_h implies that $u_h(a_i) \geq \varphi(a_i)$ for $i = 1, \ldots, m(h)$, and hence also $u_h \geq r_h u$ in Ω. This, together with the definition of v and the inequality $-\Delta u - f \geq 0$ a.e. in Ω, implies that we can write

$$(-\Delta u - f, v - u_h)_0 = (-\Delta u - f, \varphi - u_h)_{0,\Omega^-}$$

$$\leq (-\Delta u - f, \varphi - r_h\varphi)_{0,\Omega^-} \leq c(u, f)\|\varphi - r_h\varphi\|_{0,\Omega^-}$$

$$\leq c(u, f)\|\varphi - r_h\varphi\|_{0,\Omega} \leq c(u, f, \varphi)h^2.$$

The assertion of the theorem now follows from the last inequality, from (3.1′) (3.2), (3.3), and from the Friedrichs inequality. □

Let us now engage in the question of convergence of u_h to a nonregular solution. We have

Theorem 3.2. *Let $\varphi \in H^2(\Omega)$*[8] *and let us further assume that $\varphi = 0$ on Γ. Then,*

$$u_h \to u, \quad h \to 0_+ \quad \text{in } H^1(\Omega).$$

Proof. We shall verify the assumptions (3.8), (3.9) of theorem 3.1 from section 1.1.32. We already know that we can write

$$K = \varphi + K_0,$$

as well as

$$K_h = r_h\varphi + K_{0h},$$

where

$$K_{0h} = \{v_h \in V_h \mid v_h(a_i) \geq 0 \quad \forall i = 1, \ldots, m(h)\},$$

and $r_h\varphi$ is a piecewise linear Lagrangian interpolation of φ. In what follows we shall need the following result, which is presented here without proof:

$$\forall v_0 \in K_0 \; \exists \omega_n \in C_0^\infty(\Omega), \; \omega_n \geq 0 \text{ such that } \omega_n \to v_0,$$

$$n \to \infty \quad \text{in } H^1(\Omega). \tag{3.4}$$

Let $v_0 \in K_0$ and $\epsilon > 0$ be arbitrary. Then (3.4) implies the existence of $\omega_{n_0} \in C_0^\infty(\Omega)$, $\omega_{n_0} \geq 0$, such that

$$\|v_0 - \omega_{n_0}\|_1 \leq \frac{1}{2}\epsilon. \tag{3.5}$$

---

[8] This assumption is not necessary; $\varphi \in C(\bar{\Omega})$ is sufficient.

As $\omega_{n_0} \in C_0^\infty(\Omega)$, we can construct its piecewise linear Lagrangian interpolation $r_h\omega_{n_0}$. Evidently, $r_h\omega_{n_0} \in K_{0h}$ and for $h > 0$ sufficiently small,

$$\|\omega_{n_0} - r_h\omega_{n_0}\|_1 \leq \frac{1}{2}\epsilon.$$

This together with (3.5) and the triangle inequality

$$\|v_0 - r_h\omega_{n_0}\|_1 \leq \|v_0 - \omega_{n_0}\|_1 + \|\omega_{n_0} - r_h\omega_{n_0}\|_1$$

yields the existence of a sequence $w_h \in K_{0h}$ such that

$$w_h \to v_0, \quad h \to 0_+ \quad \text{in } H^1(\Omega).$$

Now let $v \in K$. Then,

$$v = \varphi + v_0, \quad v_0 \in K_0.$$

Set

$$v_h = r_h\varphi + w_h \in K_h,$$

where $w_h \in K_{0h}$ is such a sequence that

$$w_h \to v_0, \quad h \to 0_+ \quad \text{in } H^1(\Omega). \tag{3.6}$$

(We have just proved that such a sequence exists.) The triangle inequality

$$\|v - v_h\|_1 \leq \|\varphi - r_h\varphi\|_1 + \|v_0 - w_h\|_1,$$

(3.6), and the convergence

$$\|\varphi - r_h\varphi\|_1 \to 0, \quad h \to 0_+,$$

imply that $v_h \to v$, $h \to 0_+$ in $H^1(\Omega)$. Thus, (3.8) is verified. It remains to verify (3.9); that is, the validity of the implication

$$v_h \in K_h, \quad v_h \rightharpoonup v, \quad h \to 0_+ \quad \text{in } H^1(\Omega) \Rightarrow v \in K.$$

Evidently, it is sufficient to show that $v \geq \varphi$ a.e. in Ω. Let us again write

$$v_h = r_h\varphi + w_h, \quad w_h \in K_{0h}.$$

As the embedding of $H^1(\Omega)$ into $L^2(\Omega)$ is completely continuous (Nečas (1967)), it follows from $v_h \rightharpoonup v$, $h \to 0_+$ in $H^1(\Omega)$ that $v_h \to v$, $h \to 0_+$ in $L^2(\Omega)$. At the same time $r_h\varphi \to \varphi$, $h \to 0_+$ in $L^2(\Omega)$, and hence necessarily

$$w_h \to v - \varphi, \quad h \to 0_+ \quad \text{in } L^2(\Omega).$$

Consequently, we can choose a subsequence of $\{w_h\}$ such that (keeping the notation of the original sequence)

$$w_h \to v - \varphi, \quad h \to 0_+, \quad \text{a.e. in } \Omega.$$

Since $w_h \geq 0$ in Ω for all $h > 0$, we also have $v - \varphi \geq 0$ in Ω. Now the assertion on the convergence of u_h to u follows from theorem 3.1, section 1.1.32. □

1.2.4 Solution of the Dual Problem by the Finite Element Method

1.2.41. Approximation of the Dual Formulation of the Problem with an Inner Obstacle. In this section we will study the approximation of the dual formulation of the problem with an inner obstacle. To this end we will again use the Ritz method.

Let $Q_{0h}^- \subset Q_0^-$, $h \in (0,1)$, be "finite-dimensional approximations" of the convex set Q_0^- and set

$$Q_{fh}^- = \bar{q} + Q_{0h}^-,$$

where $\bar{q} \in [L^2(\Omega)]^2$ is a particular solution of the equation

$$\operatorname{div} \bar{q} = -f \quad \text{in } \Omega.$$

Evidently, $Q_{fh}^- \subset Q_f^-$ for every $h \in (0,1)$.

By an *approximation of the dual formulation* of the problem with an inner obstacle, we mean the problem to find $q_h^* \in Q_{fh}^-$ such that

$$S(q_h^*) \leq S(q_h) \quad \forall q_h \in Q_{fh}^-. \tag{4.1}$$

We have

Theorem 4.1. *For every $h > 0$ there exists exactly one solution of problem* (1.4), *and*

$$\|q^* - q_h^*\|_0^2 \leq \{(q^* - q_h, \nabla\varphi)_0 + (q^* - q_h^*, q^* - q_h)_0 + (q^*, q_h - q^*)_0\} \tag{4.2}$$

holds for any $q_h \in Q_{fh}^-$.

Proof. The existence and the uniqueness of the solution of (4.1) is a consequence of the fact that S is convex and coercive on a convex closed subset Q_{fh}^-. The inequality (4.2) is just the transcription of (3.10′) from section 1.1.32 for our particular case and notation. We also use the inclusion $Q_{fh}^- \subset Q_f$ $\forall h \in (0,1)$. □

Remark 4.1. Let the reader not be confused by the fact that for estimating the error $\|q^* - q_h^*\|_0$ we use the relations (3.10) or, as the case may be (3.10′) from section 1.1.32, which were formally used to estimate the errors of the approximations of the primal variational formulations. What is substantial is that the dual variational formulation of our problem has the same character as the problem which was formulated in a general setting in section 1.1.32, that is, the problem of minimization of a quadratic functional on a convex closed subset of a Hilbert space.

1.2.42. Construction of the Sets Q_{fh}^- and Their Approximate Properties. To be able to solve problem (4.1), we have to specify the choice of the sets Q_{0h}^-. For this purpose, we will use the finite element method. The construction will be similar to that used in section 1.1.42 for the approximation of the dual formulation of problem $\mathcal{P}_2$.

Let Ω be a bounded polygonal domain, $\{\mathcal{T}_h\}$, $h \to 0_+$ a *regular* system of triangulations of Ω. Let us define

$$Q_{0h}^- = \{q \,|\, q|_T \in (P_1(T))^2 \quad \forall T \in \mathcal{T}_h,\; (q\cdot\nu)_T + (q\cdot\nu)_{T'} = 0$$

$$\forall x \in T \cap T',\; \operatorname{div} q \le 0 \quad \text{in } \Omega\}.$$

The elements of the set Q_{0h}^- are vector functions, both of whose components are linear functions on each triangle $T \in \mathcal{T}_h$. Moreover, the condition of continuity of flows when passing from T to an adjacent triangle T' is fulfilled. The last condition, namely div $q \le 0$, guarantees that $Q_{0h}^- \subset Q_0^-$ for every $h \in (0,1)$. The reader easily verifies that $q \in Q_{0h}^-$ if and only if the inequality

$$l_1(\beta_2^e + \beta_3^e) + l_2(\beta_4^e + \beta_5^e) + l_3(\beta_1^e + \beta_6^e) \le 0 \tag{4.3}$$

holds for each triangle $T_e \in \mathcal{T}_h$ (see Haslinger (1979)), where l_i, β_j^e, $i = 1,2,3,$, $j = 1,\dots,6$ have the same meaning as in section 1.1.42. Moreover, conditions of the form

$$\beta_i + \beta_k = 0 \tag{4.4}$$

are added, which express the continuity of the flows on the interelement boundaries.

Problem (4.1) transcribed into the finite dimension is analogous to the problem of the type given in (4.33), section 1.1.42, with the only difference being that set $\mathcal{B}$ is given by the constraints (4.3), (4.4).

Let us define the mappings Π_T and r_h in the same way as in section 1.1.42. Let

$$\mathcal{R}^-(\Omega) = \{q \in [H^1(\Omega)]^2 \,|\, \operatorname{div} q \le 0 \quad \text{in } \Omega\}.$$

We can show that

$$\|q - r_h q\|_0 \le ch^j |q|_j \quad \forall q \in [H^j(\Omega)]^2,\; j = 1,2; \tag{4.5}$$

$$\|\operatorname{div} q - \operatorname{div} r_h q\|_0 \le ch |\operatorname{div} q|_1 \quad \forall q \in [H^1(\Omega)]^2;\; \operatorname{div} q \in H^1(\Omega), \tag{4.6}$$

and moreover, $r_h \in \mathcal{L}(\mathcal{R}^-(\Omega), Q_{0h}^-)$ (see Haslinger (1979)).

1.2.43. A Priori Error Bound of the Approximation of the Dual Formulation. Before estimating the rate of convergence of q_h^* to q^*, we will

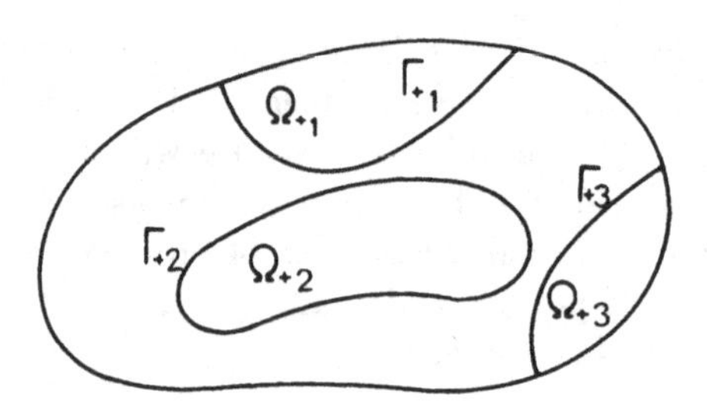

Figure 7

rearrange formula (4.2). Taking into account the fact that $q^* = \nabla u$, where $u \in H_0^1(\Omega)$ is a solution of (1.1), and *assuming* that div $q^* = \Delta u \in L^2(\Omega)$, i.e., $q^* \in Q_f^-(\text{div}, \Omega)$, we can write

$$\begin{aligned}(q^* - q_h, \nabla\varphi - q^*)_0 &= (q^* - q_h, \nabla(\varphi - u))_0 \\ &= (\text{div}(q^* - q_h), u - \varphi)_0 \quad \forall q_h \in Q_{fh}^-.\end{aligned}$$

Further, let us write

$$q^* = \bar{q} + q^0,$$

$$q_h = \bar{q} + \tilde{q}_h \quad \forall q_h \in Q_{fh}^-,$$

where $\tilde{q}_h \in Q_{0h}^-$. As $q^* \in Q_f^-(\text{div}, \Omega)$ and $f \in L^2(\Omega)$, we have $q^0 \in Q_0^-(\text{div}, \Omega)$ and moreover, q^0 is a solution of problem (1.8). The identity $Q_0^-(\text{div}, \Omega) = Q_f^-(\text{div}, \Omega)$ for $f \equiv 0$ in Ω, together with (4.2) and the relations $q^* = \bar{q} + q^0$, $q_h^* = \bar{q} + \tilde{q}_h^0$, $\tilde{q}_h^0 \in Q_{0h}^-$ yields

$$\begin{aligned}\|q^* - q_h^*\|_0^2 = \|q^0 - \tilde{q}_h^0\|_0^2 &\le (q^* - q_h^*, q^0 - \tilde{q}_h)_0 \\ &+ (\text{div}(q^0 - \tilde{q}_h), u - \varphi)_0 \quad \forall \tilde{q}_h \in Q_{0h}^-. \end{aligned} \tag{4.7}$$

The symbol Ω_+ has been used to denote the set of such $x \in \Omega$ for which $u(x) > \varphi(x)$. In the following, let us assume

$$\Omega_+ = \bigcup_{t=1}^{p} \Omega_{+t}, \quad \Omega_{+s} \cap \Omega_{+r} = \emptyset, \quad r \neq s, \tag{4.8}$$

where Ω_{+t}, $t = 1, \ldots, p$, are domains with sufficiently smooth parts of the boundary $\Gamma_{+t} \cap \Omega$ (see Figure 7).

Theorem 4.2. *Let* $q^0 \in [H^j(\Omega)]^2$, $u - \varphi \in H^k(\Omega_{+t})$ *and* $\partial^{k-1}(u - \varphi)/\partial\nu^{k-1} = 0$ *on* $\Gamma_{+t} \cap \Omega$, div $q^0 \in H^m(\Omega)$, $j, k = 1, 2$, $m = 0, 1$, $t = 1, \ldots, p$. *Moreover, let* (4.8) *hold. Then*

$$\|q^* - q_h^*\|_0 \le ch^{(k+m)/2} \left[\sum_{t=1}^{p} \|u - \varphi\|_{k,\Omega_{+t}} |\text{div}\, q^0|_{m,\Omega_t^{2h}} \right]^{1/2} + 0(h^j), \tag{4.9}$$

where $\Omega_t^{\eta h} = \{ x \in \Omega \,|\, \mathrm{dist}(x, \Gamma_{+t}) < \eta h,\ \eta > 0\}$ *and c is a positive constant independent of h.*

Proof. To prove (4.9) we shall use the relation (4.7). Put $\tilde{q}_h = r_h q^0$, where r_h has the same meaning as in the previous section. This choice is justified by $r_h q^0 \in Q_{0h}^-$. Then (4.5) implies

$$|(q^* - q_h^*, q^0 - r_h q^0)_0| \le \frac{1}{2}\|q^* - q_h^*\|_0^2 + \frac{1}{2}\|q^0 - r_h q^0\|_0^2$$

$$= \frac{1}{2}\|q^* - q_h^*\|_0^2 + 0(h^{2j}). \tag{4.10}$$

Now let us write

$$(\mathrm{div}(q^0 - r_h q^0), u - \varphi)_0 = \sum_{T \in \mathcal{T}_h} (\mathrm{div}(q^0 - r_h q^0), u - \varphi)_{0,T}.$$

If $T \subset \Omega_{+t}$ for some $t = 1, \ldots, p$, then

$$\mathrm{div}\ q^* = \mathrm{div}(\nabla u) = -f \quad \text{in } T$$

by virtue of $(1.2)_2$. Hence,

$$\mathrm{div}\ q^0 = -\mathrm{div}\,\bar{q} - f = f - f = 0 \quad \text{in } T.$$

Taking into account the definition of $r_h q^0$ and particularly (4.25), section 1.1.42, we obtain

$$\mathrm{div}(r_h q^0)|_T = 0 \quad \text{in } T,$$

and consequently

$$(\mathrm{div}(q^0 - r_h q^0), u - \varphi)_{0,T} = 0. \tag{4.11}$$

If $T \in \mathcal{T}_h$ is such that $T \subset \Omega \dot{-} \cup_{t=1}^p \bar{\Omega}_{+t}$, then $u = \varphi$ a.e. in T and (4.11) again holds.

Let I be the system of all $T \in \mathcal{T}_h$ which fulfill $T \dot{-} \Omega_{+t} \neq \emptyset$ but $T \subset \Omega_{+t}$. Set $\Omega_t^{+h} = \Omega_t^h \cap \Omega_{+t}$. Then, in virtue of the facts proven above, we can write

$$|(\mathrm{div}(q^0 - r_h q^0, u - \varphi)_{0,\Omega}| \le \sum_{T \in I} \int_T |u - \varphi| \cdot |\mathrm{div}\ q^0 - \mathrm{div}\ r_h q^0| ds$$

$$\le \sum_{t=1}^p \int_{\Omega_t^{+h}} |u - \varphi| \cdot |\mathrm{div}\ q^0 - \mathrm{div}\ r_h q^0| dx. \tag{4.12}$$

If $u - \varphi \in H^k(\Omega_{+t})$, $\partial^{k-1}(u - \varphi)/\partial \nu^{k-1} = 0$ on $\Gamma_{+t} \cap \Omega$, then (see Babuška (1970))

$$\|u - \varphi\|_{0,\Omega_t^{+h}} \le c h^k \|u - \varphi\|_{k,\Omega_{+t}}. \tag{4.13}$$

Now (4.6) yields

$$\|\operatorname{div} q^0 - \operatorname{div} r_h q^0\|^2_{0,\Omega_t^{+h}} \leq \sum_{T\cap\Omega_t^{+h}\neq\emptyset} \|\operatorname{div} q^0 - \operatorname{div} r_h q^0\|^2_{0,T}$$

$$\leq ch^{2m}|\operatorname{div} q^0|^2_{m,\Omega_t^{2h}}.$$

This estimate together with (4.10), (4.12), and (4.13) implies (4.9). □

Now we will study the convergence of q_h^* to q^* under slightly little weaker conditions on the smoothness of the solution u. In what follows we assume that $\Omega \subset \mathbf{R}^2$ is a *simply* connected polygonal domain.[9] First we will prove two auxiliary results.

Lemma 4.1. *Let $q \in Q_0^-(\operatorname{div},\Omega)$ and let $\tilde{\Omega} \supset \bar{\Omega}$ be a simply connected domain. Then there exists a function $\pi q \in L^2(\tilde{\Omega})$ with the following properties:*

$$\pi q = q \quad \text{in } \Omega, \tag{4.14}$$

$$\operatorname{div}(\pi q) \in L^2(\tilde{\Omega}), \tag{4.15}$$

$$\operatorname{div}(\pi q) \leq 0 \quad \text{in } \tilde{\Omega}. \tag{4.16}$$

Proof. As $q \in H(\operatorname{div},\Omega)$, we can define a "flow" $q\cdot\nu \in H^{-1/2}(\Gamma)$ by the formula (see theorem 1.3, section 1.1.11)

$$(q,\nabla\varphi)_0 + (\operatorname{div} q,\varphi)_0 = \langle q\cdot\nu,\varphi\rangle \quad \forall q \in C_0^\infty(\tilde{\Omega}).$$

Let w be a solution of the elliptic boundary value problem

$$-\Delta w = g \quad \text{in } \tilde{\Omega}\setminus\Omega,$$

$$w = 0 \quad \text{on } \partial\tilde{\Omega},$$

$$\frac{\partial w}{\partial\bar{\nu}} = q\cdot\bar{\nu} \quad \text{on } \partial\Omega,$$

where $g \in L^2(\tilde{\Omega})$ is a given nonnegative function and $\bar{\nu} = -\nu$ is the unit vector of the inner normal to $\partial\Omega$. Let us put

$$\pi q = \begin{matrix} q & \text{in } \Omega, \\ \nabla w & \text{in } \tilde{\Omega}\setminus\Omega. \end{matrix}$$

This immediately implies (4.14).

[9]This assumption is merely technical. A modification of our proof makes it possible to extend theorem 4.3 to the case of multiply connected sets.

Let $\varphi \in C_0^\infty(\tilde{\Omega})$ be arbitrary. Then

$$\int_{\tilde{\Omega}} \pi q \cdot \nabla\varphi\, dx = \int_{\tilde{\Omega}\setminus\Omega} \pi q \cdot \nabla\varphi\, dx + \int_\Omega q \cdot \nabla\varphi\, dx$$

$$= \int_{\tilde{\Omega}\setminus\Omega} \nabla w \cdot \nabla\varphi\, dx - \int_\Omega \operatorname{div} q \cdot \varphi\, dx + \langle q \cdot \nu, \varphi\rangle$$

$$= \int_{\tilde{\Omega}\setminus\Omega} (-\Delta w)\varphi\, dx + \langle\frac{\partial w}{\partial \bar{\nu}}, \varphi\rangle - \int_\Omega \operatorname{div} q\varphi\, dx + \langle q \cdot \nu, \varphi\rangle = \int_{\tilde{\Omega}} G \cdot \varphi\, dx,$$

where

$$G = \begin{matrix} \operatorname{div} q \in L^2(\Omega) & \text{in } \Omega, \\ \Delta w \in L^2(\tilde{\Omega}\setminus\Omega) & \text{in } \tilde{\Omega}\setminus\Omega. \end{matrix}$$

Hence, $G = \operatorname{div}(\pi q) \in L^2(\tilde{\Omega})$ and (4.16) is fulfilled. □

Lemma 4.2. *The set $Q_0^- \cap [C^\infty(\bar{\Omega})]^2$ is dense in $Q_0^-(\operatorname{div}, \Omega)$ in the norm of the space $H(\operatorname{div}, \Omega)$.*

Proof. Let $q \in Q_0^-(\operatorname{div}, \Omega)$ be arbitrary, let $\pi q = ((\pi q)_1, (\pi q)_2)$ be its extension to $\tilde{\Omega} \supset \Omega$ introduced in the previous lemma. Let us denote by $(\pi q)_h = ((\pi q)_{1h}, (\pi q)_{2h})$ the regularization of the function πq (see Nečas (1967) and section 1.1.63), that is,

$$(\pi q)_{jh}(x) = \int_{\tilde{\Omega}} (\pi q)_j(y)\omega(x-y,h)dy, \quad h > 0, \quad x, y \in \tilde{\Omega}, \quad j = 1, 2.$$

It is well known that $(\pi q)_h \in [C^\infty(\bar{\Omega})]^2$ and also

$$\|(\pi q)_h - \pi q\|_{0,\tilde{\Omega}} \to 0, \quad h \to 0_+. \tag{4.17}$$

For $h > 0$ sufficiently small and $x \in \Omega$ we can write

$$\operatorname{div}(\pi q)_h(x) = -\int_{\tilde{\Omega}} \pi q(y) \cdot \nabla_y \omega(x-y,h)dy$$

$$= \int_{\tilde{\Omega}} \operatorname{div} \pi q(y)\omega(x-y,h)dy \le 0,$$

virtue of the nonnegativeness of the regularization kernel ω. Here ∇_y stands for the gradient calculated with respect to the variable y. At the same time

$$\operatorname{div}(\pi q)_h = (\operatorname{div} \pi q)_h \quad \text{in } \Omega,$$

and similar to (4.17) we derive

$$\|\operatorname{div}(\pi q)_h - \pi q\|_{0,\Omega} \to 0, \quad h \to 0_+.$$

This together with (4.17) implies that the sequence $(\pi q)_h$ approximates q with an arbitrary accuracy in the norm of the space $H(\text{div},\Omega)$. □

As an immediate consequence we obtain

Theorem 4.3. *Let the solution u of problem* (1.1) *fulfill* $\Delta u \in L^2(\Omega)$. *Then*

$$q_h^* \to q^*, \quad h \to 0_+ \quad \textit{in } [L^2(\Omega)]^2.$$

Proof. To prove the convergence we use theorem 3.1 and remark 3.5, section 1.1.32, setting $K = Q_f^-$, $K_h = Q_{fh}^-$, $\tilde{K} = Q_f^-(\text{div},\Omega)$. Let $q \in Q_f^-(\text{div},\Omega)$. Then

$$q = \bar{q} + \tilde{q}, \quad \tilde{q} \in Q_0^-.$$

Since $f \in L^2(\Omega)$, we have $\tilde{q} \in Q_0^-(\text{div},\Omega)$ as well, and the previous lemma guarantees the existence of such $\tilde{q}_n \in Q_0^-(\text{div},\Omega) \cap [C^\infty(\bar{\Omega})]^2$ that

$$\tilde{q}_n \to \tilde{q}, \quad n \to \infty \quad \text{in } [L^2(\Omega)]^2 \quad (\text{even in } H(\text{div},\Omega)). \tag{4.18}$$

Now, for every $\tilde{q}_n$ we can construct a function $r_h\tilde{q}_n \in Q_{0h}^-$ which satisfies (see (4.5))

$$\|\tilde{q}_n - r_h\tilde{q}_n\|_0 \le ch^2|\tilde{q}_n|_2. \tag{4.19}$$

Let us define

$$q_{hn} = \bar{q} + r_h\tilde{q}_n \in Q_{fh}^-.$$

By virtue of the triangle inequality

$$\|q - q_{hn}\|_0 = \|\tilde{q} - r_h\tilde{q}_n\| \le \|\tilde{q} - \tilde{q}_n\|_0 + \|\tilde{q}_n - r_h\tilde{q}_n\|_0,$$

and of (4.18), (4.19), we conclude that

$$\|q_{hn} - q_0\|_0 \to 0, \quad n \to \infty, \quad h \to 0_+.$$

This verifies (3.8′), section 1.1.32; (1.38) of the same section is automatically fulfilled, as $Q_{fh}^- \subset Q_f^-$ for every $h > 0$. □

1.2.5 Solution of the Mixed Formulation by the Finite Element Method

Throughout this section we assume that $\varphi \in H_0^1(\Omega)$ and that the solution u of problem (1.1) satisfies $\Delta u \in L^2(\Omega)$. Then, $q^* = \nabla u \in H(\text{div},\Omega)$ and we will use the Lagrangian $\tilde{\mathcal{H}}$ for the approximation, which was defined by (2.7) (see remark (2.1).)

Let $\Omega \subset \mathbf{R}^2$ be a bounded polygonal domain, $\{\mathcal{T}_h\}$, $h \to 0_+$ a regular system of triangulations of $\bar{\Omega}$. Each $\mathcal{T}_h$ will be associated with sets

$$Q_h = \{q_h \in [L^2(\Omega)]^2 \,|q_h|_T \in [P_1(T)]^2 \quad \forall T \in \mathcal{T}_h, \ \text{div}\, q_h \in L^2(\Omega)\},$$

$$L_h = \{v_h \in L^2(\Omega) \,|\, v_h|_T \in P_0(T) \quad \forall T \in \mathcal{T}_h\};$$

further, we define $\Lambda_h = \{v_h \in L_h \,|\, v_h \geq 0 \text{ in } \Omega\}$.

The sets $Q_H \subset H(\operatorname{div}, \Omega)$, $L_h \subset L^2(\Omega)$ are evidently finite-dimensional approximations of the spaces mentioned, and $\Lambda_h \subset L^2_+(\Omega)$ for every $h \in (0,1)$.

By *an approximation of the mixed formulation of the problem with an inner obstacle,* we mean the problem of finding a saddle point (q_h^*, v_h^*) of the Lagrangian $\tilde{\mathcal{H}}$ in $Q_h \times \Lambda_h$:

$$\tilde{\mathcal{H}}(q_h^*, v_h) \leq \tilde{\mathcal{H}}(q_h^*, v_h^*) \leq \tilde{\mathcal{H}}(q_h, v_h^*) \quad \forall q_h \in Q_h, \quad \forall v_h \in \Lambda_h, \tag{5.1}$$

or equivalently, taking into account the linearity of Q_h,

$$\text{find} \quad (q_h^*, v_h^*) \in Q_h \times \Lambda_h \quad \text{such that}$$

$$\begin{aligned} (q_h^*, q_h)_0 + (\varphi, \operatorname{div} q_h)_0 + (v_h^*, \operatorname{div} q_h)_0 = 0 \quad \forall q_h \in Q_h, \\ (\operatorname{div} q_h^* + f, v_h - v_h^*)_0 \leq 0 \quad \forall v_h \in \Lambda_h. \end{aligned} \tag{5.1'}$$

In the following we will study the relation between the approximation (q_h^*, v_h^*) and the solution of the mixed formulation $(q^*, v^*) = (\nabla u, u - \varphi)$. Using the notation of section 1.1.52, we set $V = H(\operatorname{div}, \Omega)$, $L = L^2(\Omega)$, $\Lambda = L^2_+(\Omega)$, $V_h = Q_h$. However, the results of this section are not directly applicable, as the quadratic form a, which in our case has the form $(q,q)_0$, is not $H(\operatorname{div}, \Omega)$-elliptic. Therefore, we will only sketch how to obtain the assertion on the convergence of (q_h^*, v_h^*) to (q^*, v^*).

The following lemma plays the crucial role.

Lemma 5.1. *There exists a constant $\tilde{\beta} > 0$, independent of $h > 0$ and such that*

$$\sup_{Q_h} \frac{(\operatorname{div} q_h, v_h)_0}{\|q_h\|_{H(\operatorname{div}, \Omega)}} \geq \tilde{\beta} \|v_h\|_{0,\Omega} \quad \forall v_h \in L_h.$$

Proof. Let $\tilde{\Omega} \supset \Omega$ be a domain with a smooth boundary. We extend the given function $v_h \in L_h$ by zero to $\tilde{\Omega}$ (keeping its original notation). Let us consider the boundary value problem

$$\Delta w = v_h \quad \text{in } \tilde{\Omega},$$

$$w = 0 \quad \text{on } \partial\tilde{\Omega}.$$

It is known (Nečas (1967)) that $w \in H^2(\tilde{\Omega}) \cap H_0^1(\tilde{\Omega})$, and there exists a positive constant c such that

$$\|w\|_{2,\tilde{\Omega}} \leq c\|v_h\|_{0,\tilde{\Omega}} = c\|v_h\|_{0,\Omega}. \tag{5.2}$$

Set $q = \nabla w$. Then $q \in H(\text{div}, \tilde{\Omega})$ and

$$\|q\|_{H(\text{div},\Omega)} \leq \|q\|_{H(\text{div},\tilde{\Omega})} \leq c\|v_h\|_{0,\Omega}$$

by virtue of (5.2) and the definition of the norm in $H(\text{div}, \Omega)$. Now (4.5) and (5.2) imply

$$\|r_h q\|_{0,\Omega} \leq c\|q\|_{1,\Omega} = c\|\nabla w\|_{1,\Omega} \leq c\|w\|_{2,\tilde{\Omega}} \leq c\|v_h\|_{0,\Omega}. \tag{5.3}$$

Further, from the definition of the mapping r_h (see section 1.1.42) we obtain

$$(\text{div}\, r_h q, v_h)_0 = (\text{div}\, q, v_h)_0 \quad \forall v_h \in L_h. \tag{5.4}$$

As div $r_h q \in P_0$, it follows that div $r_h q$ is the orthogonal projection of div q to the space of piecewise constant functions. Thus, $\|\text{div}\, r_h q\|_{0,\Omega} \leq \|\text{div}\, q\|_{0,\Omega}$, which together with (5.3) yields

$$\|r_h q\|_{H(\text{div},\Omega)} \leq c\|v_h\|_{0,\Omega} \quad \forall v_h \in L_h.$$

Hence and from (5.4), we obtain

$$\begin{aligned}\sup_{q_h} \frac{(\text{div}\, q_h, v_h)_0}{\|q_h\|_{H(\text{div},\Omega)}} &\geq \frac{(\text{div}\, r_h q, v_h)_0}{\|r_h q\|_{H(\text{div},\Omega)}} \\ &\geq \frac{1}{c} \frac{(\text{div}\, q, v_h)_0}{\|v_h\|_{0,\Omega}} = \frac{1}{c}\|v_h\|_0,\end{aligned}$$

taking into account the definition of q. □

The following theorem is an immediate consequence of the previous lemma and remark 5.5, section 1.1.52.

Theorem 5.1. *For every $h > 0$ there exists exactly one solution (q_h^*, v_h^*) of problem* (5.1).

The crucial point of the proof of convergence of (q_h^*, v_h^*) to (q^*, v^*) is the assertion on boundedness of $\{q_h^*\}$, $\{v_h^*\}$. Let

$$\tilde{Q}_{fh}^- = \{q \in Q_h \mid (\text{div}\, q_h + f, v_h)_0 \leq 0 \quad \forall v_h \in \Lambda_h\}.$$

It is easily seen that $q_h \in \tilde{Q}_{fh}^-$ if and only if $q_h \in Q_h$ and div $(q_h|_T)$ is less or equal to the mean value of f on T. The convex set $\tilde{Q}_{fh}^-$ is an approximation (an external one, in general) of the set $Q_f^-(\text{div}, \Omega)$.

Lemma 5.2. *There is a positive constant c independent of h and such that*

$$\|q_h^*\|_0 \leq c, \quad \|v_h^*\|_0 \leq c \quad \forall h \in (0, 1).$$

Proof. Λ_h being a cone with its vertex at θ, we conclude from $(5.1')_3$:

$$(\operatorname{div} q_h^* + f, v_h^*)_0 = 0 \tag{5.5}$$

$$(\operatorname{div} q_h^* + f, v_h)_0 \leq 0 \quad \forall v_h \in \Lambda_h \Longleftrightarrow q_h^* \in \tilde{Q}_{fh}^-. \tag{5.6}$$

Restricting $(5.1')_2$ to the functions $q_h \in \tilde{Q}_{fh}^-$ and taking into account (5.5), (5.6) as well as the inclusion $\varphi \in H_0^1(\Omega)$, we conclude

$$(q_h^*, q_h - q_h^*)_0 - (\nabla\varphi, q_h - q_h^*)_0 \geq 0 \quad \forall q_h \in \tilde{Q}_{fh}^-. \tag{5.7}$$

It is clear that there is a number $r > 0$ and a sequence $\{\bar{q}_h\}$, $\bar{q}_h \in \tilde{Q}_{fh}^-$, such that

$$\|\bar{q}_h\|_0 \leq r \quad \forall h \in (0,1)$$

(it suffices to put $\bar{q}_h = r_h q$, where $q \in Q_f^- \cap [H^1(\Omega)]^2$). Substituting in (5.7) the function $\bar{q}_h$ for q_h, we obtain

$$\|q_h^*\|_0 \leq c(r, \varphi) \quad \forall h \in (0,1).$$

Lemma 5.1. implies

$$\tilde{\beta}\|v_h^*\|_0 \leq \sup_{Q_h} \frac{(\operatorname{div} q_h, v_h^*)_0}{\|q_h\|_{H(\operatorname{div},\Omega)}}. \tag{5.8}$$

From $(5.1)_2$ we obtain

$$\frac{(\operatorname{div} q_h, v_h^*)_0}{\|q_h\|_{H(\operatorname{div},\Omega)}} \leq (\|q_h^*\|_0 + \|\varphi\|_0) \quad \forall q_h \in Q_h.$$

This inequality, together with the boundedness of $\{q_h^*\}$ and with (5.8), completes the proof of the lemma. □

Proceeding analogously to the proof of theorem 5.3, section 1.1.52, we could prove

$$q_h^* \to q^*, \quad h \to 0_+ \quad \text{in } [L^2(\Omega)]^2,$$
$$v_h^* \to v^*, \quad h \to 0_+ \quad \text{in } L^2(\Omega).$$

If the solution u of problem (1.1) is sufficiently smooth, it is also possible to determine the rate of convergence of (q_h^*, v_h^*) to (q^*, v^*) (see Brezzi, Hager, and Raviart (1979)).

Chapter 2

One-Sided Contact of Elastic Bodies

An important application of the theory of variational inequalities has been found in the problem of one-sided contact of elastic bodies. This problem was formulated in a simplified form by Signorini as early as 1933 for the specific case of one-sided contact of one body with a perfectly rigid and smooth foundation. The Signorini problem was studied by Fichera (1964, 1972), who gave the proof of existence and regularity of the weak solution, and discussed the problem of nonuniqueness of solution.

If we suppose that friction occurs on the surface of contact, and that this friction is governed, for example, by Coulomb's Law, we obtain the so-called Signorini problem with friction. This problem for a long time remained open as concerns the existence of its solution (see Duvaut and Lions (1972)). The proof of existence has only recently been given in a paper by Nečas, Jarušek, and Haslinger (1980). In this paper, existence can be proven provided the coefficient of friction has a compact support in the zone of contact.

In this chapter, we first introduce the formulation and an approximate solution of contact problems for two elastic bodies without friction. In addition to formulations in displacements, we will also study dual variational problems—that is, formulations in stresses. Further, we will study problems with friction of Coulomb's type. In this case, we will give the existence proof for the Signorini problem with friction, as well as iteration algorithms of an approximate solution.

Throughout the chapter, approximate solutions are defined via discretization by the finite element method with piecewise linear functions on the triangulation of the given domains.

2.1 Formulation of Contact Problems

The one-sided contact of elastic bodies has been analyzed in a number of papers of technical rather than mathematical nature (see Chan and Tuba (1971); Cowry and Seireg (1971); Francavilla and Zienkiewicz (1975); Fredriksson (1976)). However, the authors of these papers have given no formulation of the continuous problem, instead starting directly from the finite-dimensional formulation by the finite element method.

In order to be able to analyze the approximate solutions, we first introduce definitions of solutions of the continuous contact problems. To this end we shall use variational inequalities, proceeding similarly to Fichera (1964), Panagiotopulos (1975), Duvaut (1976), Kikuchi and Oden (1979), and others.

For the sake of simplicity, let us consider contact *without friction*. First, we give the "local" classical formulation—that is, a system of differential equations and boundary conditions. Then we introduce the global–variational formulation and prove that both the classical and variational formulations are in a certain sense equivalent.

Throughout the chapter, let us consider:

— the plane problem,

— two bounded bodies,

— the theory of small strain,

— the linear generalized Hooke's Law for generally inhomogeneous, anisotropic materials,

— zero initial values for both stress and strain,

— the constant temperature field.

Let the elastic bodies occupy bounded domains $\Omega', \Omega'' \subset \mathbf{R}^2$ with Lipschitz boundaries. The superscript of one or two dashes will indicate in the sequel the correspondence to the body Ω' or Ω'', respectively.

Let $x = (x_1, x_2)$ be Cartesian coordinates. We look for the vector field of displacements $u = (u_1, u_2)$ on the set $\Omega' \cup \Omega''$, that is, $u' = (u'_1, u'_2)$ on Ω' and $u'' = (u''_1, u''_2)$ on Ω'', and for the corresponding tensor field of strain

$$e_{ij}(u) = \frac{1}{2}\left(\frac{\partial u_i}{\partial x_j} + \frac{\partial u_j}{\partial x_i}\right), \quad i, j = 1, 2. \tag{1.1}$$

The stress tensor is determined by the generalized Hooke's Law

$$\tau_{ij} = c_{ijkm} e_{km}, \quad i, j = 1, 2, \tag{1.2}$$

where a repeated index always means summation over the numbers 1,2. Let the coefficients c_{ijkm} be bounded and measurable functions of $x \in \Omega' \cup \Omega''$,

$$c_{ijkm} = c_{jikm} = c_{kmij}, \tag{1.3}$$

and let there exist a positive constant c_0 such that

$$c_{ijkm}(x) e_{ij} e_{km} \geq c_0 e_{ij} e_{ij} \tag{1.4}$$

holds for all symmetric matrices e_{ij} and almost all $x \in \Omega' \cup \Omega''$.

The stress tensor satisfies the equations of equilibrium

$$\frac{\partial \tau_{ij}}{\partial x_j} + F_i = 0, \quad i = 1, 2, \tag{1.5}$$

where F_i are components of the vector of the body forces.

We assume that the body Ω' is fixed by its part Γ_u,

$$u = 0 \quad \text{on } \Gamma_u \subset \partial\Omega'. \tag{1.6}$$

On some parts of the boundaries the surface load is given, that is,

$$\tau_{ij}^M n_{ij}^M = P_i^M \quad \text{on } \Gamma_\tau^M \subset \partial\Omega^M, \quad M = {}' , {}''; \quad i = 1, 2, \tag{1.7}$$

where n^M denotes the outer unit normal to $\partial\Omega^M$, while P_i^M are the components of the surface load.

Let the conditions of the "classical" (two-sided) contact be fulfilled on a part $\Gamma_0 \subset \partial\Omega''$:

$$u_n = 0, \quad T_t = 0 \quad \text{on } \Gamma_0 \subset \partial\Omega'', \tag{1.8}$$

where

$$u_n = u_i n_i, \quad T_t = \tau_{ij} n_j t_i, \quad t = (t_1, t_2) = (-n_2, n_1)$$

are the normal component of the displacement and the tangential component of the stress tensor, respectively.

The conditions (1.8) occur, for example, on the axis of symmetry of the given problem, and they enable us to reduce the solution of the problem to only a half of the given elastic system.

Contact can occur on the other parts of the boundary $\partial\Omega' \cup \partial\Omega''$. In the following, we will distinguish two types of contact problems: (1) with a bounded zone of contact, and (2) with an increasing zone of contact.

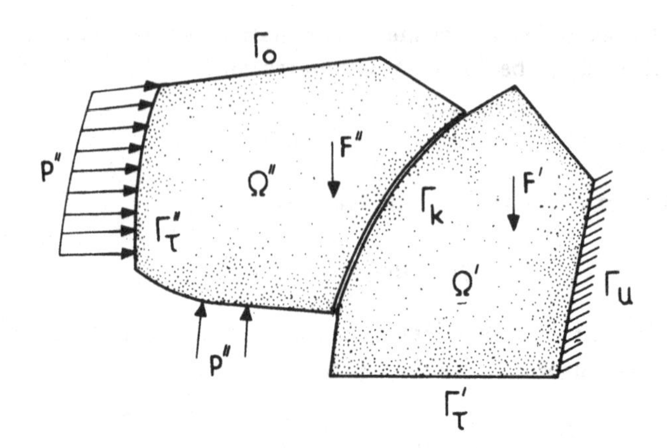

Figure 8

2.1.1 Problems with Bounded Zone of Contact

First, let us consider the case when the zone of contact during the process of deformation cannot expand beyond a certain domain, which is determined by the geometric situation in the vicinity of the set $\partial\Omega' \cap \partial\Omega''$—see Figure 8.

We define the zone of contact

$$\Gamma_K = \partial\Omega' \cap \partial\Omega'',$$

so that we have decompositions

$$\partial\Omega' = \bar{\Gamma}_u \cup \bar{\Gamma}'_\tau \cup \Gamma_K, \quad \partial\Omega'' = \bar{\Gamma}_0 \cup \bar{\Gamma}''_\tau \cup \Gamma_K, \tag{1.9}$$

where Γ_u, Γ'_τ, Γ''_τ, Γ_0 are pairwise disjoint open parts of the boundaries. Let Γ_u and Γ_K have positive measure. The other parts either have positive measure or are empty.

We say that *one-sided contact* occurs on Γ_K, if

$$u'_n + u''_n \leq 0 \quad \text{on } \Gamma_K, \tag{1.10}$$

where

$$u'_n = u'_i \cdot n'_i, \quad u''_n = u''_i \cdot n''_i, \quad n' = -n''.$$

Let us briefly indicate how to derive condition (1.10), which essentially represents the *condition of nonpenetrating* of the bodies into each other:

Let us assume that before the deformation both bodies Ω' and Ω'' had contact along the whole arc Γ_K (see Figure 9). Let us identify the axis x_1

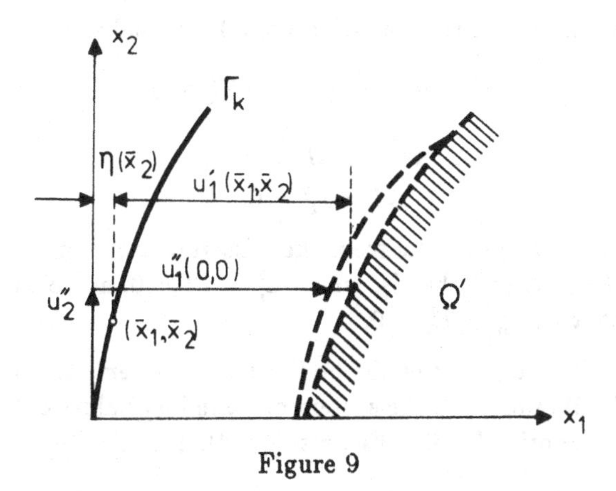

Figure 9

with the normal n'' and the axis x_2 with the tangent t'' at a certain point $0 \in \Gamma_K$. During the deformation process the points $0' \in \partial\Omega'$ and $0'' \in \partial\Omega''$ are generally displaced in a different way, nonetheless, always to that the body Ω'' cannot penetrate the body Ω'. This condition yields

$$u_1''(0,0) \leq u_1'(\bar{x}_1, \bar{x}_2) + \eta(\bar{x}_2), \tag{1.11}$$

where η is the function determined by the curve Γ_K, and $\bar{x} = (\bar{x}_1, \bar{x}_2)$ is such a point on Γ_K that $u_2'(\bar{x}_1, \bar{x}_2) + \bar{x}_2 = u_2''(0,0)$. Naturally, the point $\bar{x}$ is unknown, and thus (1.11) is too complicated a condition. This is why we simplify it by introducing "natural" hypotheses:

$$\eta(\bar{x}_2) \doteq 0, \tag{1.12}$$

$$u_1'(\bar{x}_1, \bar{x}_2) \doteq u_1'(0,0). \tag{1.13}$$

Obviously, (1.12) holds for a "flat" arc Γ_K; (1.13) is valid, for example, if the mutual displacement $|u_2' - u_2''|$ and the derivative $|\partial u_1'/\partial x_2|$ are small in a neighborhood of the point 0.

By substituting (1.12), (1.13) and $u_1'' = u_n''$, $u_1' = -u_n'$ into (1.11) we obtain the condition (1.10) for the point $0 \in \Gamma_K$.

Further, let us consider the contact forces. By the Action and Reaction Law we have

$$T_n' = T_n'', \quad T_t' = T_t'' \quad \text{on } \Gamma_K.$$

The assumption of vanishing friction implies that the tangential components also vanish. The normal components evidently cannot be tensions. Hence,

$$T_t' = T_t'' = 0, \quad T_n' = T_n'' \leq 0.$$

Altogether, we introduce the following boundary conditions on Γ_K:

$$u'_n + u''_n \leq 0, \quad T'_n = T''_n \leq 0, \tag{1.14}$$

$$(u'_n + u''_n)T'_n = 0, \tag{1.15}$$

$$T'_t = T''_t = 0. \tag{1.16}$$

The condition (1.15) results from the following argument: at the points where no contact occurs, that is, where $u'_n + u''_n < 0$, no contact force can arise either, that is, $T'_n = T''_n = 0$.

Remark 1.1. Provided one of the bodies becomes perfectly rigid, the system (1.14)–(1.16) reduces to the system of boundary values of the Signorini problem (see Signorini (1933); Fichera (1964), (1972); Duvaut and Lions (1972)).

Definition 1.1. A function u is called a classical solution of Problem $\mathcal{P}_1$ with a bounded zone of contact if it satisfies equations (1.1), (1.2), (1.5) in $\Omega' \cup \Omega''$, boundary conditions (1.6) on Γ_u, (1.7) on $\Gamma'_\tau \cup \Gamma''_\tau$, (1.8) on Γ_0, and (1.14), (1.15), (1.16) on Γ_K.

2.1.2 Problems with Increasing Zone of Contact

In some cases, the range of the contact zone may expand during the deformation process. Such a situation arises if the bodies Ω', Ω'' have smooth boundaries in the vicinity of the intersection $\partial\Omega' \cap \partial\Omega''$.

In these cases, our definition of one-sided contact would not be suitable, and requires some modifications.

As a typical case, let us consider the situation in Figure 10. Let us fix the local coordinate system (ξ, η) in such a way that the ξ-axis has the direction n'' and the η-axis has the direction of the common tangent at a certain point $P \in \partial\Omega' \cap \partial\Omega''$ at the center of the contact zone. The figure corresponds to the state before deformation. Let us now estimate those parts of $\partial\Omega'$, $\partial\Omega''$ that could come in contact during the deformation process:

$$\Gamma_K^M = \{(\xi, \eta) \mid a \leq \eta \leq b, \ \xi = f^M(\eta)\}, \quad M = \ ',''$$

f^M are continuous in the interval $[a, b]$. (This interval must be a priori estimated so as to guarantee that it contains the projection of the possible contact zone.)

Analogous to the derivation of condition (1.10), we obtain the *condition of nonpenetrating* in the form

$$u''_\xi - u'_\xi \leq \epsilon(\eta) \quad \forall \eta \in [a, b], \tag{1.17}$$

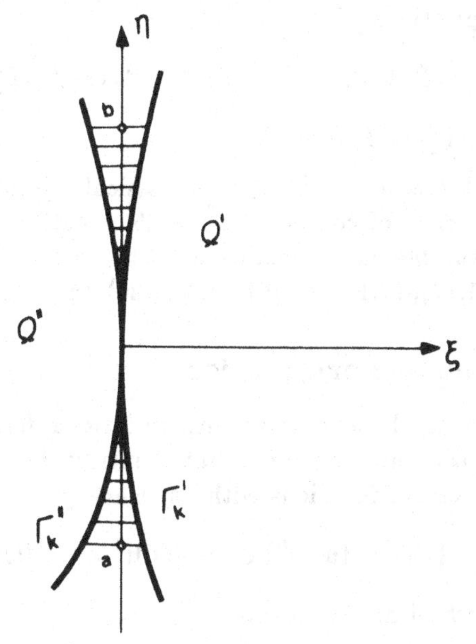

Figure 10

where $\epsilon(\eta) = f'(\eta) - f''(\eta)$ is the distance of both boundaries before deformation and u'_ξ, u''_ξ are the ξ-components of the displacement vectors.

Similarly to the preceding section, we further derive the following conditions:

$$-T'_\xi(\cos\alpha')^{-1} = T''_\xi(\cos\alpha'')^{-1} \leq 0, \tag{1.18}$$

$$T'_\eta = T''_\eta = 0, \tag{1.19}$$

$$(u''_\xi - u'_\xi - \epsilon)T''_\xi = 0, \tag{1.20}$$

which are valid for all points of $\Gamma'_K \cup \Gamma''_K$ with the same η-coordinate, $\eta \in [a, b]$. Here we use the notation

$$(\cos\alpha^M)^{-1} = \left(1 + \left(\frac{df^M}{d\eta}\right)^2\right)^{1/2}, \quad M = {}', {}'',$$

α^M being the angle between the η-axis and the tangent to Γ^M_K.

Condition (1.20) results by the following argument: if no contact occurs, that is, if $u''_\xi - u'_\xi - \epsilon < 0$, then the contact forces vanish: $T''_\xi = T'_\xi = 0$. Condition (1.19) represents an approximation of the condition of zero friction—it neglects the projections $T^M_\xi \sin\alpha^M$.

Let the decompositions

$$\partial\Omega' = \bar{\Gamma}_u \cup \bar{\Gamma}'_\tau \cup \Gamma'_K, \quad \partial\Omega'' = \bar{\Gamma}_0 \cup \bar{\Gamma}''_\tau \cup \Gamma''_K$$

be valid with $\Gamma^M_\tau \cap \Gamma^M_K = \emptyset$, $M = {}',{}''$.

Definition 1.2. A function u is called a classical solution of problem $\mathcal{P}_2$ with an increasing zone of contact if u fulfills equations (1.1), (1.2) and (1.5) in $\Omega' \cup \Omega''$, the boundary conditions (1.6) on Γ_u, (1.7) on $\Gamma'_\tau \cup \Gamma''_\tau$, (1.8) on Γ_0, and (1.17), (1.18), (1.19), and (1.20) on $\Gamma'_K \cup \Gamma''_K$.

2.1.3 Variational Formulations

Problems $\mathcal{P}_1$ and $\mathcal{P}_2$ can be associated with variational formulations on the basis of the principle of minimum of potential energy. Let us first introduce the space of displacement functions with finite energy

$$\mathcal{H}^1(\Omega) = \{u \mid u = (u', u'') \in [H^1(\Omega')]^2 \times [H^1(\Omega'')]^2\},$$

and the space of virtual displacements

$$V = \{u \in \mathcal{H}^1(\Omega) \mid u' = 0 \text{ on } \Gamma_u, \ u''_n = 0 \text{ on } \Gamma_0\}.$$

Further, we denote by $\mathcal{H}^k(\Omega)$, $k \geq 0$ integer, the space

$$[H^k(\Omega')]^2 \times [H^k(\Omega'')]^2,$$

that is,

$$u \in \mathcal{H}^k(\Omega) \Longleftrightarrow u^M = u|_{\Omega^M} \in [H^k(\Omega^M)]^2, \quad M = {}',{}''.$$

The norm and seminorm in $\mathcal{H}^k(\Omega)$ are introduced as usual,

$$|||u|||^2_{k,\Omega} = \|u'\|^2_{k,\Omega'} + \|u''\|^2_{k,\Omega''},$$

$$[[u]]^2_{k,\Omega} = [u']^2_{k,\Omega'} + [u'']^2_{k,\Omega''},$$

where $\|u^M\|_{k,\Omega^M}$ and $[u^M]_{k,\Omega^M}$, $M = {}',{}''$ denote the norm and the seminorm, respectively, in $[H^k(\Omega^M)]^2$. If there is no danger of misunderstanding, we shall omit the symbol Ω, writing only $|||u|||_k$, $[[u]]_k$.

Let us define the functional of potential energy

$$\mathcal{L}(v) = \frac{1}{2}A(v,v) - L(v), \tag{1.21}$$

where

$$A(u,v) = \int_\Omega c_{ijkm} e_{ij}(u) e_{km}(v)\,dx, \quad \Omega = \Omega' \cup \Omega'', \tag{1.22}$$

$$L(v) = \int_{\Omega} F_i v_i dx + \int_{\Gamma_\tau} P_i v_i ds, \quad \Gamma_\tau = \Gamma'_\tau \cup \Gamma''_\tau. \tag{1.23}$$

The set of admissible displacements K for problem $\mathcal{P}_1$ with a bounded zone of contact is introduced by

$$K = \{v \in V \mid v'_n + v''_n \leq 0 \quad \text{on } \Gamma_K\}. \tag{1.24}$$

Definition 1.3. A function $u \in K$ is called a weak (variational) solution of problem $\mathcal{P}_1$ with a bounded zone of contact, if

$$\mathcal{L}(u) \leq \mathcal{L}(v) \quad \forall v \in K. \tag{1.25}$$

Theorem 1.1. *Every classical solution of problem $\mathcal{P}_1$ is its weak solution. If a weak solution is sufficiently smooth, then it is a classical solution as well.*

Proof. 1^0 Let u be a classical solution. Then $\tau_{ij}(u) = c_{ijkm} e_{km}(u)$ fulfills (1.5). Multiplying (1.5) by a test function $w \in V$ and integrating over Ω' and Ω'', we obtain

$$\begin{aligned} 0 &= \int_{\Omega} \left(-\tau_{ij}(u) \frac{\partial w_i}{\partial x_j} + F_i w_i \right) dx + \int_{\partial\Omega' \cup \partial\Omega''} \tau_{ij}(u) n_j w_i ds \\ &= -A(u, w) + L(w) + \int_{\Gamma_0} (T_n(u) w_n + T_t(u) w_t) ds \\ &+ \int_{\Gamma_K} [T'_n(u) w'_n + T'_t(u) w'_t + T''_n(u) w''_n + T''_t(u) w''_t] ds. \end{aligned}$$

The integral over Γ_0 vanishes, as $w_n = 0$ and $T_t(u) = 0$ on Γ_0. Using further (1.14), (1.16), we find that

$$A(u, w) - L(w) = \int_{\Gamma_K} T'_n(u)(w'_n + w''_n) ds.$$

Let $v \in K$ and put $w = v - u$. At the points with $u'_n + u''_n < 0$, we have $T'_n(u) = 0$ (see (1.15)). At the points with $u'_n + u''_n = 0$ we have

$$w'_n + w''_n = v'_n + v''_n \leq 0, \quad T'_n(u) \leq 0.$$

Hence, the integral over Γ_K is nonnegative and

$$A(u, v - u) - L(v - u) \geq 0 \quad \forall v \in K. \tag{1.26}$$

By theorem 1.1.1 we know that (1.25) and (1.26) are equivalent, since both K and L are convex. Hence, u is a weak solution of $\mathcal{P}_1$.

2^0 Let u be a sufficiently smooth solution of $\mathcal{P}_1$. Thus, u fulfills the variational inequality (1.26). Integrating (1.26) by parts and denoting $v - u = w$, we can write

$$0 \le A(u,w) - L(w) = -\int_\Omega \left(\frac{\partial \tau_{ij}(u)}{\partial x_j} + F_i\right) w_i dx$$

$$-\int_{\Gamma_\tau} P_i w_i ds + \int_{\partial\Omega' \cup \partial\Omega''} (T_n w_n + T_t w_t) ds. \tag{1.27}$$

Let us choose $v = u \pm w$, where $w_i \in C_0^\infty(\Omega^M)$, $M = {}',{}''$, so that $v \in K$ and (1.27) imply the equations of equilibrium (1.5).

Let $v = u \pm w$ and let the support of the traces of functions w_i on $\partial\Omega' \cup \partial\Omega''$ belong to $\Gamma'_\tau \cup \Gamma''_\tau$. Then again $v \in K$, and (1.27) yields

$$0 = \int_{\Gamma_\tau} (T_i - P_i) w_i ds.$$

Hence, conditions (1.7) on $\Gamma'_\tau \cup \Gamma''_\tau$ follow. Conditions (1.6) and (1.8) are fulfilled by virtue of the definition of $u \in K$.

Let $v = u \pm w$, where the support of the trace of w_i belongs to Γ_0 and $w_n = 0$ on Γ_0. Then, $v \in K$ and (1.27) yield

$$0 = \int_{\Gamma_0} T_t w_t ds.$$

This implies the latter condition from (1.8). Thus, we have derived from (1.27)

$$0 \le \int_{\Gamma_K} (T'_n w'_n + T'_t w'_t + T''_n w''_n + T''_t w''_t) ds. \tag{1.28}$$

Let us now choose a function w such that

$$w'_n = -w''_n = \pm\psi, \quad w'_t = w''_t = 0 \quad \text{on } \Gamma_K.$$

Then,

$$0 = \int_{\Gamma_K} (T'_n - T''_n)\psi \, ds \Rightarrow T'_n = T''_n \quad \text{on } \Gamma_K. \tag{1.29}$$

Further, if we choose

$$w'_n = w''_n = 0, \quad w''_t = 0, \quad w'_t = \pm\psi \quad \text{on } \Gamma_K,$$

then

$$0 = \int_{\Gamma_K} T'_t \psi \, ds \Rightarrow T'_t = 0 \quad \text{on } \Gamma_K.$$

Analogously we derive that $T_t'' = 0$ on Γ_K. It remains to verify conditions (1.14), (1.15). To this end choose a function w such that

$$w_n'' = 0, \quad w_n' = \psi \le 0 \quad \text{on } \Gamma_K.$$

Then $v = u + w \in K$ and (1.28) yield

$$0 \le \int_{\Gamma_K} T_n' \psi \, ds \quad \forall \psi \le 0.$$

Hence, $T_n' \le 0$ on Γ_K. By virtue of (1.29) we also derive (1.14).

Let $u_n' + u_n'' < 0$ at a point $x \in \Gamma_K$. Then there exists a smooth function $\psi \ge 0$ on Γ_K such that $\psi(x) > 0$ and $u_n' + u_n'' + \psi \le 0$ on Γ_K. There exists $w \in V$ such that $w_n' = \psi$, $w_n'' = 0$ on Γ_K, hence $v = u + w \in K$. Condition (1.28) together with the inequality $T_n' \le 0$ on Γ_K implies

$$0 \le \int_{\Gamma_K} T_n' \psi \, ds \Rightarrow T_n'(x) = 0,$$

hence, (1.15) holds. □

Let us now consider problem P_2 with an increasing zone of contact. Define the set K_ϵ of admissible displacements by

$$K_\epsilon = \{v \in V \mid v_\xi'' - v_\xi' \le \epsilon \quad \text{for a.e. } \eta \in [a, b]\}.$$

Definition 1.4. A function $u \in K_\epsilon$ is called a weak (variational) solution of problem P_2 with an increasing zone of contact, if

$$\mathcal{L}(u) \le \mathcal{L}(v) \quad \forall v \in K_\epsilon. \tag{1.30}$$

Theorem 1.2. *Every classical solution of problem P_2 is its weak solution as well. If a weak solution of problem P_2 is sufficiently smooth, then it is a classical solution as well.*

Proof. 1^0 Let u be a classical solution. Multiplying the equations (1.5) by a function $w \in V$ and integrating by parts, we obtain

$$0 = -A(u, w) + L(w) + \int_{\Gamma_K'} (T_\xi' w_\xi' + T_\eta' w_\eta') ds'$$

$$+ \int_{\Gamma_K''} (T_\xi'' w_\xi'' + T_\eta'' w_\eta'') ds''.$$

Here we have also used boundary conditions (1.6), (1.7), and (1.8). By virtue of (1.19) we may write

$$A(u, w) - L(w) = \int_a^b [T_\xi' w_\xi' (\cos \alpha')^{-1} + T_\xi'' w_\xi'' (\cos \alpha'')^{-1}] d\eta.$$

Further, we make use of the relations

$$T_\xi'' u_\xi'' = T_\xi''(u_\xi' + \epsilon),$$

$$T_\xi'' w_\xi'' = T_\xi''(v_\xi'' - u_\xi'') = T_\xi''(v_\xi'' - u_\xi' - \epsilon),$$

which follow from (1.20) for an arbitrary $w = u - v$. Applying (1.18), we can write for every $v \in K_\epsilon$:

$$T_\xi'(\cos\alpha')^{-1}(v_\xi' - u_\xi') + T_\xi''(\cos\alpha'')^{-1}(v_\xi'' - u_\xi' - \epsilon)$$

$$= T_\xi''(\cos\alpha'')^{-1}(v_\xi'' - v_\xi' - \epsilon) \geq 0$$

on the interval $[a,b]$. Hence, we conclude $u \in K_\epsilon$ and

$$A(u, v-u) - L(v-u) \geq 0 \quad \forall v \in K_\epsilon, \tag{1.31}$$

which is a variational inequality equivalent to (1.30), since both K_ϵ and $\mathcal{L}$ are convex.

2^0 Let $u \in K_\epsilon$ be a sufficiently smooth weak solution of problem $\mathcal{P}_2$. Integrating in (1.31) by parts, we obtain in the same way as in the proof of theorem 1.1 that u fulfills the equations (1.5) and the boundary conditions (1.7), (1.8).

Denoting $v - u = w$, we then have

$$0 \leq \sum_{M=',''} \int_{\Gamma_K^M} (T_\xi^M w_\xi^M + T_\eta^M w_\eta^M)\, ds^M. \tag{1.32}$$

Let us choose w, satisfying

$$w_\xi' = w_\xi'' = \pm\psi, \quad w_\eta' = w_\eta'' = 0$$

on the interval $[a,b]$. Then we can write

$$0 = \int_a^b \psi[T_\xi'(\cos\alpha')^{-1} + T_\xi''(\cos\alpha'')^{-1}]\eta,$$

which implies

$$-T_\xi'(\cos\alpha')^{-1} = T_\xi''(\cos\alpha'')^{-1}. \tag{1.33}$$

Now let

$$w_\xi' = w_\xi'' = 0, \quad w_\eta'' = 0, \quad w_\eta' = \pm\psi.$$

This yields

$$0 = \int_{\Gamma_K'} \psi T_\eta'\, ds' \Rightarrow T_\eta' = 0,$$

and similarly we find that $T''_\eta = 0$.

Choosing w such that

$$w' = 0, \quad w''_\xi = \psi \leq 0,$$

we conclude that

$$0 \leq \int_{\Gamma''_K} \psi T''_\xi \, ds'' \quad \forall \psi \leq 0 \Rightarrow T''_\xi \leq 0 \quad \text{on } \Gamma''_K.$$

This together with (1.33) implies (1.18).

It is still necessary to verify condition (1.20). Let us assume that

$$u''_\xi - u'_\xi < \epsilon$$

at a point $\bar{\eta} \in [a,b]$. Then there is a smooth function $\psi \geq 0$ on the interval $[a,b]$, such that $\psi(\bar{\eta}) > 0$ and

$$u''_\xi - u'_\xi + \psi \leq \epsilon \quad \forall \eta \in [a,b].$$

There exists a function $w \in V$ such that $w''_\xi = \psi$ on Γ''_K, $w'_\xi = 0$ on Γ'_K. Then $v = u + w \in K_\epsilon$. As $T''_\xi \leq 0$ on Γ''_K, we find from (1.32) that

$$0 \leq \int_{\Gamma''_K} \psi T''_\xi \, ds'' \Rightarrow T''_\xi(\bar{\eta}) = 0$$

and the condition (1.20) is satisfied. □

2.2 Existence and Uniqueness of Solution

In this section we discuss the conditions guaranteeing the existence and uniqueness of a weak solution of problems $\mathcal{P}_1$ and $\mathcal{P}_2$.

2.2.1 Problem with Bounded Zone of Contact

First we introduce the subspace of displacements of rigid bodies

$$\mathbf{R} = \{z \in \mathcal{H}^1(\Omega) \mid z = (z', z''), \ z_1^M = a_1^M - b^M x_2,$$

$$z_2^M = a_2^M + b^M x_1, \quad M = {}',{}''\},$$

where $a_i^M \in \mathbf{R}^1$ and $b^M \in \mathbf{R}^1$, $i = 1, 2$, are arbitrary.

Evidently $e_{ij}(z) = 0$ for every $z \in R$, and hence

$$A(v, z) = 0 \quad \forall z \in R.$$

Conversely, if $p \in \mathcal{H}^1(\Omega)$, $e_{ij}(p) = 0 \ \forall i,j$, then $p \in R$. (Proof is found in Hlaváček, Nečas (1981).)

Lemma 2.1. *Let there exist a weak solution of problem $\mathcal{P}_1$. Then*

$$L(y) \leq 0 \quad \forall y \in K \cap R. \tag{2.1}$$

Proof. A weak solution u fulfills condition (1.26). Substituting there $v = u + y$ with $y \in K \cap R$, then $v \in K$, and

$$0 = A(u, y) \geq L(y). \qquad \square$$

Theorem 2.1. *Let $V \cap \mathbf{R} = \{0\}$ or*

$$L(z) \neq 0 \quad \forall z \in V \cap R \dot{-} \{0\}. \tag{2.2}$$

Then there is at most one weak solution of problem $\mathcal{P}_1$.

Proof. Let u^1, u^2 be two weak solutions. Using (1.26) we can write

$$A(u^1, u^2 - u^1) \geq L(u^2 - u^1),$$

$$A(u^2, u^1 - u^2) \geq L(u^1 - u^2).$$

The sum of these inequalities yields

$$A(u^1 - u^2,\ u^2 - u^1) \geq 0.$$

Denoting $u^1 - u^2 = z$, we have $A(z, z) \leq 0$. Now the condition (1.4) implies that $e_{ij}(z) = 0 \ \forall i, j$, hence $z \in \mathbf{R} \cap V$. If $\mathbf{R} \cap V = \{0\}$, then $z = 0$ and the solution is unique.

If $z \neq 0$, denote $u^2 = u$, $u^1 = u + z$. Then

$$A(u, z) = A(z, z) = 0,$$

$$\mathcal{L}(u) = \mathcal{L}(u + z) \Rightarrow L(u) = L(u + z) \Rightarrow L(z) = 0,$$

which contradicts assumption (2.2). Hence, again $z = 0$. $\square$

Example 2.1. Let Γ_0 consist of line segments parallel to the x_1-axis (see Figure 8). Then

$$V \cap \mathbf{R} = \{z \mid z' = (0,0),\ \ z'' = (a, 0),\ \ a \in \mathbf{R}^1\}.$$

Assume that $n_1'' \geq 0$ almost everywhere on Γ_K and that there is $x \in \Gamma_K$ with $n_1''(x) > 0$. Then,

$$K \cap \mathbf{R} = \{y \mid y' = (0,0),\ \ y'' = (a, 0),\ \ a \leq 0\}.$$

Indeed, $y \in K \cap \mathbf{R} \subset V \cap \mathbf{R}$,

$$y'_n + y''_n = an''_1 \leq 0 \quad \text{on } \Gamma_K \Longleftrightarrow a \leq 0.$$

Lemma 2.1 now imlies that a weak solution exists only if

$$V''_1 = \int_{\Omega''} F''_1 dx + \int_{\Gamma''_\tau} P''_1 ds \geq 0.$$

Indeed, substituting $y \in K \cap \mathbf{R}$ in condition (2.1) we obtain

$$0 \geq L(y) = aV''_1 \quad \forall a \leq 0.$$

Theorem 2.1 implies that if $V''_1 \neq 0$, then there exists at most one weak solution. Indeed, for $z \in V \cap R \dot{-} \{0\}$ we have

$$L(z) = aV''_1, \quad a \neq 0,$$

and if $V''_1 \neq 0$, then $L(z) \neq 0$.

Let us now present a general result on the existence of a weak solution of problem $\mathcal{P}_1$.

Define the set of "two-sided" admissible displacements of rigid bodies

$$\mathbf{R}^* = \{z \in K \cap \mathbf{R} \,|\, z \in \mathbf{R}^* \Rightarrow -z \in \mathbf{R}^*\}.$$

We immediately see that

$$\mathbf{R}^* = \{z \in V \cap \mathbf{R} \,|\, z'_n + z''_n = 0 \quad \text{on } \Gamma_K\}. \tag{2.3}$$

Theorem 2.2. *Let*

$$L(y) \leq 0 \quad \forall y \in K \cap \mathbf{R}, \tag{2.4}$$

$$L(y) < 0 \quad \forall y \in K \cap \mathbf{R} \dot{-} \mathbf{R}^*. \tag{2.5}$$

Then there exists a weak solution u of problem $\mathcal{P}_1$. Any other solution $\tilde{u}$ can be written in the form $\tilde{u} = u + y$, where $y \in V \cap \mathbf{R}$ is such a function that $u + y \in K$, $L(y) = 0$.

Proof. Can be obtained on the basis of a general abstract theorem following Fichera (1972) (see theorem 1.II, ibid.). □

Nonuniqueness of the solution is an obstacle to the numerical analysis of the given problem. Besides, when proving the convergence of the approximate method (see section 2.3) we will require the functional $\mathcal{L}$ of potential energy to be coercive on the set K of admissible functions. Therefore, in section 2.2 and 2.3 we restrict our considerations to the cases where the dimension of the space of virtual displacements of rigid bodies is at most one. This will enable us to define a contact problem with a unique solution, which, moreover, will exhibit the required coerciveness.

Remark 2.1. The case of more-dimensional spaces of virtual displacements of rigid bodies will be studied in the dual variational formulation (i.e., in terms of stresses) (see section 2.4).

Theorem 2.3. *Denote* $V \cap \mathbf{R} = \mathbf{R}_V$. *Assume that*

$$K \cap \mathbf{R} = \mathbf{R}_V, \tag{2.6}$$

$$L(y) = 0 \quad \forall y \in \mathbf{R}_V. \tag{2.7}$$

Further, let

$$V = H \oplus \mathbf{R}_V$$

be the orthogonal decomposition of the space V *(with respect to an arbitrary inner product). Then* $\mathcal{L}$ *is coercive on* H *(that is,* $\mathcal{L}(v) \to +\infty$ *for* $|||v|||_1 \to \infty$, $v \in H$*); there exists a unique solution* $\hat{u} \in \hat{K}$ *of the problem*

$$\mathcal{L}(\hat{u}) \leq \mathcal{L}(z) \quad \forall z \in \hat{K}, \quad \hat{K} = K \cap H; \tag{2.8}$$

every weak solution of problem $\mathcal{P}_1$ *can be written in the form* $u = \hat{u} + y$, *where* $\hat{u} \in \hat{K}$ *is the solution of problem* (2.8) *and* $y \in \mathbf{R}_V$; *if* $\hat{u} \in \hat{K}$ *is the solution of* (2.8), *then* $u = \hat{u} + y$, y *being an arbitrary element from* $\mathbf{R}_V$, *represents a weak solution of problem* $\mathcal{P}_1$.

Remark 2.2. The assumption (2.6) can be fulfilled only if $\dim \mathbf{R}_V \leq 1$. Indeed, $\dim \mathbf{R}_V \leq 3$ and the case $\dim \mathbf{R}_V = 2$ is impossible. Thus, let us consider the case $\dim \mathbf{R}_V = 3$, that is, $\Gamma = \emptyset$ and

$$\mathbf{R}_V = \{y = (y', y'') \mid y' = 0, \; y_1'' = a_1 - bx_2, \; y_2'' = a_2 + bx_1\},$$

where a_1, a_2, b are arbitrary real constants. Consequently, the body Ω'' is entirely free. Since the set $K \cap \mathbf{R} \subset \mathbf{R}_V$ is subjected to the condition $y_n'' \leq 0$ on Γ_K, we have $K \cap \mathbf{R} \neq \mathbf{R}_V$, which contradicts (2.6).

An example satisfying $\dim \mathbf{R}_V = 1$ and (2.6) is shown in Figure 11. In this case $a_2 = b = 0$, a_1 is arbitrary. If the force resultant satisfies $V_1'' = 0$, then $L(y) = a_1 V_1''$ for all $a_1 \in \mathbf{R}^1$ and (2.7) holds as well.

Another example is that of both Γ_0 and Γ_K being parts of concentric circumferences. Then the rigid body Ω'' can only rotate. If the resultant moment satisfies

$$M = \int_{\Omega''} (x_1 F_2'' - x_2 F_1'') dx + \int_{\Gamma_\tau''} (x_1 P_2'' - x_2 P_1'') ds = 0,$$

then $L(y) = bM = 0 \; \forall b \in \mathbf{R}^1$ and again (2.7) holds.

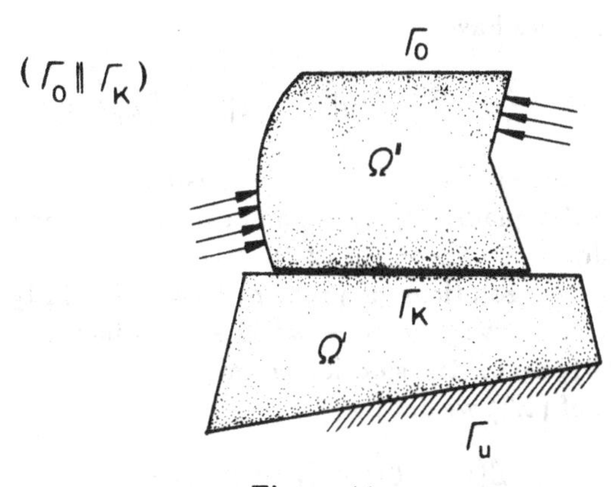

Figure 11

Remark 2.3. From the point of view of numerical methods, it is suitable to introduce in V the following types of inner product (see Hlaváček, Nečas (1981), Chapter 7). Let dim $\mathbf{R}_V = 1$. Set

$$(u, v)_V = \int_\Omega e_{ij}(u)e_{ij}(v)dx + p(u)p(v),$$

where p is a continuous linear functional on V such that

$$\{y \in \mathbf{R}_V \mid p(y) = 0\} = \{0\}. \tag{2.9}$$

For example, if (see Figure 8)

$$\mathbf{R}_V = \{y = (y', y'') \mid y' = 0,\ y_1'' = a \in \mathbf{R}^1,\ y_2'' = 0\},$$

then we can choose

$$p(v) = \int_{\Gamma_1} v_1'' ds, \tag{2.10}$$

where $\Gamma_1 \subset \partial\Omega''$, mes $\Gamma_1 > 0$. Then (see Hlaváček, Nečas (1970), Part I, remark 4)

$$H = V \ominus \mathbf{R}_V = \{v \in V \mid p(v) = 0\}. \tag{2.11}$$

Proof of Theorem 2.3. 1^0 Every $v \in H$ satisfies Korn's inequality (see Hlaváček, Nečas (1970), Part I, remark 3.4)

$$c_1 |||v|||_1 \leq |v|, \tag{2.12}$$

where $|||\cdot|||_1$ is the norm in $\mathcal{H}^1(\Omega)$ and

$$|v|^2 = \int_\Omega e_{ij}(v)e_{ij}(v)dx. \tag{2.13}$$

Then, for all $v \in H$ we have

$$\mathcal{L}(v) \geq \frac{1}{2} c_0 |v|^2 - L(v) \geq c |||v|||_1^2 - \|L\| \, |||v|||_1,$$

which implies the coerciveness of $\mathcal{L}$ on the subspace H.

2^0 Since $\mathcal{L}$ is also quadratic, convex and the set $\hat{K}$ is convex and closed, there exists a solution $\hat{u}$ of problem (2.8).

Let u^2, u^1 be two solutions of problem (2.8). Similarly to the proof of theorem 2.1, we derive $z = u^1 - u^2 \in \mathbf{R}_V$. Since $z \in H$, we have $z \in \mathbf{R}_V \cap H = \{0\}$, hence, the solution is unique.

3^0 By virtue of (2.7) we obtain

$$\mathcal{L}(v) = \mathcal{L}(v + y) \quad \forall y \in \mathbf{R}_V. \tag{2.14}$$

Moreover, for the orthogonal projection $P_H : V \to H$ we have

$$P_H(K) = K \cap H. \tag{2.15}$$

Indeed, let $v \in K$. By virtue of (2.3) and (2.6) we obtain

$$P_H v = v - P_{R_V} v, \quad \mathbf{R}^* = \mathbf{R}_V,$$

$$(P_H v)'_n + (P_H v)''_n = v'_n + v''_n \leq 0 \quad \text{on } \Gamma_K,$$

hence $P_H v \in K \cap H$. The converse inclusion

$$K \cap H = P_H(K \cap H) \subset P_H(K)$$

is trivial.

Now let u be a weak solution of problem $\mathcal{P}_1$. Using (2.14), we can write

$$\mathcal{L}(P_H v) = \mathcal{L}(P_H v + P_{R_V} v) = \mathcal{L}(v) \quad \forall v \in V.$$

Further, $P_H u \in K \cap H$ by (2.15),

$$\mathcal{L}(P_H u) = \mathcal{L}(u) \leq \mathcal{L}(v) = \mathcal{L}(P_H v) \quad \forall v \in K,$$

and (2.15) implies that $P_H u$ is a solution of problem (2.8). Since this problem has a single solution, we have $P_H u = \hat{u}$, $u = \hat{u} + y$, $y \in \mathbf{R}_V$.

4^0 Let $u = \hat{u} + y$, where $y \in \mathbf{R}_V$ is arbitrary. Then $u \in K$ by virtue of (2.3), and (2.14) implies

$$\mathcal{L}(u) = \mathcal{L}(\hat{u}) \leq \mathcal{L}(z) \quad \forall z \in \hat{K}. \tag{2.16}$$

Let $v \in K$. Then (2.14) together with the decomposition

$$v = P_H v + P_{R_V} v$$

imply

$$\mathcal{L}(z) = \mathcal{L}(v) \tag{2.17}$$

for $z = P_H v \in P_H(K) = K \cap H = \hat{K}$.

Finally, (2.16) and (2.17) yield

$$\mathcal{L}(u) \leq \mathcal{L}(v) \quad \forall v \in K.$$

Theorem 2.4. *Assume that*

$$\mathbf{R}^* = \{0\}, \quad \mathbf{R}_V \neq \{0\}, \tag{2.18}$$

$$L(y) \neq 0 \quad \forall y \in \mathbf{R}_V \dot{-} \{0\}, \tag{2.19}$$

and let either $K \cap \mathbf{R} = \{0\}$ *or*

$$K \cap \mathbf{R} \neq \{0\}, \tag{2.20}$$

$$L(y) < 0 \quad \forall y \in K \cap \mathbf{R} \dot{-} \{0\}. \tag{2.21}$$

Then $\mathcal{L}$ *is coercive on* K, *and there exists a unique weak solution of problem* $\mathcal{P}_1$.

Remark 2.4. Assumption (2.19) can be fulfilled only if dim $\mathbf{R}_V \leq 1$. Indeed, for $\Gamma_0 = \emptyset$, dim $\mathbf{R}_V = 3$ (cf. remark 2.2), the identity

$$L(y) = a_1 V_1'' + a_2 V_2'' + bM'' = 0$$

holds for every vector (a_1, a_2, b) orthogonal to (V_1'', V_2'', M'') in the space $\mathbf{R}^3$. (Here V_1'', V_2'', M'' stand for the components of the force and moment resultants of the load acting on the body Ω'' (see example 2.1 and remark 2.2). An example satisfying dim $\mathbf{R}_V = 1$, (2.18), and (2.20) is in Figure 12. Another example with dim $\mathbf{R}_V = 1$, which satisfies (2.18) and $K \cap \mathbf{R} = \{0\}$, is shown in Figure 13.

Let Γ_0 be parallel to the x_1-axis and let $V_1'' > 0$. Then

$$y \in \mathbf{R}_V \dot{-} \{0\} \Rightarrow L(y) = a_1 V_1'' \neq 0,$$

hence the condition (2.19) is fulfilled. When the situation corresponds to that shown in Figure 12, then it is also easy to verify (2.21).

Proof of Theorem 2.4. 1^0 Let us first consider the case $K \cap \mathbf{R} = \{0\}$. We shall use the following *Abstract Theorem 1* (see Nečas (1975), theorem 2.2):

Let $|u|$ be a seminorm in a Hilbert space H with a norm $\|u\|$. Let us define a subspace

$$\mathcal{R} = \{u \in H \mid |u| = 0\}.$$

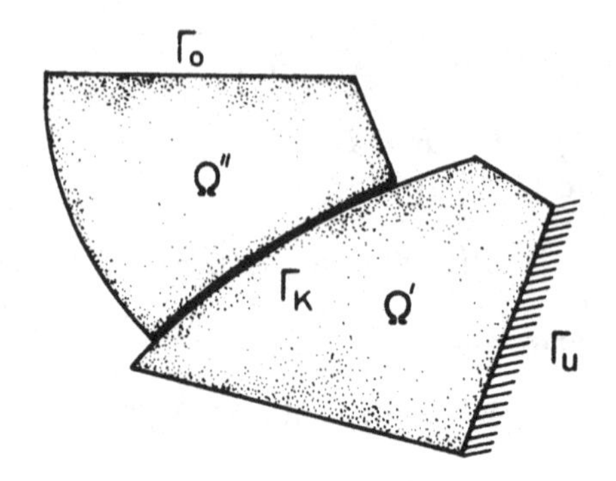

Figure 12

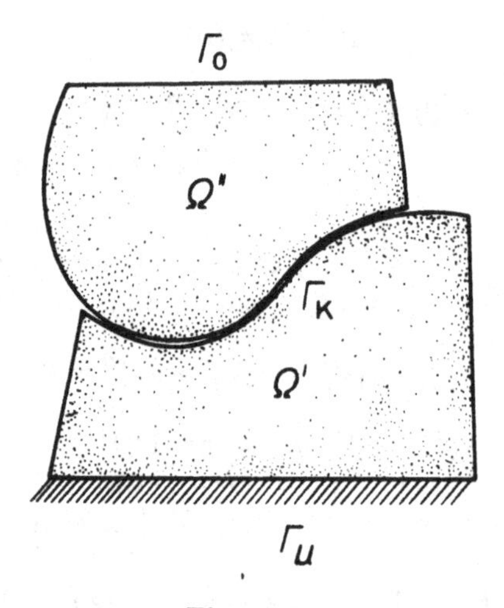

Figure 13

Assume dim $\mathcal{R} < \infty$ and

$$C_1\|u\| \le |u| + \|P_{\mathcal{R}}u\| \le C_2\|u\| \quad \forall u \in H, \tag{2.22}$$

where $P_{\mathcal{R}}$ is the orthogonal projection to $\mathcal{R}$.

Let K be a closed convex subset H containing the origin, $K \cap \mathbf{R} = \{0\}$. Let $\beta : H \to \mathbf{R}^1$ be a penalty functional whose Gateaux differential satisfies

$$D\beta(tu, v) = tD\beta(u, v) \quad \forall t > 0, \quad u, v \in H,$$

and let

$$\beta(u) = 0 \Longleftrightarrow u \in K.$$

Then

$$|u|^2 + \beta(u) \ge c\|u\|^2 \quad \forall u \in H. \tag{2.23}$$

We apply this abstract theorem to our case with $H = V$, $\mathcal{R} = V \cap \mathbf{R} = \mathbf{R}_V$, defining $|v|$ according to (2.13) and

$$\beta(u) = \frac{1}{2}\int_{\Gamma_K} [(u_n' + u_n'')^+]^2 ds.$$

In order to verify (2.22), we use an inequality of Korn's type (see Hlaváček and Nečas (1970)) and the decomposition

$$V = Q \oplus \mathbf{R}_V.$$

Thus, we obtain for all $u \in V$ the inequality

$$\begin{aligned} |||u|||_1^2 = |||P_Q u|||_1^2 + |||P_{R_V} u|||_1^2 &\le C|P_Q u|^2 + |||P_{R_V} u|||_1^2 \\ &= C|u|^2 + |||P_{R_V} u|||_1^2, \end{aligned}$$

which implies the left-hand part of (2.22). The right-hand part is obvious.

Now (2.23) implies

$$|u|^2 \ge C|||u|||_1^2 \quad \forall u \in K,$$

which easily yields that $\mathcal{L}$ is coercive on K. Hence, there exists a weak solution u of problem $\mathcal{P}_1$.

If u_1, u_2 are two weak solutions, then proceeding in the same way as in the proof of theorem 2.1, we obtain

$$y = u^1 - u^2 \in \mathbf{R}_V,$$

$$\mathcal{L}(u^1) = \mathcal{L}(u^2) \Rightarrow L(u^1) = L(u^2) \Rightarrow L(y) = 0.$$

By assumption (2.19), we conclude $y = 0$.

2^0 Let us consider the case (2.20), (2.21). We shall use *Abstract Theorem 2* (see Nečas (1975), theorem 2.3):

Let the assumptions of abstract theorem 1 be fulfilled except $K \cap \mathcal{R} = \{0\}$ (that is, we assume that $K \cap \mathcal{R} \neq \{0\}$). Further, let f be a continuous linear functional on H, such that

$$f(y) < 0 \quad \forall y \in K \cap \mathcal{R} \dot{-} \{0\}.$$

Then

$$|u|^2 + \beta(u) - f(u) \geq C_1\|u\| - C_2 \quad \forall u \in H. \tag{2.24}$$

We can apply this abstract theorem with the same H, $\mathcal{R}$, $|\cdot|$, β as in 1^0, putting in addition

$$f(v) = L(v).$$

Then (2.24) implies the coerciveness of $\mathcal{L}$ on K. The existence and uniqueness of the weak solution is then proved in the same way as above. □

Remark 2.5. The simplest case is the so-called coercive case with $V \cap \mathbf{R} = \{0\}$. Then an inequality of Korn's type holds, namely,

$$|||v|||_1 \leq C|v| \quad \forall v \in V,$$

so that $\mathcal{L}$ is coercive on the whole space V. The rest of the proof of existence and uniqueness is easy.

2.2.2 Problem with Increasing Zone of Contact

Let us again consider the case when the space of virtual displacements of the rigid bodies has the dimension one. We first introduce a result analogous to theorem 2.3.

Theorem 2.5. *Denote*

$$K_0 = \{v \in V \mid v''_\xi - v'_\xi \leq 0 \quad \forall \eta \in [a, b]\},$$

and assume that

$$\mathbf{R}_V = K_0 \cap \mathbf{R}, \tag{2.25}$$

$$L(y) = 0 \quad \forall y \in \mathbf{R}_V. \tag{2.26}$$

Let $V = H \oplus \mathbf{R}_V$ be the orthogonal decomposition of the space V (with respect to an arbitrary inner product).

Then $\mathcal{L}$ is coercive on H; there exists a unique solution $\hat{u} \in \hat{K}_\epsilon$ of problem

$$\mathcal{L}(\hat{u}) \leq \mathcal{L}(z) \quad \forall z \in K_\epsilon \cap H \equiv \hat{K}_\epsilon; \tag{2.27}$$

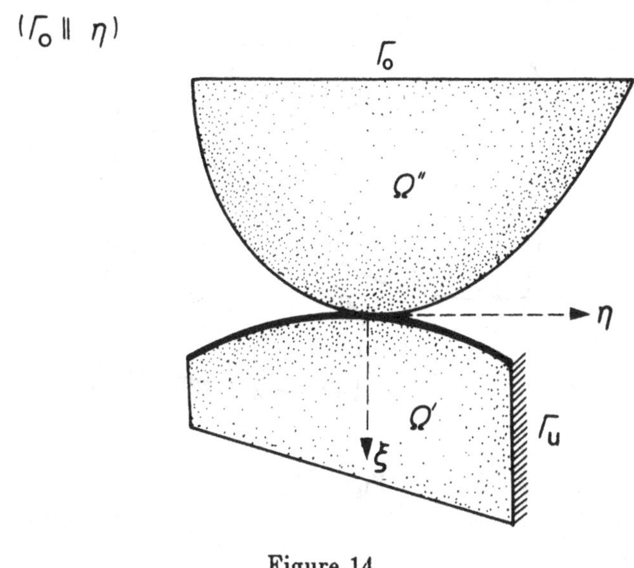

Figure 14

every weak solution of problem $\mathcal{P}_2$ can be written in the form $u = \hat{u} + y$, where $\hat{u}$ solves (2.27) and $y \in \mathbf{R}_V$; if $\hat{u} \in \hat{K}_\epsilon$ is a solution of (2.27), then $u = \hat{u}+y$, where y is an arbitrary element of $\mathbf{R}_V$, represents a weak solution of problem $\mathcal{P}_2$.

Remark 2.6. By the same argument as in remark 2.2, it follows that (2.25) can hold only if $\dim \mathbf{R}_V \leq 1$.

The case when condition (2.25) is fulfilled is in Figure 14. Then

$$\mathbf{R}_V = \{y = (y', y'') \mid y' = 0,\ y'' = (a, 0),\ a \in R^1\}$$

and (2.26) holds provided $V_1'' = 0$.

Remark 2.7. The choice of a suitable inner product in the space V can be made by the method suggested in remark 2.3.

Proof of Theorem 2.5. Analogous to that of theorem 2.3. □

Theorem 2.6. *Let us assume that Γ_0 consists of line segments parallel to the x_1-axis, $\cos(\xi, x_1) > 0$ (see Figure 15) and*

$$V_1'' = \int_{\Omega''} F_1'' dx + \int_{\Gamma_\tau''} P_1'' ds > 0. \tag{2.28}$$

Then $\mathcal{L}$ is coercive on K_ϵ and there exists a unique solution of problem $\mathcal{P}_2$.

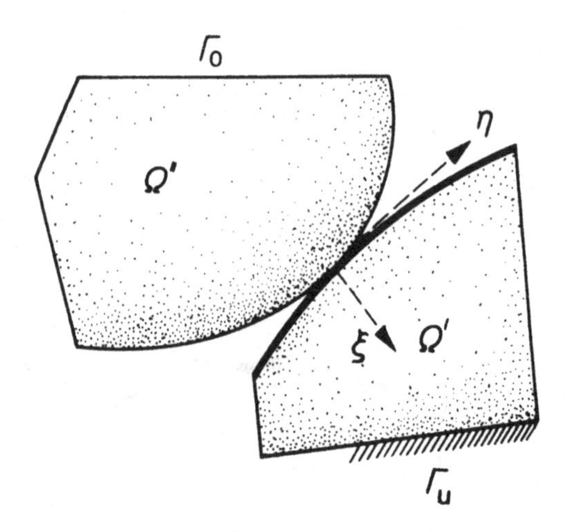

Figure 15

Proof. Let us set

$$p_0(v) = \int_a^b (v''_\xi - v'_\xi)\,d\eta,$$
$$V_p = \{v \in V \mid p_0(v) = 0\}.$$

Then

$$\mathbf{R} \cap V_p = \{0\}. \tag{2.29}$$

Indeed,

$$\mathbf{R} \cap V_p \subset \mathbf{R}_V = \{z = (z', z'') \mid z' = 0,\ z'' = (c, 0),\ c \in \mathbf{R}^1\}.$$

The identity $p_0(v) = 0$ yields

$$0 = \int_a^b z''_\xi\,d\eta = c \int_a^b \cos(\xi, x_1)\,d\eta \Rightarrow c = 0.$$

By means of (2.29) we can prove an inequality of Korn's type (see Hlaváček, Nečas (1970)):

$$|v| \geq C|||v|||_1 \quad \forall v \in V_p. \tag{2.30}$$

Let $v \in V$. Define $y \in \mathbf{R}_V$ by the relations

$$y' = 0,\quad y''_1 = p_0(v)d^{-1},\quad y''_2 = 0,$$

with

$$d = \int_a^b \cos(\xi, x_1)\,d\eta.$$

It is easily verified that the difference $Pv = v - y$ satisfies

$$p_0(Pv) = p_0(v) - p_0(y) = p_0(v) - \int_a^b p_0(v) d^{-1} \cos(\xi, x_1) d\eta = 0,$$

hence, $Pv \in V_p$.

With the help of (2.30) we can write

$$\mathcal{L}(v) = \frac{1}{2} A(Pv, Pv) - L(Pv) - L(y) \geq C_1 |||Pv|||_1^2 - C_2 |||Pv|||_1 - y_1'' V_1''. \tag{2.31}$$

When $|||v|||_1 \to \infty$ then at least one of the norms $|||Pv|||_1$, $|||y|||_1$ tends to infinity. Moreover,

$$v \in K_\epsilon \Rightarrow p_0(v) \leq \int_a^b \epsilon \, d\eta < +\infty, \tag{2.32}$$

$$|||y|||_1 = |y_1''| \left(\int_{\Omega''} dx \right)^{1/2} = |p_0(v)| d^{-1} (\text{mes}\, \Omega'')^{1/2}. \tag{2.33}$$

1^0 Let $|||y|||_1 \to \infty$. Then (2.32) together with (2.33) imply $-p_0(v) \to +\infty$, and hence $-y_1'' \to \infty$. As

$$C_1 |||Pv|||_1^2 - C_2 |||Pv|||_1 \geq C_3 > -\infty,$$

we conclude from (2.31) and (2.28) that $\mathcal{L}(v) \to +\infty$.

2^0 Let $|||Pv|||_1 \to +\infty$. Then (2.32), (2.28) yield

$$\mathcal{L}_1(Pv) = C_1 |||Pv|||_1^2 - C_2 |||Pv|||_1 \to +\infty,$$

$$\mathcal{L}_2(y) = -y_1'' V_1'' = -p_0(v) d^{-1} V_1'' \geq -d^{-1} V_1'' \int_a^b \epsilon \, d\eta > -\infty.$$

By virtue of (2.31) we have

$$\mathcal{L}(v) \geq \mathcal{L}_1(Pv) + \mathcal{L}_2(y) \to +\infty.$$

Thus, we have proved that $\mathcal{L}$ is coercive on K_ϵ.

Since K_ϵ is a closed convex subset of V and the functional $\mathcal{L}$ is convex and continuous on V, a solution of problem $\mathcal{P}_2$ exists.

Uniqueness is a consequence of condition (2.28). Indeed, we first prove—as in the proof of theorem 2.1—that the two solutions u^1 and u^2 differ from each other by an element $z \in \mathbf{R}_V$ with $L(z) = 0$. On the other hand, however, $L(z) = cV_1''$, $c \in \mathbf{R}^1$. Condition (2.28) implies $c = 0$, that is, $z = 0$.

2.3 Solution of Primal Problems by the Finite Element Method

In this section we will study approximations of contact problems by the finite element method. We will describe a construction of finite-dimensional approximations of the set of admissible displacements, which will be used for defining approximate solutions. This will be done first for the contact problems with a bounded zone of contact, the contact zone being given first by a piecewise linear curve, then by a smooth one. Subsequently, we will deal in the same way with problems with an increasing zone of contact. Further, we will discuss the mutual relation of the approximate and the exact solution. We will also find the rate of convergence, provided the exact solution is sufficiently smooth.

2.3.1 Approximation of the Problem with a Bounded Zone of Contact

Let us consider problem $\mathcal{P}_1$, using the symbols V, K, $\mathcal{L}$, A, and L in the same sense as in section 2.1.3.

I. First, let us assume that Ω', $\Omega'' \subset \mathbf{R}^2$ are bounded domains with *polygonal* boundaries $\partial\Omega'$, $\partial\Omega''$. In this case, we can write $\bar{\Gamma}_K$ in the form

$$\bar{\Gamma}_K = \bigcup_{i=1}^{m} \bar{\Gamma}_{K,i},$$

where $\bar{\Gamma}_{K,i}$ is a closed line segment with an initial point A_i and an end-point A_{i+1}. Let $\mathcal{T}_h'$, $\mathcal{T}_h''$ denote triangulations of polygonal domains Ω' and Ω''. Here we observe the current rules, which were formulated in section 1.1.31. Naturally, we assume that both $\mathcal{T}_h'$ and $\mathcal{T}_h''$ are consistent with the respective decompositions of the boundaries $\partial\Omega'$ and $\partial\Omega''$. Moreover, the nodes lying on Γ_K *belong to both the triangulations.* The pair $\{\mathcal{T}_h', \mathcal{T}_h''\}$ defines a decomposition of the set $\Omega = \Omega' \cup \Omega''$. More frequently, we will use a simpler notation, namely, $\mathcal{T}_h = \{\mathcal{T}_h', \mathcal{T}_h''\}$. $\mathcal{T}_h$ is said to be *regular* if both $\mathcal{T}_h'$, $\mathcal{T}_h''$ are regular. We associate every triangulation $\mathcal{T}_h$ with a finite dimensional space V_h, given by

$$V_h = \{v_h \in [C(\bar{\Omega}')]^2 \times [C(\bar{\Omega}'')]^2 \cap V \,|\, v_{h|T} \in [P_1(T)]^2 \;\; \forall T \in \mathcal{T}_h\}. \quad (3.1)$$

Let a_j^i, $j = 1, \ldots, m_i$ be the vertices of $\mathcal{T}_h$ lying on $\bar{\Gamma}_{K,i}(a_1^i \equiv A_i,\ a_{m_i}^i \equiv A_{i+1})$, $i = 1, \ldots, m$, and let n^i be the unit vector of the outer normal of the side $\Gamma_{K,i}$ with respect to Ω'. Let us define

$$K_h = \{v_h \in V_h \,|\, n^i \cdot (v_h' - v_h'')(a_j^i) \le 0,\ i = 1, \ldots, m,\ j = 1, \ldots, m_i\}. \quad (3.2)$$

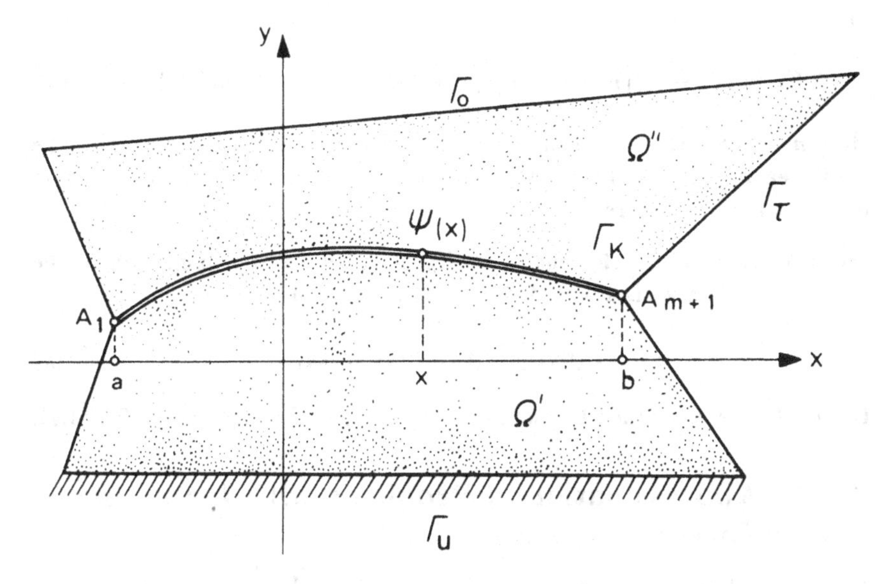

Figure 16

The reader easily proves

Lemma 3.1. *$K_h \subset K$ for every $h \in (0,1)$.*

II. Let us consider the case of sets Ω', Ω'' with *more general* boundaries than those studied in I. For the sake of simplicity, let us restrict ourselves to the case when only Γ_K is curved. Let ψ be a continuous concave (or convex) curve defined on $[a,b]$ (see Figure 16), whose graph coincides with $\bar{\Gamma}_K$. On $\bar{\Gamma}_K$ let us choose $m+1$ points $A_1, \ldots A_{m+1}$ such that $A_1 = (a, \psi(a))$, $A_{m+1} = (b, \psi(b))$. Let A_i, $A_{i+1} \in \bar{\Gamma}_K$, $S \in \Omega^M$, $M = ','',$. A *curved element* T is the closed set bounded by the line segments SA_i, SA_{i+1} and by the arc A_iA_{i+1}. The minimal inner angle of the straight triangle $A_iA_{i+1}S$ will be called the minimal inner angle of the curve element T. We say that an ordered pair $\mathcal{T}_h = \{\mathcal{T}_h', \mathcal{T}_h''\}$ is a *triangulation* of the set $\bar{\Omega}$, if $\mathcal{T}_h^M$, $M = ','',$ is a triangulation of Ω^M, $M = ','',$ which consists of the one hand of the curved elements along the part Γ_K, and on the other hand of straight triangular elements inside Ω^M, $M = ','',$. Let h denote the maximum of diameters and θ the minimal inner angle of all elements $T \in \mathcal{T}_h$. We introduce, as usual, the notion of a regular system of triangulations. We define

$$V_h = \{v_h \in [(C(\bar{\Omega}')]^2 \times [C(\bar{\Omega}'')]^2 \cap V \mid v_{h|T} \in [P_1(T)]^2 \ \ \forall T \in \mathcal{T}_h\} \quad (3.3)$$

and

$$K_h = \{v_h \in V_h \mid n \cdot (v_h' - v_h'')(A_i) \leq 0 \quad \forall i = 1, \ldots, m+1\}. \tag{3.4}$$

Here n denotes the vector of the outer (with respect to Ω') normal. The reader easily verifies that in this case the inclusion $K_h \subset K$ $\forall h \in (0,1)$ does not hold any more.

Definition 3.1. An element $u_h \in K_h$ is called an approximation of the contact problem with a bounded zone of contact, if

$$\mathcal{L}(u_h) \leq \mathcal{L}(v_h) \quad \forall v_h \in K_h, \tag{$\mathcal{P}_{1h}$}$$

the set K_h being defined by one (and only one) of the formulae (3.2), (3.4).

2.3.2 Approximation of Problems with Increasing Zone of Contact

Let us now describe the approximation of problem $\mathcal{P}_2$, considering the same decomposition of the boundaries $\partial\Omega'$, $\partial\Omega''$ as in section 2.1.2.

For the sake of simplicity, let us suppose in the sequel that only Γ_K', Γ_K'' are curved and that the functions f', f'' describing these arcs (see section 2.1.2) are twice continuously differentiable on $[a,b]$. Curved elements T are defined in the same way as in II of the preceding section. For the construction of the finite dimensional space of functions on T we use the technique due to Zlámal (1973).

Let $\hat{T}$ be a triangle with vertices (0,0), (1,0), (0,1). Let A_i, $A_{i+1} \in \Gamma_K'$, $S \in \Omega'$ (say), and let $x = \varphi(s)$, $y = \psi(s)$, $s \in [0,1]$, $\varphi, \psi \in C^2([0,1])$, be parametric equations of the arc A_iA_{i+1}. The symbol T stands for the curved element determined by the points A_i, A_{i+1}, S. It is known that—provided the diameter of T is not big—there exists a one to one mapping $F_T : \mathbf{R}^2 \to \mathbf{R}^2$ of $\hat{T}$ onto T and, moreover, this mapping is continuously differentiable in each of its variables.

Let $\hat{P} = P_1(\hat{T})$ be the set of linear polynomials defined on $\hat{T}$. The corresponding set of functions defined on the curved element $T = F_T(\hat{T})$ is the set

$$P(T) = \{p \mid \exists \hat{p} \in \hat{P}, \ \ p = \hat{p} \circ F_T^{-1}\}, \tag{3.5}$$

where F_T^{-1} is the inverse mapping of F_T.

The triangulation $\mathcal{T}_h = \{\mathcal{T}_h', \mathcal{T}_h''\}$ of the set $\bar{\Omega}$ consists on the one hand of the curved elements along Γ_K', Γ_K'' and, on the other hand, of the inner triangular elements. The elements along Γ_K', Γ_K'' are constructed in the following way: let $\{C_j\}_{j=1}^m$ be a partition of $[a,b]$, $C_1 = a$, $C_m = b$. The

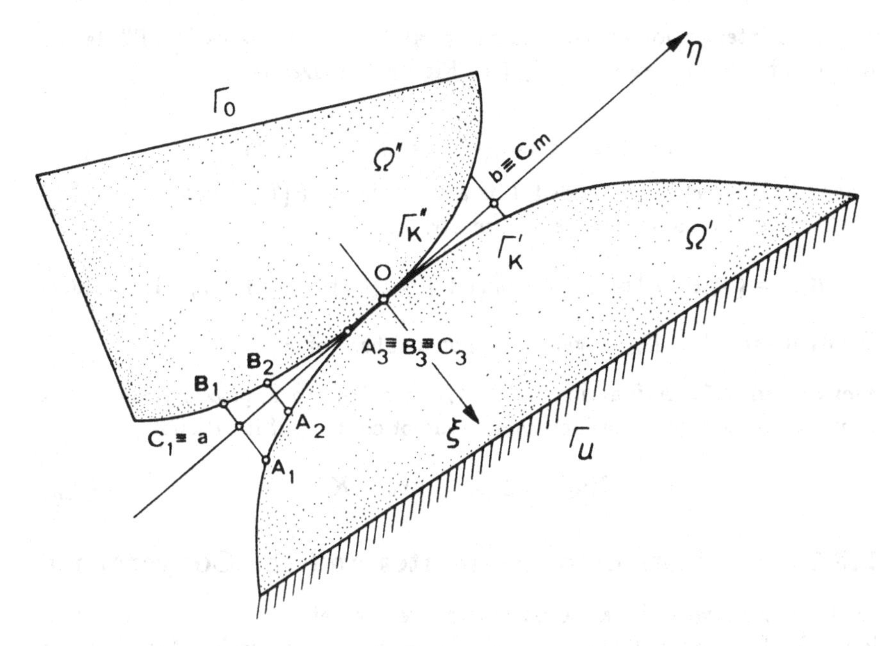
Γ_0
Ω''
$b \equiv C_m$
η
Γ_K''
Γ_K'
Ω'
O
$A_3 \equiv B_3 \equiv C_3$
B_2
B_1
$C_1 \equiv a$
A_2
A_1
ξ
Γ_u

Figure 17

points of intersection of perpendiculars at C_j with the arcs Γ'_K, Γ''_K let be denoted by A_j, B_j, respectively (see Figure 17). Define

$$V_h = \{v_h \in V \mid v_{h|T} \in [P(T)]^2 \quad \forall T \in \mathcal{T}_h\}, \tag{3.6}$$

where $P(T) = P_1(T)$ provided T is a triangle, or $P(T)$ is defined by (3.5) in the case of a curved element. Finally, let

$$K_{\epsilon h} = \{v_h \in V_h \mid v''_{h\xi}(B_j) - v'_{h\xi}(A_j) \leq \epsilon(C_j), \quad j = 1, \ldots, m\}. \tag{3.7}$$

We easily see that $K_{\epsilon h}$ is generally not a subset of K_ϵ.

Definition 3.2. A function $u_h \in K_{\epsilon h}$ is called an approximation of the contact problem with an increasing zone of contact, if it satisfies

$$\mathcal{L}(u_h) \leq \mathcal{L}(v_h) \quad \forall v_h \in K_{\epsilon h}. \tag{$\mathcal{P}_{2h}$}$$

2.3.3 A Priori Error Estimates and the Convergence

In this section we will derive estimates of error between u and u_h for problems $\mathcal{P}_1$, $\mathcal{P}_2$, provided the solution u is *sufficiently smooth*. Moreover, for problem $\mathcal{P}_1$ with Ω', Ω'' polygonal (i.e., case I), as well as for problem $\mathcal{P}_2$, we will prove convergence of u_h to u even for a nonsmooth solution u. To this end we will use the result of section 1.1.32. First, let us transcribe the relations (3.10), (3.10') of the just mentioned section, using the symbols introduced in this chapter. We have

$$c_0|u - u_h|^2 = A(u - u_h, u - u_h) \leq L(u - v_h) + L(u_h - v) + A(u_h - u, v_h - u)$$
$$+ A(u, v - u_h) + A(u, v_h - u) \quad \forall v_h \in K_h, \; v \in K, \tag{3.8}$$

and, provided $K_h \subset K$,

$$c_0|u - u_h|^2 \leq A(u - u_h, u - u_h) \leq L(u - v_h) + A(u_h - u, v_h - u)$$
$$+ A(u, v_h - u) \quad \forall v_h \in K_h. \tag{3.9}$$

Here we have also used properties (1.3) and (1.4) of the coefficients of Hooke's Law (the symbol $|\cdot|$ is defined by (2.13)).

2.3.31. Bounded Zone of Contact. When studying the approximation of problem $\mathcal{P}_1$, we will consistently distinguish between cases I and II. From the viewpoint of the error estimate, both cases essentially differ in one point. In case I, the convex sets K_h represent the *inner* approximation of K, while in case II, K_h are generally not subsets of K, which makes it necessary to use the more complicated relation (3.8) in order to establish an error estimate.

2.3.311. Polygonal Domains. Let us consider the case of bounded polygonal domains Ω', Ω'', with K_h defined by (3.2). It is evident that sufficient conditions for the existence and uniqueness of solution of problem $\mathcal{P}_{1h}$ are those which guarantee the existence and uniqueness of the continuous problem $\mathcal{P}_1$. This is a consequence of the fact that the coerciveness of $\mathcal{L}$ on K automatically guarantees the coerciveness of $\mathcal{L}$ on each K_h (since $K_h \subset K$ $\forall h \subset (0,1)$) and moreover, $\mathbf{R}$ (the space of displacements of rigid bodies, introduced in section 2.2.1) is part of V_h. We have

Theorem 3.1. *Let there exist solutions u, u_h of problems $\mathcal{P}_1$, $\mathcal{P}_{1h}$, respectively, and let $u \in \mathcal{H}^2(\Omega) \cap K$, $u', u'' \in [W^{1,\infty}(\Gamma_{K,i})]^2$, $i = 1, \dots, m$,*[1] *$T_n'(u) = T_n''(u) \in L^\infty(\Gamma_K)$. Further, let us assume that the set of points at which the change of $u_n' - u_n'' < 0$ to $u_n' - u_n'' = 0$ occurs, is finite. Then*

$$|u - u_h| \le ch\{[[u]]_{2,\Omega}^2 + \sum_{i=1}^{m} \|T_n(u)\|_{\infty,\Gamma_{Ki}} \cdot$$

$$(|u'|_{1,\infty,\Gamma_{Ki}} + |u''|_{1,\infty,\Gamma_{Ki}})\}^{1/2}. \tag{3.10}$$

provided the system $\{\mathcal{T}_h\}$, $h \to 0_+$, is regular.

Proof. As $K_h \subset K$ for all $h \in (0,1)$, we will estimate the error by (3.9). For $v_h \in K_h$ we obtain by integration by parts that

$$L(u - v_h) + A(u, v_h - u) = (T_n(u), (v_{hn}' - u_{hn}') - (v_{hn}'' - u_{hn}''))_{0,\Gamma_K},$$

and hence (3.9) can be written in the form

$$c_0|u - u_h|^2 \le A(u_h - u, v_h - u) + (T_n(u), (v_{hn}' - u_{hn}')$$

$$- (v_{hn}'' - u_{hn}''))_{0,\Gamma_K} \quad \forall v_h \in K_h. \tag{3.11}$$

Here $T_n(u)$ is the common value of $T_n'(u)$, $T_n''(u)$ on Γ_K and $(\cdot,\cdot)_{0,\Gamma_K}$ is the inner product in $L^2(\Gamma_K)$. For the function v_h we choose the piecewise linear Lagrangian interpolation of the function u, which we denote by $r_h u = (r_h u', r_h u'')$. Its definition immediately implies that $r_h u \in K_h$, since

$$n^i \cdot (r_h u' - r_h u'')(a_j^i) = n^i \cdot (u' - u'')(a_j^i) \le 0.$$

---

[1] The symbol $u' \in W^{k,\infty}(\Gamma_{K,i})$ means that there exists the $(k-1)$-st derivative of the trace u' in the direction $\Gamma_{K,i}$, and this derivative is an absolutely continuous function of the parameter of the side $\Gamma_{K,i}$. Moreover, the k-th derivative, which exists a.e. on $\Gamma_{K,i}$, is bounded and measurable on $\Gamma_{K,i}$. The symbols $\|\cdot\|_{k,\infty,\Gamma_{Ki}}$ and $|\cdot|_{k,\infty,\Gamma_{Ki}}$ stand for the norm and seminorm, respectively, in $W^{k,\infty}(\Gamma_{Ki})$. For $k = 0$ we simply write $\|\cdot\|_{\infty,\Gamma_{Ki}}$ and $|\cdot|_{\infty,\Gamma_{Ki}}$, respectively.

By using the classical results concerning the interpolation of functions by piecewise linear polynomials we obtain

$$|A(u_h - u, r_h u - u)| \le \frac{1}{2}|u - u_h|^2 + c[[u - r_h u]]_1^2$$

$$\le \frac{1}{2}|u - u_h|^2 + ch^2[[u]]_2^2. \tag{3.12}$$

Let

$$\Gamma_{K,i}^- = \{x \in \Gamma_{K,i} \mid (u_n' - u_n'')(x) < 0\},$$
$$\Gamma_{K,i}^0 = \{x \in \Gamma_{K,i} \mid (u_n' - u_n'')(x) = 0\}.$$

Set $\mathcal{U}_i = (r_h u' - r_h u'') \cdot n^i$ on $\Gamma_{K,i}$. It is easily seen that $\mathcal{U}_i$ is a piecewise linear Lagrangian interpolation of the function $u_n' - u_n''$ on $\Gamma_{K,i}$. Let $s_j^i = a_j^i a_{j+1}^i$ be the side of a triangle which lies on $\Gamma_{K,i}$. Exactly one of the following three possibilities may occur concerning its location with respect to $\Gamma_{K,i}^-$, $\Gamma_{K,i}^0$:

(i) $s_j^i \subseteq \Gamma_{K,i}^0$. In this case $u_n' - u_n'' = 0$ on s_j^i, and thus

$$\mathcal{U}_i \equiv 0 \quad \text{on } s_j^i \tag{3.13}$$

as well.

(ii) $s_j^i \subseteq \Gamma_{K,i}^-$. Then (1.15) implies that

$$T_n(u) = 0 \quad \text{on } s_j^i. \tag{3.14}$$

(iii) Finally, s_j^i may contain in its interior points from $\Gamma_{K,i}^-$ as well as those from $\Gamma_{K,i}^0$. Let $\mathcal{T}_i$ denote the set of those $s_j^i \subseteq \Gamma_{K,i}$ for which this last case occurs. With regard to the choice of v_h, and by virtue of (3.13), (3.14) and the definition of $\mathcal{U}_i$, we obtain that the right-hand side of (3.11) can be written in the form

$$\sum_{i=1}^{m} \sum_{s_j^i \in \mathcal{T}_i} (T_n(u), \mathcal{U}_i - (u_n' - u_n''))_{0,s_j^i}. \tag{3.15}$$

Using the assumptions of the theorem on the smoothness of $T_n(u)$ and $u_n' - u_n''$ on Γ_K, we conclude

$$|(T_n(u), \mathcal{U}_i - (u_n' - u_n''))_{0,s_j^i}| \le h\|T_n(u)\|_{\infty,s_j^i} \|\mathcal{U}_i - (u_n' - u_n'')\|_{\infty,s_j^i}$$

$$\le ch^2\|T_n(u)\|_{\infty,s_j^i} |u_n' - u_n''|_{1,\infty,s_j^i}$$

$$\le ch^2\|T_n(u)\|_{\infty,s_j^i} (|u'|_{1,\infty,s_j^i} + |u''|_{1,\infty,s_j^i}), \tag{3.16}$$

using in addition the fact that $\mathcal{U}_i$ is a piecewise linear interpolation of $u'_n - u''_n$ on $\Gamma_{K,i}$. As the set of points on Γ_K at which the change from $u'_n - u''_n < 0$ to $u'_n - u''_n = 0$ occurs is finite, the number of elements of the set $\mathcal{T}_i$ is bounded from above independently of h. This together with (3.11), (3.12), (3.15), and (3.16) yields the estimate (3.10). □

Remark 3.1. The existence of solutions u, u_h of problems $\mathcal{P}_1$, $\mathcal{P}_{1h}$, respectively, is presumed. Sufficient conditions of existence of solutions were formulated in section 2.2. Let us point out that the uniqueness of solution for any one of the problems *is not required.*

Remark 3.2. If the solution u is merely supposed to fulfill $u \in \mathcal{H}^2(\Omega) \cap K$, $T_n(u) \in L^2(\Gamma_K)$, then it is possible to show

$$|u - u_h| = 0(h^{3/4}), \quad h \to 0_+. \tag{3.17}$$

Remark 3.3. In the coercive case, when Korn's inequality holds on the space V, we can write the *norm* in $\mathcal{H}^1(\Omega)$ on the left hand side of (3.10) and (3.17). At the same time, $\mathcal{P}_1$ and $\mathcal{P}_{1h}$ have unique solutions u and u_h, respectively.

Since the smoothness conditions imposed on the solutions are relatively strict, we will, in the following, study the convergence of u_h to the solution u without any additional assumptions on its smoothness. To this end, we first prove the following auxiliary result.

Lemma 3.2. *Assume that* $\bar{\Gamma}_K \cap \bar{\Gamma}_u = \emptyset$, $\bar{\Gamma}_K \cap \bar{\Gamma}_0 = \emptyset$, *and let there exist only a finite number of boundary points of* $\bar{\Gamma}_\tau \cap \bar{\Gamma}_K$, $\bar{\Gamma}_u \cap \bar{\Gamma}_\tau$, $\bar{\Gamma}_\tau \cap \bar{\Gamma}_0$. *Then the set*

$$\mathcal{M} \equiv K \cap [C^\infty(\bar{\Omega}')]^2 \times [C^\infty(\bar{\Omega}'')]^2$$

is dense in K *with respect to the norm of the space* $\mathcal{H}^1(\Omega)$.

Proof. Let $u \in K$ be an arbitrarily chosen but fixed function. Consider a system of open sets $\{B_i\}_{i=0}^r$ covering $\bar{\Omega}' \cup \bar{\Omega}''$ and satisfying

$$B_0 \subset \Omega', \quad B_1 \subset \Omega'',$$

$$\bar{\Gamma}_K \subset \bigcup_{j=2}^{k} B_j \quad (k < r),$$

$$\bar{\Gamma}_K \cap B_i \neq \emptyset \Longleftrightarrow 2 \le i \le k.$$

We say that a point $P \in \partial\Omega' \cup \partial\Omega''$ is a singular point if P is either a vertex of the polynomial boundary or an element of one of the intersections

$$\bar{\Gamma}_K \cap \bar{\Gamma}_\tau, \quad \bar{\Gamma}_u \cap \bar{\Gamma}_\tau, \quad \bar{\Gamma}_0 \cap \bar{\Gamma}_\tau.$$

In the following we will assume that each B_j contains at most one singular point. Let $\{\varphi_i\}_{i=0}^r$ be the corresponding partition of unity, that is,

$$\varphi_i \in C_0^\infty(B_i), \quad 0 \le \varphi_i \le 1, \quad \sum_{i=0}^{r} \varphi_i(x) = 1 \quad \forall x \in \bar{\Omega}' \cup \bar{\Omega}'',$$

and let

$$u^j = u\varphi_j, \quad j = 0, \dots, r.$$

Evidently supp $u^j \in B_j$, $u^j \in \mathcal{H}^1(\Omega)$, $\sum_{j=0}^r u^j = u$. Now, for each u^j we construct an infinitely differentiable function from K, which is close (in the norm of $\mathcal{H}^1(\Omega)$) to the function u^j. To this end we divide the sets B_i, $i = 0, \dots, r$ into several groups.

Group 1. Let $j \le k$ and let B_j contain no singular point. Let us introduce a local Cartesian coordinate system (ξ, η) so that the axis ξ coincides with Γ_K, and the axis η with the unit vector of the outer normal to Γ_K. Then we can write (omitting the index j):

$$\Gamma_K \cap B = \{(\xi, \eta) \mid |\xi| < \xi_0, \quad \eta = 0\},$$

$$u^M = u_\xi^M e_\xi + u_\eta^M e_\eta, \quad M = {}', {}''$$

(e_ξ, e_η are the unit vectors along the ξ- and η-axes, respectively), and moreover,

$$u_n' - u_n'' = u_\eta' - u_\eta'' \le 0 \quad \text{on } \Gamma_K. \tag{3.18}$$

Let us now continue the function u_η' to $B \cap \Omega''$ and u_η'' to $B \cap \Omega'$ so that the resulting functions, denoted respectively by Eu_η', Eu_η'', are even with respect to η. Let $R_\mathcal{H} Eu_\eta'$ stand for the regularization of the function Eu_η':

$$\mathbf{R}_\mathcal{H} Eu_\eta'(x) = \int_B \omega_\mathcal{H}(x - x') Eu_\eta'(x') dx', \quad x' = (\xi', \eta'), \tag{3.19}$$

where $\omega_\mathcal{H}(x, \mathcal{H})$ is the usual regularization kernel (see section 1.1.63, Chapter 1). It can be shown (see Nečas (1967)) that there exists a function $v \in H^1(B)$ such that

$$v \le 0 \text{ in } B, \quad \text{supp } v \subset B,$$

$$v = u_\eta' - u_\eta'' < 0 \quad \text{on } \Gamma_K.$$

Then we can write

$$Eu_\eta' - Eu_\eta'' = v + z, \tag{3.20}$$

where $z \in H^1(B)$ and the restriction $z|_{\Omega^M} \in H_0^1(B \cap \Omega^M)$, $M = {}^{\prime},{}^{\prime\prime}$. The regularized function $\mathbf{R}_{\varkappa} v$ evidently fulfills

$$\mathbf{R}_{\varkappa} v \leq 0 \ \text{ on } \Gamma_K, \quad \mathbf{R}_{\varkappa} v \in C_0^{\infty}(B),$$

$$\mathbf{R}_{\varkappa} v \to v \ \text{ in } H^1(B) \ \text{ for } \varkappa \to 0_+.$$

Since $z|_{\Omega^M} \in H_0^1(B \cap \Omega^M)$, $M = {}^{\prime},{}^{\prime\prime}$, there exist functions $z_{\varkappa}^M \in C_0^{\infty}(B \cap \Omega^M)$ such that

$$z_{\varkappa}^M \to z|_{\Omega^M}, \quad \varkappa \to 0_+, \quad M = {}^{\prime},{}^{\prime\prime}$$

in the norm $H^1(B \cap \Omega^M)$. Altogether, we have

$$\mathbf{R}_{\varkappa} v + z_{\varkappa} \leq 0 \ \text{ on } \Gamma_K, \quad z_{\varkappa} = (z'_{\varkappa}, z''_{\varkappa}) \in C_0^{\infty}(B),$$

$$\mathbf{R}_{\varkappa} v + z_{\varkappa} \to v + z \ \text{ in } \ H^1(B). \tag{3.21}$$

Finally, let us set

$$u'_{\eta\varkappa} = \mathbf{R}_{\varkappa} E u'_{\eta}|_{\Omega'}$$

$$u''_{\eta\varkappa} = [\mathbf{R}_{\varkappa} E u'_{\eta} - (\mathbf{R}_{\varkappa} v + z_{\varkappa})]|_{\Omega''}.$$

Then, (3.19), (3.20), and (3.21) imply

$$u_{\eta\varkappa}^M \to u_{\eta}^M, \quad \varkappa \to 0_+ \ \text{ in } \ H^1(B \cap \Omega^M). \tag{3.22}$$

Moreover,

$$u'_{\eta\varkappa} - u''_{\eta\varkappa} = \mathbf{R}_{\varkappa} v + z_{\varkappa} \leq 0 \quad \text{on } \Gamma_K. \tag{3.23}$$

The components u_{ξ}^M can be regularized directly.

Group 2. Let $j \leq k$ and let B_j contain a vertex $P \in \Gamma_K$. In the following we shall use a skew coordinate system. Let e^1, e^2 and n^1, n^2 be the tangent and the normal vectors with respect to $\partial\Omega'$ (see Figure 18). Then we can write (again omitting the index j):

$$u' = \sum_{p=1}^{2} \frac{u'^{(p)}}{e^p \cdot n^p} e^p,$$

where $u'^{(p)} = u' \cdot n^p$. Evidently

$$u'^{(p)} = u'_n \ \text{ on } \ \Gamma^{(p)}, \quad p = 1, 2.$$

The same decomposition holds for u'' as well, and

$$u''^{(p)} = -u''_n \ \text{ on } \ \Gamma^{(p)}, \quad p = 1, 2.$$

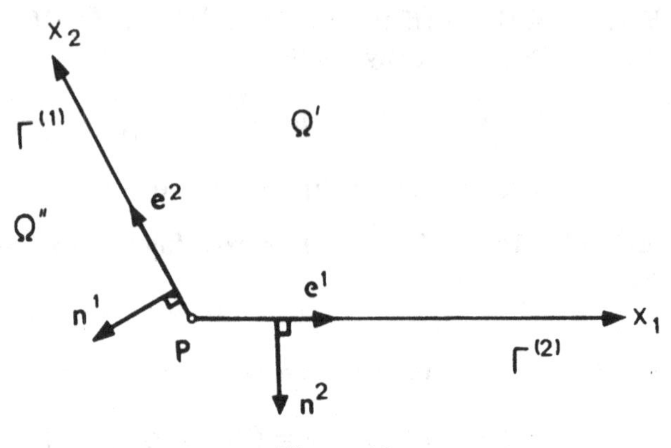

Figure 18

Altogether,

$$u_n' + u_n'' = u'^{(p)} - u''^{(p)} \leq 0 \quad \text{on} \quad \Gamma^{(p)}.$$

Let us first consider the component $u^{M(2)}$. By a Lipschitzian mapping T we map the set $B \cap \Omega'$ to the upper halfplane $\{(\xi, \eta) \mid \eta > 0\}$, with $T\Gamma^{(2)}$ and $T\Gamma^{(1)}$ coinciding with the positive and negative halfaxes ξ, respectively. We continue the function

$$\hat{u}^{M(2)}(\xi, \eta) = u^{M(2)}(T^{-1}(\xi, \eta))$$

to the lower halfplane $\{(\xi, \eta) \mid \eta < 0\}$ so as to obtain an even function in the variable η. Set

$$Eu^{M(2)}(x) = E\hat{u}^{M(2)}(T(x)).$$

Regularization yields $\mathbf{R}_{\varkappa} Eu'^{(2)} \in C_0^\infty(B)$.

Now let us continue the function

$$\hat{U} = \hat{u}'^{(2)} - \hat{u}''^{(2)} \leq 0 \quad \text{on } T\Gamma^{(2)}$$

from the positive halfaxis ξ to the negative one so as to obtain an even function $E\hat{U}$ in ξ. Then there is a function $\hat{v} \in H^1(TB)$ such that $\hat{v} = E\hat{U}$ on the ξ-axis, $\hat{v} \leq 0$ in TB, supp $\hat{v} \subset TB$. Put

$$v(x) = \hat{v}(Tx).$$

Evidently $v \in H^1(B)$, supp $v \subset B$, $v \leq 0$ in B, $v = u'^{(2)} - u''^{(2)}$ on $\Gamma^{(2)}$. Hence, we can write

$$Eu'^{(2)} - Eu''^{(2)} = v + z,$$

where $z \in H^1(B)$, supp $z \subset B$, $z = 0$ on $\Gamma^{(2)}$. Regularizing the function v we obtain

$$\mathbf{R}_{\varkappa} v \leq 0 \quad \text{on } \Gamma, \quad \mathbf{R}_{\varkappa} v \to v, \quad \varkappa \to 0_+, \quad \text{in } H^1(B). \tag{3.24}$$

As $z = 0$ on $\Gamma^{(2)}$, we can find a function $w \in H^1(B)$, supp $w \subset B$, such that

$$w = z \quad \text{on} \quad \Gamma^{(1)} \cup \Gamma^{(2)},$$

and further,

$$w = 0$$

in a certain "angular neighborhood" $|\theta| < \theta_0$ of $\Gamma^{(2)}$. Let us introduce a function w_λ, $\lambda \in \mathbf{R}^1$, $\lambda > 0$ by the relation

$$w_\lambda(x) = w(x + \lambda e^1).$$

For $\varkappa < C\lambda$ with $C > 0$ we have the identity $\mathbf{R}_{\varkappa} w_\lambda = 0$ on $\Gamma^{(2)}$, and

$$\|\mathbf{R}_{\varkappa} w_\lambda - w\|_1 \leq \|\mathbf{R}_{\varkappa} w_\lambda - w_\lambda\|_1 + \|w_\lambda - w\|_1 \to 0, \quad \lambda \to 0_+. \tag{3.25}$$

Let us write the function z in the form

$$z = w + z_0, \quad z_0|_{\Omega^M} \in H_0^1(B \cap \Omega^M), \quad M = {}',{}''.$$

The definition of z_0 implies the existence of $z_{0\varkappa}^M \in C_0^\infty(B \cap \Omega^M)$ such that

$$\mathbf{R}_{\varkappa} w_\lambda + z_{0\varkappa} = 0 \quad \text{on } \Gamma^{(2)}, \quad z_{0\varkappa} = (z'_{0\varkappa}, z''_{0\varkappa}),$$

$$\mathbf{R}_{\varkappa} w_\lambda + z_{0\varkappa} \to w + z_0 = z, \quad \lambda \to 0_+ \quad \text{in} \quad H^1(B). \tag{3.26}$$

Set

$$u_{\varkappa}^{'(2)} = \mathbf{R}_{\varkappa} E u^{'(2)}|_{\Omega'},$$

$$u_{\varkappa}^{''(2)} = [\mathbf{R}_{\varkappa} E u^{'(2)} - (\mathbf{R}_{\varkappa} v + \mathbf{R}_{\varkappa} w_\lambda + z_{0\varkappa})]|_{\Omega''}.$$

Then (3.24), (3.25), and (3.26) imply

$$u_{\varkappa}^{M(2)} \to u^{M(2)} \quad \text{in} \quad H^1(B \cap \Omega^M) \tag{3.27}$$

and

$$u_{\varkappa}^{'(2)} - u_{\varkappa}^{''(2)} = \mathbf{R}_{\varkappa} v + \mathbf{R}_{\varkappa} w_\lambda + z_{0\varkappa} \leq 0 \quad \text{on} \quad \Gamma^{(2)}.$$

The components $u^{M(1)}$ can be analyzed analogously. As the Cartesian coordinates w_k of a vector w can be expressed in the form

$$w_k = a_1 w^{(1)} + a_2 w^{(2)}$$

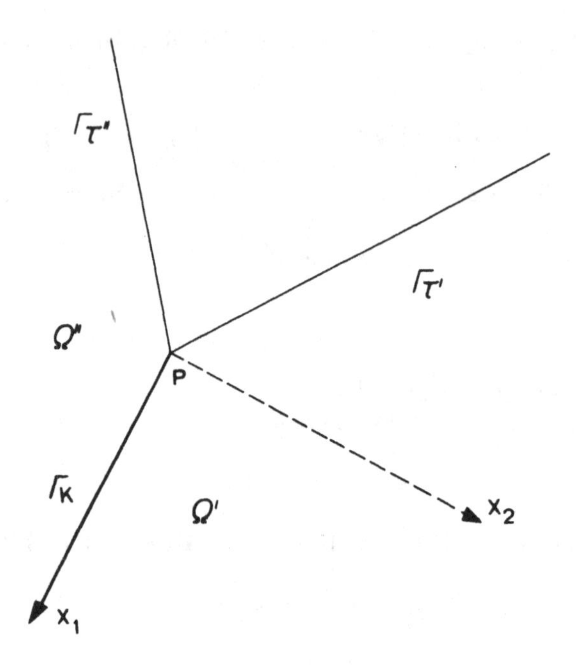

Figure 19

with given constants a_1, a_2, we have

$$\|w_k\|_1^2 \le C \sum_{p=1}^{2} \|w^{(p)}\|_1^2, \quad k = 1, 2.$$

Thus, setting

$$u_{\varkappa}^M = \sum_{p=1}^{2} \frac{u_{\varkappa}^{M(p)}}{e^p \cdot n^p} e^p, \quad M = {}' , {}'' ,$$

we conclude from (3.27) that

$$\|u_{\varkappa}^j - u^j\|_1 \to 0, \quad \lambda \to 0_+, \quad \varkappa \le C\lambda.$$

Group 3. Let $j \le k$ and let B_j contain a point $P \in \Gamma_K \cap \bar{\Gamma}_\tau$. We introduce a new Cartesian coordinate system with its origin at P, whose x_1-axis coincides with Γ_K (see Figure 19). In that case

$$u_n' + u_n'' = -u_2' + u_2'' \le 0 \quad \text{on } \Gamma_K.$$

We proceed in the same way as in Group 2. The components corresponding to u_2', u_2'' are $u'^{(2)}$, $u''^{(2)}$, the half-ray corresponding to Γ_K is

$\Gamma^{(2)}$. The components u_1', u_1'' can be regularized arbitrarily, for they are subjected to no boundary condition.

Group 4. Let B_j contain a point $P \in \bar{\Gamma}_0 \cap \bar{\Gamma}_\tau$, which is possibly a vertex at the same time. If we place the local Cartesian coordinate system so that the x_1-axis coincides with Γ_0, then

$$u_n'' = \pm u_2'' = 0 \quad \text{on } \Gamma_0.$$

Again there exists a function $v \in H^1(B_j \cap \Omega'')$ such that $\operatorname{supp} v \subset B_j$, $v = u_2''$ on $\partial\Omega''$ and $v = 0$ in a certain "angular neighborhood" of Γ_0. Let us define the function $v_\lambda(x)$ by

$$v_\lambda(x) = v(x + \lambda)$$

with a suitable chosen vector $\lambda \in \mathbf{R}^2$. Then, its regularization $\mathbf{R}_\varkappa v_\lambda$ obviously satisfies $\mathbf{R}_\varkappa v_\lambda = 0$ on Γ_0, $\mathbf{R}_\varkappa v_\lambda \to v$ in $H^1(B_j)$ for $\varkappa < c|\lambda|$, $\lambda \to 0_+$. As we can write

$$u_2'' = v + z,$$

where $z \in H_0^1(B_j \cap \Omega'')$, we obtain that $u_{2\varkappa}'' = 0$ on Γ_0 as well, and

$$u_{2\varkappa}'' = \mathbf{R}_\varkappa v_\lambda + z_\lambda \to u_2'' \ \text{ in } H^1(B_j \cap \Omega'').$$

Here, $z_\varkappa \in C_0^\infty(B_j \cap \Omega'')$ satisfy $z_\varkappa \to z$, $\varkappa \to 0_+$ in $H^1(B_j \cap \Omega'')$.

Group 5. Let B_j contain a point $P \in \bar{\Gamma}_u \cap \bar{\Gamma}_\tau$. Then, the same reasoning as that used in the previous section for u_2'' will be applied to the individual components u_k', $k = 1, 2$.

The other cases, when $B_j \cap \partial\Omega' \subset \Gamma_u$, $B_j \cap \partial\Omega'' \subset \Gamma_0$, $B_j \cap \partial\Omega^M \subset \Gamma_\tau$, as well as the regularization of functions that are defined in B_0, B_1, are easy (again it suffices to use classical results on the density, see Nečas (1967)). Finally, if we set

$$u_\varkappa^M = \sum_{j=0}^{r^M} u_\varkappa^{M_j}, \quad M = {}',{}'',$$

we conclude on the basis of the above results that

$$|||u_\varkappa - u|||_1 \to 0, \quad \varkappa \to 0_+, \quad |\lambda| \to 0_+, \quad \varkappa < c|\lambda|.$$

Moreover, $u_\varkappa$ are infinitely differentiable in $\bar{\Omega}$ and fulfill all the boundary conditions that appear in the definition of K. $\square$

The following convergence result is an immediate consequence of the previous lemma and of remark 3.9, section 1.1.32.

Theorem 3.2. *Let $\mathcal{L}$ be coercive on K and let $\mathcal{P}_1$ have exactly one solution u. Further, let us assume that all the assumptions of lemma 3.2 are fulfilled. Then,*

$$|||u - u_h|||_1 \to 0, \quad h \to 0_+$$

for any regular system of triangulations $\{\mathcal{T}_h\}$, $h \to 0_+$.

Proof. Since $K_h \subset K$ for all $h \in (0,1)$, it is sufficient to verify the following assertion:

$$\forall v \in K \; \exists v_h \in K_h : v_h \to v, \quad h \to 0_+ \quad \text{in the norm } \mathcal{H}^1(\Omega).$$

This result is obtained in the standard manner. First we approximate the function v with an arbitrary accuracy by a function $w \in \mathcal{M}$ and for the function w we construct its piecewise linear Lagrange interpolation $r_h w$ over the given triangulation $\mathcal{T}_h$. If $w \in K$, then $r_h w \in K_h$. The rest of the assertion of the theorem is an immediate consequence of remark 3.9, section 1.1.32. □

2.3.312. Curved Contact Zone. As was already said above, the a priori error estimate in this case will be much more complicated than was the case with polygonal domains, since K_h generally are *not* subsets of K. Moreover, the conditions that guarantee the existence of solution of $\mathcal{P}_1$ cannot be automatically transferred to its approximation $\mathcal{P}_{1h}$. If $\mathcal{L}$ is coercive on K, then it need not be coercive on K_h (again due to the fact that generally, $K_h \not\subset K$). Before we proceed to the study of the error itself, we present some results that will be needed in what follows.

Lemma 3.3. *Let $Q \subset \mathbf{R}^2$ be a bounded convex domain whose boundary ∂Q is twice continuously differentiable. Let $\{\mathcal{T}_h\}$, $h \to 0_+$, be a strongly (α, β)-regular system of triangulations of $\bar{Q}$, $\beta = 2$, with the longest straight sides of the triangles $T \in \mathcal{T}_h$ not longer than the longest chord connecting the endpoints of the arcs $A_i A_{i+1}$ of the curved elements of $T \in \mathcal{T}_h$. Then*

$$\|u - r_h u\|_{0,\partial Q} \le c h^{3/2} \|u\|_{2,Q} \tag{3.28}$$

holds for each $u \in H^2(Q)$, where $r_h u$ means the piecewise linear Lagrangian interpolation of u over $\mathcal{T}_h$ and c is a positive constant independent of u, h.

Proof. See Nitsche (1971). □

Lemma 3.4 (Inverse Inequality). *Let p be a linear function, which is defined on $[a,b]$ $(-\infty < a < b < \infty)$. Then*

$$\|p\|_{1,[a,b]} \le c(b-a)^{-1/2} \|p\|_{1/2,[a,b]}, \tag{3.29}$$

where $\|\cdot\|_{1,[a,b]}$ *and* $\|\cdot\|_{1/2,[a,b]}$ *denote the norms in the spaces* $H^1([a,b])$ *and* $H^{1/2}([a,b])$*, respectively, and* c *is a positive constant independent of* p, a, b.

Proof. Follows directly from the definition of norms in the corresponding Sobolev spaces (see Nečas (1967)). □

Lemma 3.5. *Let an arc* A_iA_{i+1} *form the curved side of a boundary element* $T \in \mathcal{T}_h$. *Let* $v \in P_1(T)$ *and let* T_h *be the triangle resulting by connecting the points* A_iA_{i+1} *by a segment. Then*

$$\|v\|^2_{1,\Delta(T,T_h)} \leq ch\|v\|^2_{1,T_h},$$

where $\Delta(T, T_h) \equiv (T \dot{-} T_h) \cup (T_h \dot{-} T)$ *and* c *is a positive constant independent of* v, h.

Proof. See Fix and Strang (1973). □

The main result of this section is

Theorem 3.3. *Let problems* $\mathcal{P}_1$, $\mathcal{P}_{1h}$ *have solutions* u, u_h*, respectively. Let* $u \in \mathcal{H}^2(\Omega) \cap K$, $T_n(u) \in L^2(\Gamma_K)$ *and let the norms* $|||u_h|||_1$ *remain bounded. Assume that a system of triangulations* $\{\mathcal{T}_h\}$, $h \to 0_+$, *fulfills all the requirements formulated in lemma 3.3. Finally, let the function* ψ *describing* Γ_K *be three times continuously differentiable on* $[a,b]$. *Then*

$$|u - u_h| \leq c(u)h^{3/4}, \quad h \to 0_+. \tag{3.30}$$

Proof. To establish estimate (3.30) we use relation (3.8). Similar to the proof of theorem 3.1, using integration by parts, we can write (3.8) in the form

$$\begin{aligned} c_0|u - u_h|^2 \leq A(u_h - u, v_h - u) &+ (T_n(u), (v_n' - u_{hn}') - (v_n'' - u_{hn}''))_{0,\Gamma_K} \\ &+ (T_n(u), (v_{hn}' - u_n') - (v_{hn}'' - u_n''))_{0,\Gamma_K} \\ &\forall v \in K,\ v_h \in K_h. \end{aligned}$$

The first and third terms on the right-hand side of the above inequality are estimated in the same way as in the proof of theorem 3.1. We set $v_h = r_h u \in K_h$, that is, v_h is the piecewise linear Lagrangian interpolation of the exact solution u on $\mathcal{T}_h$. Taking into account our choice of v_h, we obtain

$$\begin{aligned} |A(u_h - u, r_h u - u)| &\leq \frac{1}{2}|u_h - u|^2 + c|||u - r_h u|||_1^2 \\ &\leq \frac{1}{2}|u_h - u|^2 + ch^2|||u|||_2^2, \end{aligned} \tag{3.31}$$

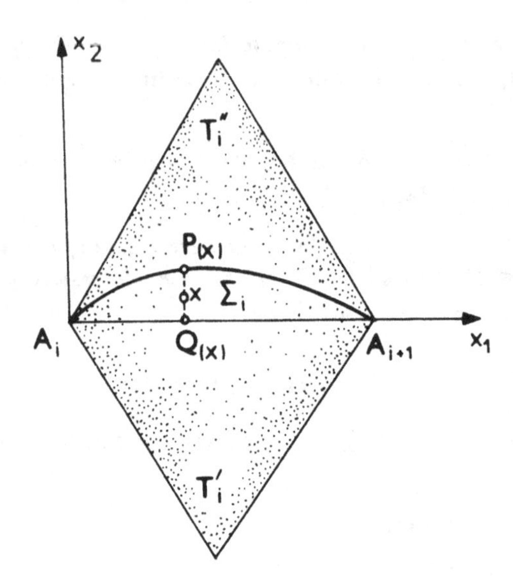

Figure 20

$$\begin{aligned}(T_n(u), (r_h u' - u')\cdot n - (r_h u'' - u'')\cdot n)_{0,\Gamma_K} \\ \le c(\|r_h u' - u'\|_{0,\Gamma_K} + \|r_h u'' - u''\|_{0,\Gamma_K}) \\ \le ch^{3/2}|||u|||_{2,\Omega},\end{aligned}$$

by virtue of the Hölder inequality, the inclusion $T_n(u) \in L^2(\Gamma_K)$ and (3.28).[2]

The most complicated estimate is that of the term

$$(T_n(u), (v'_n - u'_{hn}) - (v''_n - u''_{hn}))_{0,\Gamma_K}, \quad v \in K. \tag{3.32}$$

In the following we will construct the function $v \in K$ so as to make expression (3.32) small.

Let $T'_i \in \mathcal{T}'_h$, $T''_i \in \mathcal{T}''_h$ be two adjacent curved elements, with $A_i A_{i+1}$ their common part (see Figure 20).

We choose the Cartesian coordinate system (x_1, x_2) with its origin at the point A_i and the x_1-axis coinciding with the straight line $\overline{A_i A_{i+1}}$. Let

[2]The following consideration is necessary for making possible the use of lemma 3.3: Let $\tilde{\Omega}'$ be a convex domain, whose boundary $\partial\tilde{\Omega}'$ is twice continuously differentiable and $\partial\tilde{\Omega}' \supset \Gamma_K$. Let $Eu' \in [H^2(\mathbf{R}^2)]^2$ be the continuous Calderon extension (see Nečas (1967)) of the function $u' \in [H^2(\Omega')]^2$. Then, in accordance with (3.28) we have $\|u' - r_h u'\|_{0,\Gamma_K} \le \|Eu' - r_h(Eu')\|_{0,\partial\tilde{\Omega}'} \le ch^{3/2}\|Eu'\|_{2,\tilde{\Omega}} \le ch^{3/2}\|u'\|_{2,\Omega}$. An anlogous argument can be used for $\|u'' - r_h u''\|_{0,\Gamma_K}$.

Σ_i be the closed set bounded by the arc $A_iA_{i+1} \equiv s_i \subset \Gamma_K$ and the segment $\overline{A_iA_{i+1}}$. Let $x \in \Sigma_i$. The symbols $P(x)$, $Q(x)$ will denote the points of intersection of the line perpendicular to $\overline{A_iA_{i+1}}$ through the point x, with the arc s_i and the segment $\overline{A_iA_{i+1}}$, respectively. Let us extend an arbitrary function $v \in [P_1(T_i'')]^2$ to $T_i'' \cup \Sigma_i$ in the usual way:

$$Ev \in [P_1(t_i'' \cup \Sigma_i)]^2, \quad Ev|_{T_i''} = v.$$

For the sake of simplicity of notation, we use the symbol v for the extended function as well.

Let us now define functions $\mathcal{U}_h$, $\tilde{\mathcal{U}}_h$ on $\cup_i\Sigma_i$ by

$$\mathcal{U}_h(x) = (u_h'(x) - u_h''(x)) \cdot n(P(x)),$$

$$\tilde{\mathcal{U}}_h(x) = (u_h'(Q(x)) - u_h''(Q(x))) \cdot n(P(x)) \equiv (\tilde{u}_h' - \tilde{u}_h'')(x) \cdot n(P(x)),$$

where

$$\tilde{u}_h'(x) \equiv u_h'(Q(x)), \quad \tilde{u}_h''(x) \equiv u_h''(Q(x)), \quad x \in \Sigma_i.$$

Evidently,

$$\mathcal{U}_h(x) = \tilde{\mathcal{U}}_h(x) \quad \text{provided } x \in \bigcup_{i=1}^{m} \overline{A_iA_{i+1}}.$$

Let $\Phi_i(x)$, $x \in \overline{A_iA_{i+1}}$, denote the linear Lagrangian interpolation of the function $\mathcal{U}_h$ on $\overline{A_iA_{i+1}}$ and let us define a function $\tilde{\Phi}$ on $\cup_i\Sigma_i$ by

$$\tilde{\Phi}(x) = \Phi_i(Q(x)), \quad x \in \Sigma_i.$$

From the definition of $\tilde{\Phi}$ we see that $\tilde{\Phi} \le 0$ on Γ_K. Let us estimate $\|\tilde{\Phi} - \mathcal{U}_h\|_{0,\Gamma_K}$. The triangular inequality implies

$$\|\tilde{\Phi} - \mathcal{U}_h\|_{0,\Gamma_K} \le \|\tilde{\Phi} - \tilde{\mathcal{U}}_h\|_{0,\Gamma_K} + \|\tilde{\mathcal{U}}_h - \mathcal{U}_h\|_{0,\Gamma_K}. \tag{3.33}$$

First, let us estimate the second term on the right-hand side of (3.33):

$$\|\tilde{\mathcal{U}}_h - \mathcal{U}_h\|_{0,\Gamma_K}^2 = \sum_{i=1}^{m} \|\tilde{\mathcal{U}}_h - \mathcal{U}_h\|_{0,s_i}^2$$

$$\le 2\left(\sum_{i=1}^{m} \|u_h' - \tilde{u}_h'\|_{0,s_i}^2 + \sum_{i=1}^{m} \|u_h'' - \tilde{u}_h''\|_{0,s_i}^2\right).$$

Let τ be the value of the parameter of the arc s_i corresponding to the point $P(x) = (P_1(x), P_2(x))$, and let us denote $Q_1(x) = x_1$. For $M = {}',{}''$, we have

$$u_{hj}^M - \tilde{u}_{hj}^M = \int_0^{P_2(x)} \frac{\partial}{\partial x_2}(u_{hj}^M - \tilde{u}_{hj}^M)dx_2 = \int_0^{P_2(x)} \frac{\partial}{\partial x_2} u_{hj}^M dx_2, \quad j = 1, 2.$$

Integrating this identity and using the Fubini theorem, we obtain

$$\|u_{hj}^M - \tilde{u}_{hj}^M\|_{0,s_i}^2 \le ch^2 |u_{hj}^M|_{1,\Sigma_i}^2, \quad j = 1,2.$$

Finally, from this inequality and lemma 3.5, we conclude

$$\|\tilde{\mathcal{U}}_h - \mathcal{U}_h\|_{0,\Gamma_K}^2 \le ch^2 \left(\sum_{i=1}^m |u_h'|_{1,\Sigma_i}^2 + \sum_{i=1}^m |u_h''|_{1,\Sigma_i}^2 \right) \le ch^3 \|u_h\|_{1,\Omega}^2. \quad (3.34)$$

Let us estimate

$$\|\tilde{\Phi} - \tilde{\mathcal{U}}_h\|_{0,\Gamma_K}^2 = \sum_{i=1}^m \|\tilde{\Phi} - \tilde{\mathcal{U}}_h\|_{0,s_i}^2.$$

Evidently,

$$\begin{aligned} \tilde{\Phi}(\tau) - \tilde{\mathcal{U}}_h(\tau) &= \int_0^{Q_1(\tau)} \frac{d}{dx_1}(\Phi_i(x_1,0) - \tilde{\mathcal{U}}_h(x_1,0))dx_1 \\ &+ \int_0^{P_2(x)} \frac{d}{dx_2}(\Phi_i(Q_1(x),x_2) - \tilde{\mathcal{U}}_h(Q_1(\tau),x_2))dx_2 \\ &= \int_0^{Q_1(\tau)} \frac{d}{dx_1}(\Phi_i(x_1,0) - \tilde{\mathcal{U}}_h(x_1,0))dx_1. \end{aligned}$$

Since $\psi \in C^3([a,b])$, we have $\tilde{\mathcal{U}}_h \in H^2(\overline{A_iA_{i+1}})$. Hence,

$$|\tilde{\Phi}(\tau) - \tilde{\mathcal{U}}_h(\tau)|^2 \le ch|\Phi_i - \tilde{\mathcal{U}}_h|_{1,\overline{A_iA_{i+1}}}^2 \le ch^3 |\tilde{\mathcal{U}}_h|_{2,\overline{A_iA_{i+1}}}^2. \quad (3.35)$$

The definition of $\tilde{\mathcal{U}}_h$, together with the inclusion $\tilde{u}_h'$, $\tilde{u}_h'' \in P_1(\overline{A_iA_{i+1}})$, yields

$$|\tilde{\mathcal{U}}_h|_{2,\overline{A_iA_{i+1}}}^2 \le c[\|u_h'\|_{1,\overline{A_iA_{i+1}}}^2 + \|u_h''\|_{1,\overline{A_iA_{i+1}}}^2].$$

Using this inequality, (3.35), lemma 3.4 applied to $\overline{A_iA_{i+1}}$, and the definition of the strong (α,β)-regularity of $\{\mathcal{T}_h\}$, $h \to 0_+$, we conclude

$$\begin{aligned} \|\tilde{\Phi} - \tilde{\mathcal{U}}_h\|_{0,s_i}^2 &\le ch^4(\|u_h'\|_{1,\overline{A_iA_{i+1}}}^2 + \|u_h''\|_{1,\overline{A_iA_{i+1}}}^2) \\ &\le ch^3(\|u_h'\|_{1/2,\overline{A_iA_{i+1}}}^2 + \|u_h''\|_{1/2,\overline{A_iA_{i+1}}}^2). \end{aligned} \quad (3.36)$$

Summing (3.36) for $i = 1,\dots,m$ we obtain

$$\|\tilde{\Phi} - \tilde{\mathcal{U}}_h\|_{0,\Gamma_K}^2 \le ch^3(\|u_h'\|_{1/2,\Gamma_h}^2 + \|u_h''\|_{1/2,\Gamma_h}^2), \quad (3.37)$$

where $\Gamma_h = \overline{\cup_{i=1}^m A_iA_{i+1}}$ is a polygonal approximation of Γ_K. Using now the theorem on traces and lemma 3.5, we infer

$$\|u_h'\|_{1/2,\Gamma_h}^2 \le c\|u_h'\|_{1,\Omega' \dot{-} \cup_i \Sigma_i}^2 \le c\|u_h'\|_{1,\Omega'}^2,$$

$$\|u_h''\|^2_{1/2,\Gamma_h} \le c\|u_h''\|^2_{1,\Omega''\cup\cup_i\Sigma_i} \le c\|u_h''\|^2_{1,\Omega''}.$$[3]

Using these estimates, (3.33), (3.34), and (3.37), we conclude

$$\|\tilde{\Phi} - \mathcal{U}_h\|_{0,\Gamma_K} \le ch^{3/2}\|u_h\|_{1,\Omega}. \tag{3.38}$$

Now, let $v = (v', v'') \in V$ satisfy $v'' \equiv 0$ on Ω'' and $v' \cdot n = \tilde{\Phi}$ on Γ_K. Then,

$$v' \cdot n - v'' \cdot n = v' \cdot n = \tilde{\Phi} \le 0 \quad \text{on } \Gamma_K,$$

and consequently, $v \in K$. If a function v constructed in this way is substituted in estimate (3.32), we can write

$$(T_n(u), (v_n' - u_{hn}') - (v_n'' - u_{hn}''))_{0,\Gamma_K} = (T_n(u), \tilde{\Phi} - \mathcal{U}_h)_{0,\Gamma_K} \le ch^{3/2}\|u_h\|_{1,\Omega}. \tag{3.39}$$

Then, estimate (3.30) is a consequence of (3.31), (3.39), and of the fact that $|||u_h|||_1$ remain bounded. □

Remark 3.4. In the coercive case, when Korn's inequality holds on the whole space V, all the assumptions of the previous theorem are obviously fulfilled. $\mathcal{P}_1$ and $\mathcal{P}_{1h}$ have exactly one solution u and u_h, respectively. Moreover, as K_h is a convex cone with its vertex at 0, we have

$$\alpha|||u_h|||_1^2 \le A(u_h, u_h) = L(u_h) \le c|||u_h|||_1,$$

which implies boundedness of the sequence of norms $|||u_h|||_1$. Finally, on the left-hand side of (3.30) we can write the norm in $\mathcal{H}^1(\Omega)$ instead of $|\cdot|$.

Remark 3.5. The situation is considerably more complicated in the semi-coercive case when only

$$c_0|v|^2 \le A(v, v) \quad \forall v \in V$$

holds. A sufficient condition for the sequence $|||u_h|||_1$, $h \in (0, 1)$, to be bounded is that $\mathcal{Y}$ be coercive on $\cup_{h>0}K_h$ and $\overline{\cup_{h>0}K_h} = K$, the closure being taken with respect to the norm of $\mathcal{H}^1(\Omega)$ (see remark 3.9, section 1.1.32).

In section 2.2 we have formulated sufficient conditions guaranteeing the coerciveness of $\mathcal{Y}$ on $\mathcal{K}$. In some special cases, these results guarantee the coerciveness of $\mathcal{Y}$ on $\cup_{h>0}K_h$.

---

[3] The constant c in the estimates

$$\|u_h'\|_{1/2,\Gamma_h} \le c\|u_h'\|_{1,\Omega' \dot{-} \cup_i\Sigma_i},$$

$$\|u_h''\|_{1/2,\Gamma_h} \le c\|u_h''\|_{1,\Omega''\cup\cup_i\Sigma_i},$$

generally depends on h. Nevertheless, it can be shown that for $h > 0$ sufficiently small, c can be estimated from above independently of h.

For example, let Γ_K contain a segment I, and let us define

$$K_I = \{v \in V \mid v'_n - v''_n \leq 0 \text{ on } I\}.$$

Then, evidently the convex sets K_h defined by means of (3.4) satisfy the inclusion $K_h \subset K_I$ for all $h \in (0,1)$. Since $\cup_{h>0} K_h \subset K_I$, the coerciveness of $\mathcal{Y}$ on this union follows for example from the coerciveness of $\mathcal{Y}$ on K_I.

The coerciveness of $\mathcal{Y}$ on $\cup_{h>0} K_h$ can be studied even in more complicated cases (see Haslinger (1979), where these problems are studied for some semicoercive cases of the Signorini problem). We do not intend to discuss the problem of density of $\cup_{h>0} K_h$ in K here. Nonetheless, let us mention that the corresponding density result can be obtained by modifying the proof of lemma 3.2.

Remark 3.6. If we want to establish the convergence of the approximate solutions u_h to a nonregular solution u, then with regard to the fact that $K_h \not\subset K$ we have to verify the implication

$$v_h \in K_h, \quad v_h \rightharpoonup v, \quad h \to 0_+ \quad \text{(weakly) in } V \Rightarrow v \in K.$$

For the Signorini problem, this was accomplished in Haslinger (1979). (See also lemma 3.7 of the next section.)

2.3.32. Increasing Zone of Contact. Let us consider problem $\mathcal{P}_2$ and its approximation $\mathcal{P}_{2h}$, defined in section 2.3.2. We will study the rate of convergence of u_h to u, provided u is sufficiently smooth. Then we will prove the convergence of u_h to a nonregular solution u. To this end, we will make use of the results of section 2.1.2 and 2.2.2.

Theorem 3.4. *Let $\mathcal{P}_2$, $\mathcal{P}_{2h}$ have solutions u, u_h, respectively. Let $u \in \mathcal{H}^2(\Omega) \cap K_\epsilon$, $u'_\xi \in W^{1,\infty}(\Gamma'_K)$, $u''_\xi \in W^{1,\infty}(\Gamma''_K)$, $f'f'' \in C^2([a,b])$. Further, let us assume that the number of points in $[a,b]$ at which the inequality $u''_\xi - u'_\xi < \epsilon$ changes into the identity $u''_\xi - u'_\xi = \epsilon$ $(\epsilon = f'' - f')$ is finite. Then,*

$$|u - u_h| \leq c(u, f', f'')h$$

for an arbitrary regular system $\{\mathcal{T}_h\}$, $h \to 0_+$.

Proof. We proceed analogously to the proof of theorem 3.3. We start with relation (3.8). After integrating by parts we can write it in the form

$$c_0|u - u_h|^2 \leq A(u_h - u, v_h - u) + (T'_\xi(u), v'_{h\xi} - u'_\xi)_{0,\Gamma'_K}$$

$$+ (T''_\xi(u), v''_{h\xi} - u''_\xi)_{0,\Gamma''_K} + (T'_\xi(u), v'_\xi - u'_{h\xi})_{0,\Gamma'_K} + (T''_\xi(u), v''_\xi - u''_{h\xi})_{0,\Gamma''_K}$$

$$\forall v_h \in K_{\epsilon h}, \quad v \in K_\epsilon,$$

where all symbols have the same meaning as in section 2.1.2. Let the function v_h be chosen to be the corresponding P-interpolation of the exact solution, constructed by using the isoparametric technique (see Zlámal (1973)):

$$v_h = u_I, \quad \text{where}$$

$$u_I|_T = \hat{\pi}(u|_T \circ F_T) \circ F_T^{-1} \quad \forall T \in \mathcal{T}_h.$$

Here $T = F_T(\hat{T})$, and $\hat{\pi}$ is the operator of the linear Lagrangian interpolation of $\hat{T}$ (see section 2.3.2). The definition of the function u_I implies that $u_I \in K_{\epsilon h}$. The well-known approximative properties of u_I (see Zlámal (1973)) imply

$$|A(u_h - u, u_I - u)| \leq \frac{1}{2}|u_h - u|^2 + c[[u_I - u]]_1^2$$

$$\leq \frac{1}{2}|u_h - u|^2 + ch^2|||u|||_{2,\Omega}^2. \tag{3.40}$$

Let us write

$$(T'_\xi(u), u'_{I\xi} - u'_\xi)_{0,\Gamma'_K} + (T''_\xi(u), u''_{I\xi} - u''_\xi)_{0,\Gamma''_K}$$

$$= \int_a^b T_\xi(u)\{(u''_{I\xi} - u'_{I\xi}) - (u''_\xi - u'_\xi)\}d\eta,$$

where

$$T_\xi(u) \stackrel{\text{def}}{\equiv} T''_\xi(u)(\cos\alpha'')^{-1} = -T'_\xi(u)(\cos\alpha_I)^{-1}.$$

Denote

$$W_h(\eta) = u''_{I\xi}(f''(\eta), \eta) - u'_{I\xi}(f'(\eta), \eta),$$

$$\mathcal{U}(\eta) = u''_\xi(f''(\eta), \eta) - u'_\xi(f'(\eta), \eta), \quad \eta \in [a, b].$$

The functions $u'_I(f'(\eta), \eta)$, $u''_I(f''(\eta), \eta)$ are piecewise linear in the variable η on $[a, b]$ with vertices at points C_j. Since $\xi \in \mathbf{R}^2$ is a constant vector, W_h is a piecewise linear function on $[a, b]$ as well. Let us divide the interval $[a, b]$ into two disjoint subsets Γ^0, Γ^-, where

$$\Gamma^0 = \{\eta \in [a, b] \mid u''_\xi - u'_\xi = \epsilon\},$$

$$\Gamma^- = \{\eta \in [a, b] \mid u''_\xi - u'_\xi < \epsilon\}.$$

If $[C_i, C_{i+1}] \subseteq \Gamma^0$, then the function W_h is a piecewise linear Lagrangian interpolation of ϵ on $[C_i, C_{i+1}]$, and

$$\int_{C_i}^{C_{i+1}} T_\xi(u)(W_h(\eta) - \mathcal{U}(\eta))d\eta = \int_{C_i}^{C_{i+1}} T_\xi(u)(W_h - \epsilon)d\eta$$

$$\leq ch^2 |\epsilon|_{2,\overline{C_iC_{i+1}}}. \tag{3.41}$$

If $[C_i, C_{i+1}] \subseteq \bar{\Gamma}^-$, then $T_\xi(u) \equiv 0$ on $[C_i, C_{i+1}]$, and hence

$$\int_{C_i}^{C_{i+1}} T_\xi(u)(W_h(\eta) - \mathcal{U}(\eta))d\eta = 0. \tag{3.42}$$

Let $\mathcal{T}$ be the system of all $[C_i, C_{i+1}] \subseteq [a, b]$ whose interiors simultaneously contain points from both Γ^0 and Γ^-. By the assumptions of the theorem we obtain

$$\int_{C_i}^{C_{i+1}} T_\xi(u)(W_h(\eta) - \mathcal{U}(\eta))d\eta$$

$$\leq h \|T_\xi(u)\|_{\infty,\overline{C_iC_{i+1}}} \|W_h - \mathcal{U}\|_{\infty,\overline{C_iC_{i+1}}}$$

$$\leq ch^2 \|T_\xi(u)\|_{\infty,\overline{C_iC_{i+1}}} |\mathcal{U}|_{1,\infty,\overline{C_iC_{i+1}}}. \tag{3.43}$$

For the same reason as in theorem 3.1, the number of elements of $\mathcal{T}$ is bounded from above independently of h. From (3.41)–(3.43), we then obtain

$$\int_a^b T_\xi(u)[(u''_{I\xi} - u'_{I\xi}) - (u''_\xi - u'_\xi)]d\eta \leq c(u, f', f'')h^2. \tag{3.44}$$

It remains to estimate the expression

$$(T'_\xi(u), v'_\xi - u'_{h\xi})_{0,\Gamma'_K} + (T''_\xi(u), v''_\xi - u''_{h\xi})_{0,\Gamma''_K}$$

$$= \int_a^b T_\xi(u)[(v''_\xi - u''_{h\xi}) - (v'_\xi - u'_{h\xi})]d\eta, \quad v \in K_\epsilon.$$

Let us denote

$$\mathcal{U}_h(\eta) = u''_{h\xi}(f''(\eta), \eta) - u'_{h\xi}(f'(\eta), \eta), \quad \eta \in [a, b],$$

and define a function W^h on $[a, b]$ by

$$W^h(\eta) = \inf_{\eta \in [a,b]} [\mathcal{U}_h(\eta), \epsilon(\eta)].$$

From the definition it immediately follows that $W^h \leq \epsilon$ on $[a, b]$ and moreover, $W^h \in H^1([a, b])$. As

$$W^h - \mathcal{U}_h = \begin{array}{ll} 0 & \text{provided } \mathcal{U}_h \leq \epsilon, \\ \epsilon - \mathcal{U}_h & \text{otherwise,} \end{array}$$

we can write

$$\left| \int_a^b T_\xi(u)(\mathcal{U}_h - W^h) d\eta \right| \leq c \|\mathcal{U}_h - \epsilon\|_{0,\delta}, \tag{3.45}$$

where $\delta \subseteq [a,b]$ is the set of points at which $\mathcal{U}_h > \epsilon$. As $\mathcal{U}_h(C_j) \leq \epsilon(C_j)$, $j = 1, \ldots, m$, we have $\mathcal{U}_h(C_j) \leq \epsilon_I(C_j)$, where ϵ_I stands for the piecewise linear Lagrangian interpolation of ϵ on $[a,b]$. Further, $\mathcal{U}_h$ and ϵ_I are piecewise linear on $[a,b]$, and consequently, $\mathcal{U}_h \leq \epsilon_I$ on the whole $[a,b]$. This implies that (3.45) can be further estimated by

$$\left| \int_a^b T_\xi(\mathcal{U}_h - W^h) d\eta \right| \leq c\|\epsilon_I - \epsilon\|_{0,\delta}$$

$$\leq c\|\epsilon_I - \epsilon\|_{0,[a,b]} \leq ch^2 |\epsilon|_{2,[a,b]}.$$

The rest of the proof is analogous to that of theorem 3.3. Let $v = (v', v'') \in V$ fulfill $v'' = 0$ on Ω'' and $-v'_\xi = W^h$. Then, $v \in K_\epsilon$ and

$$\int_a^b T_\xi(u)[(v''_\xi - v'_\xi) - (u''_{h\xi} - u'_{h\xi})] d\eta = \int_a^b T_\xi(u)[W^h - \mathcal{U}_h] d\eta$$

$$\leq ch^2 |\epsilon|_{2,[a,b]}.$$

The assertion of the theorem follows from this estimate and from (3.40), (3.44). □

For the same reasons as in the case of the contact problem with a bounded zone of contact, we will study the convergence of the approximate solutions u_h to the nonregular solution u. To this aim we shall need two auxiliary assertions.

Lemma 3.6. *Let $f^M \in C^m((a-\delta, b+\delta))$, $m \geq 1$, $\delta > 0$, $\bar\Gamma_K^M \cap \bar\Gamma_u = \emptyset$, $\bar\Gamma_K^M \cap \bar\Gamma_0 = \emptyset$, $M = {}'\!,{}''$, and let the intersections $\bar\Gamma_u \cap \bar\Gamma_\tau$, $\bar\Gamma_0 \cap \bar\Gamma_\tau$ consist of a finite number of points. Let $v \in K_\epsilon$ fulfill the condition $v''_\xi - v'_\xi \leq f' - f''$ in $(a-\delta, b+\delta)$. Then v belongs to the closure (in the norm of $\mathcal{H}^1(\Omega)$) of the set*

$$K_\epsilon \cap [C^m(\bar\Omega')]^2 \times [C^m(\bar\Omega'')]^2.$$

Proof. Let us consider a system of open sets $\{B_i\}_{i=0}^r$ covering $\bar\Omega' \cup \bar\Omega''$ and such that $\bar B_0 \subset \Omega'$, $\bar B_1 \subset \Omega''$, $\bar\Gamma'_K \cup \bar\Gamma''_K \subset \cup_{j=2}^k B_j$ with $(\bar\Gamma'_K \cup \bar\Gamma''_K) \cap B_i \neq \emptyset \iff 2 \leq i \leq k$. Let us assume that the union of arcs (see Figure 21) $PQ' \cup PQ''$, $Q^M = (f^M(b), b)$, $M = {}'\!,{}''$, contains at most one singular point (that is, a vertex of Ω^M or a point from $\bar\Gamma_u \cap \bar\Gamma_\tau$, $\bar\Gamma_0 \cap \bar\Gamma_\tau$). In the

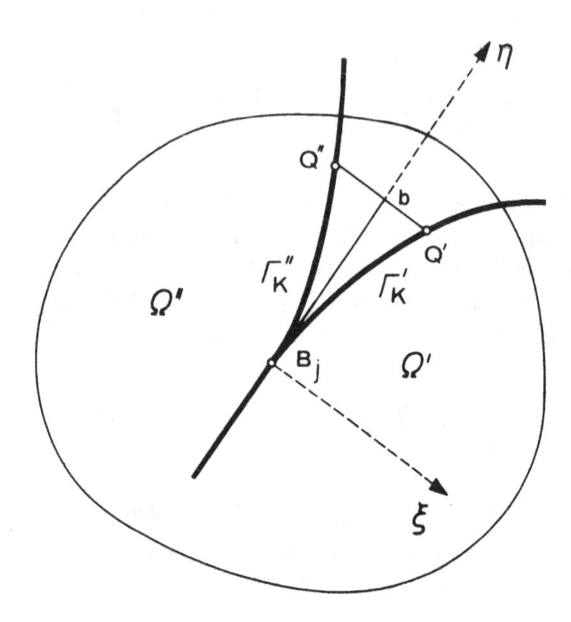

Figure 21

same way as in the proof of lemma 3.2 we put $u^j = u\varphi_j$, where $\{\varphi_j\}_{j=0}^r$ is a partition of the unit corresponding to the covering $\{B_j\}_{j=0}^r$. For every u^j we construct a smooth approximation from K_ϵ. We can proceed essentially in the same way as in the proof of lemma 3.2. The situation requiring a special analysis is shown in Figure 21.

Let us note that in this case we have $\varphi_j \equiv 1$ on $\Gamma_K' \cup \Gamma_K''$. In the following we will omit the index j. First we will map $\Omega' \cap B$ to the right halfplane ($\hat{\xi} > 0$) and $\Omega'' \cap B$ to the left halfplane ($\hat{\xi} < 0$) by means of two mappings:

$$\hat{x} = T^M x : \{\hat{\xi}^M = \xi - f^M(\eta),\ \hat{\eta}^M = \eta\}, \quad M = \ ',\,'',$$

$$\hat{x} = (\hat{\xi}, \hat{\eta}), \quad x = (\xi, \eta).$$

Let us denote $\hat{B} = T'(\bar{\Omega}' \cap B) \cup T''(\bar{\Omega}'' \cap B)$ and $\hat{u}^M(\hat{x}) = u^M((T^M)^{-1}\hat{x})$. Since

$$u_\xi''(f''(\eta), \eta) - u_\xi'(f'(\eta), \eta) - \epsilon(\eta) \le 0, \quad \eta_0 < \eta \le b, \tag{3.46}$$

we also have

$$\hat{U}(\hat{\eta}) \equiv (\hat{u}_\xi'' - \hat{u}_\xi' - \hat{\epsilon}) \le 0 \quad \text{for } \hat{\xi} = 0, \quad \eta_0 < \eta \le b.$$

Let us extend $\hat{\epsilon}$ to the interval $\hat{\eta} > b$ so that this extension fulfills $E\hat{\epsilon} \in C^m$ and so that $E\hat{\mathcal{U}}$ be nonnegative; $E\hat{\mathcal{U}} \in H^{1/2}$ and supp $E\hat{\mathcal{U}} \subset \hat{B}$.[4] Then there exists a function $\hat{v} \in H^1(\hat{B})$, $\hat{v} \leq 0$ in $\hat{B}$, $\hat{v} = E\hat{u}$ for $\hat{\xi} = 0$ and supp $\hat{v} \subset \hat{B}$.

Now let us extend the functions $\hat{u}_\xi^M$ over the $\hat{\eta}$-axis so that the resulting functions $E\hat{u}_\xi^M$ be even in $\hat{\xi}$. Then we can write

$$E\hat{u}_\xi'' - E\hat{u}_\xi' - E\hat{\epsilon} = \hat{v} + \hat{z},$$

where $\hat{z} \in H^1(\hat{B})$, $\hat{z}|_{\hat{\xi}=0} = 0$. Regularizing the functions $\hat{v}$ and $\hat{z}$ we obtain

$$(R_\varkappa \hat{v} + \hat{z}_{\hat{\varkappa}}|_{\hat{\xi}=0} \leq 0, \quad R_\varkappa \hat{v} + \hat{z}_\varkappa \to \hat{v} + \hat{z} \quad \text{in } H_1(\hat{B}).$$

Set

$$\hat{u}_{\xi\varkappa}'' = R_\varkappa E\hat{u}_\xi''|_{T''\cap\Omega''},$$

$$\hat{u}_{\xi\varkappa}' = [R_\varkappa E\hat{u}_\xi'' - R_{\hat{\varkappa}}\hat{v} - \hat{z}_{\hat{\varkappa}}]|_{T'\cap\Omega'} - E\hat{\epsilon},$$

$$u_{\xi\varkappa}'' = \hat{u}_{\xi\varkappa}''(\xi - f''(\eta), \eta), \quad u_{\xi\varkappa}' = \hat{u}_{\xi\varkappa}'(\xi - f'(\eta), \eta).$$

It is evident that $u_\varkappa^M \in C^m$, $u_{\xi\varkappa}$ fulfill (3.46) and $\|u_{\xi\varkappa}^M - u_\xi^M\|_{1,\Omega^M} \to 0$, since both T^M and $(T^M)^{-1}$ are Lipschitzian mappings. □

Lemma 3.7. *Let $\varphi \in C([a,b])$, $(-\infty < a < b < \infty)$, let $D_n : a = x_0^n < x_1^n < \ldots < x_n^n = b$ be a partition of $[a,b]$ whose norm fulfills $\nu(D_n) \to 0$, $n \to \infty$. Let $\{\psi_n\}_{n=1}^\infty$ be a sequence of piecewise linear functions with vertices at the points x_i^n, such that $\psi_n(x_i^n) \leq \varphi(x_i^n)$ for all $i = 0, \ldots, n$; $n = 1, 2, \ldots$. Finally, let $\psi_n \to \psi$ a.e. in $[a,b]$. Then $\psi \leq \varphi$ a.e. in $[a,b]$.*

Proof. Can be found in Haslinger (1977). □

The following convergence result is an immediate consequence of the preceding two lemmas.

Theorem 3.5. *Let $\mathcal{P}_2$ have exactly one solution u and let $\mathcal{Y}$ be coercive on $\overline{\cup_{h>0} K_{\epsilon h}}$. Let all the assumptions of lemma 3.6 be fulfilled with $m = 2$. Then*

$$|||u - u_h|||_1 \to 0, \quad h \to 0_+,$$

for any regular system $\{\mathcal{T}_h\}$.

Proof. As in general $K_{\epsilon h} \not\subset K_\epsilon$, we have to verify the conditions (3.8) and (3.9) of section 1.1.32. (3.8) is an immediate consequence of lemma 3.6

[4] For example, we can put $E\hat{\epsilon} = \varphi_j(f' - f'')$ provided the point $(0, b + \delta)$ lies outside of $\hat{B}$.

(see also the proof of the analogous assertion in theorem 3.2). Let us verify (3.9). Let $v_h \in K_{\epsilon h}$ be such that

$$v_h \rightarrow v, \quad h \rightarrow 0_+ \quad \text{in } \mathcal{H}^1(\Omega). \tag{3.47}$$

As the map $\gamma : V \rightarrow [L^2(\Gamma'_K)]^2 \times [L^2(\Gamma''_K)]^2$ which maps a function $v \in V$ to the traces of the individual components on Γ'_K and Γ''_K is totally continuous (see Nečas (1967)), we obtain from (3.47) that

$$v_h^M \rightarrow v^M \quad \text{in } [L^2(\Gamma_K^M)]^2, \quad M = {}',{}''.$$

Hence, we can choose subsequences $\{v'_h\}$, $\{v''_h\}$ (keeping the original notation for them), such that

$$V_h(\eta) \equiv v''_{h\xi}(f''(\eta), \eta) - v'_{h\xi}(f'(\eta), \eta) \rightarrow v''_\xi(f''(\eta), \eta)$$
$$- v'_\xi(f'(\eta), \eta) \equiv V(\eta) \quad \text{a.e. on } [a, b].$$

Since $V_h(\eta)$ is piecewise linear on $[a, b]$ and $V_h(C_i) \leq \epsilon(C_i)$ $\forall i = 1, \ldots, m$, it follows from lemma 3.7 that $V \leq \epsilon$ a.e. on $[a, b]$. Hence and from remark 3.9, section 1.1.32, we obtain the assertion of the lemma. □

In conclusion, let us say several words about the numerical realization of problems $\mathcal{P}_{1h}$, $\mathcal{P}_{2h}$. The reader can easily see that both the problems in a finite dimension lead to problems of quadratic programming: to find the minimum of the quadratic function

$$\mathcal{L}(x) = \frac{1}{2}(x, Ax) - (f, x) \tag{3.48}$$

on the closed convex subset

$$\mathcal{K}_E = \{x \in \mathbf{R}^n \mid Bx \leq d\},^5 \tag{3.49}$$

where A is an $n \times n$ stiffness matrix, $f \in \mathbf{R}^n$ the vector arising by the integration of the body and surface forces, B is generally a rectangular matrix of type $m \times n$, and $d \in \mathbf{R}^m$ is a given vector.

Let us have a more detailed look at the definition of the set $\mathcal{K}_E$. When dealing with problem $\mathcal{P}_1$ we have $d \equiv 0$, and the number of constraints depends on the number of nodes of the triangulation, which lie on Γ_K in the following manner: if Γ_K consists of a single line segment or if Γ_K is curved, then m is equal to the number of the nodes. If Γ_K is a broken line consisting of p segments ($p > 1$), then the total number of constraints is greater by $p-1$. In problem $\mathcal{P}_2$ the number of constraints equals the number

[5] We assume that the stable boundary conditions are already taken into account when constructing the stiffness matrix A.

of the points of partition in $[a,b]$ and $d = (\epsilon(C_1), \epsilon(C_2), \ldots, \epsilon(C_m))^T$. In both cases, the character of the matrix B is the same. Each row of the matrix contains at most four nonzero entries (components of the vector n or ξ), while in each column there is at most one, or in the case of $\mathcal{P}_1$ with Γ_K a broken line, at most two such entries. In the following, we will assume that each node lying on Γ_K is assigned exactly one inequality constraint (leaving the case of Γ_K piecewise linear, when there are two constraints at the vertices, to the reader). Let $\mathcal{T}_i = \{k_1^i, k_2^i, k_3^i, k_4^i\}$, $i = 1, \ldots, m$ be the quadruple of indices of nonzero entries of the i-th row of B. Introducing new variables $y = (y_1, \ldots, y_n)$ by

$$y_j = x_j, \quad j \neq k_4^i, \quad i = 1, \ldots, m,$$

$$y_{k_4^i} = \sum_{j=1}^{4} b_{ik_j^i} x_{k_j^i},$$

or, in the matrix form,

$$y = Cx \Longleftrightarrow x = C^{-1} y,$$

we can formulate the minimization problem described by (3.48), (3.49) with respect to the variable y: find the minimum of the quadratic function

$$\mathcal{L}(y) = \frac{1}{2}(y, \mathcal{A} y) - (\mathcal{F}, y)$$

on the set

$$\mathcal{K}_E = \{y \in \mathbf{R}^n, \quad y_{k_4^i} \leq d_{k_4^i}, \quad i = 1, \ldots, m\},$$

where $\mathcal{A} = (C^{-1})^T A C^{-1}$, $\mathcal{F} = (C^{-1})^T f$. The matrix $\mathcal{A}$ has the same properties as A, that is, it is symmetric and positive definite or semidefinite. The advantange of this formulation consists of the fact that its numerical solution can be carried out by the SOR method with an additional projection (see section 1.1.31). The convergence of this method is guaranteed provided that $\mathcal{A}$ (and hence A as well) is positive definite; that is, for the coercive case of the contact problem.

Another possibility is to solve problem (3.48), (3.49) by using Uzawa's algorithm and thus reduce the problem with constraints to a sequence of problems without constraints. Nonetheless, it is important that matrix A be regular again. However, we know from the above considerations that semicoercive problems leading to A singular are rather frequent in applications. This is why we will discuss them here more thoroughly.

In Chapter 1, section 1.6.1, we briefly mentioned the method of conjugate gradients, which makes it possible to find the minimum of a quadratic function with a positive semidefinite matrix A subject to constraints of the form

$$
\begin{aligned}
(a_i, x) - b_i &\le 0, \quad i \in \mathcal{T}^-,\\
(a_i, x) - b_i &= 0, \quad i \in \mathcal{T}^0,
\end{aligned} \tag{3.50}
$$

where a_i, $i \in \mathcal{T}^- \cup \mathcal{T}^0$ are linearly independent vectors from $\mathbf{R}^n$, $b_i \in \mathbf{R}^1$. Let us describe this algorithm in more detail. The proofs and additional details can be found in Pšeničnyj and Danilin (1975).

Let us first suppose that $\mathcal{T}^- = \emptyset$; that is, we seek the minimum of $\mathcal{L}$ on the set

$$
K = \{x \in \mathbf{R}^n \mid (a_i, x) - b_i = 0, \quad i = 1, \ldots, q\}. \tag{3.51}
$$

Let P be the projector operator to the space generated by the vectors $a_1, \ldots, a_q$. It can be shown that P may be written in the form

$$
P = A^*_{\mathcal{T}^0}(A_{\mathcal{T}^0} A^*_{\mathcal{T}^0})^{-1} A_{\mathcal{T}^0},
$$

where $A_{\mathcal{T}^0}$ is the matrix whose rows are formed by the vectors $a_1, \ldots, a_q$. In contact problems, when at most four components of the vector x are linked by constraints in (3.51), it is not necessary to construct P by means of its matrix representation. It is possible to directly deduce the value of Px. Let us now introduce the following algorithm:

$$
\text{let } x_0 \in K \text{ be arbitrary, } p_1 = -(I - P)\mathcal{L}'(x_0). \tag{3.52}
$$

If we know $x_k \in K$, we set

$$
x_{k+1} = x_k + \alpha_{k+1} p_{k+1}, \text{ where}
$$

$$
p_{k+1} = -(I - P)\mathcal{L}'(x_k) + \frac{\|(I - P)\mathcal{L}'(x_k)\|^2}{\|(I - P)\mathcal{L}'(x_{k-1})\|^2} p_k,
$$

$$
\alpha_{k+1} = -\frac{(\mathcal{L}'(x_k), p_{k+1})}{(p_{k+1}, A p_{k+1})}.
$$

Here, $\mathcal{L}'(x) \equiv \operatorname{grad} \mathcal{L}(x) = Ax - f$ and I stands for the unit matrix. The following lemma is valid:

Lemma 3.8. *Let there exist a unique minimum $x*$ of the quadratic function $\mathcal{L}$ on the convex set K given by (3.51). Then there is an index k, $k \le n$, such that*

$$
x_k = x*.[6]
$$

Let us now consider the case when both $\mathcal{T}^0$ and $\mathcal{T}^-$ are *nonempty*. Define

$$
\mathcal{T}(x) = \{i \mid (a_i, x) - b_i = 0, \quad i \in \mathcal{T}^0 \cup \mathcal{T}^-\}.
$$

[6] Of course, we assume that no round-off errors occur.

Choose $x_0 \in \mathbf{R}^n$ such that it satisfies the constraints (3.50). Find the set $\mathcal{T}_0 \equiv \mathcal{T}(x_0)$ and construct the projection operator $P_{\mathcal{T}_0}$ by the relation

$$P_{\mathcal{T}_0} = A^*_{\mathcal{T}_0}(A_{\mathcal{T}_0}A^*_{\mathcal{T}_0})^{-1}A_{\mathcal{T}_0}.$$

$A_{\mathcal{T}_0}$ is a rectangular matrix, whose rows are formed by the vectors a_i, $i \in \mathcal{T}_0$. Now we calculate an auxiliary vector

$$u_0 = -(A_{\mathcal{T}_0}A^*_{\mathcal{T}_0})^{-1}A_{\mathcal{T}_0}\mathcal{L}'(x_0),$$

and with its help,

$$(I - P_{\mathcal{T}_0})\mathcal{L}'(x_0) = \mathcal{L}'(x_0) + A^*_{\mathcal{L}_0}u_0.$$

We distinguish two possibilities:

1^0 $(I - P_{\mathcal{T}_0})\mathcal{L}'(x_0) = 0$. In this case, it can be shown that x_0 is a point of minimum of $\mathcal{L}$ with respect to the set

$$(a_i, x) = b_i \quad \forall i \in \mathcal{T}_0.$$

If *all* components u_0^i of the vector u_0, $i \in \mathcal{T}_0 \cap \mathcal{T}^-$ are *nonnegative*, then x_0 is the minimum of $\mathcal{L}$ subject to the constraints (3.50) as well, which means that x_0 is the required solution.

In the opposite case, when there is $j \in \mathcal{T}_0 \cap \mathcal{T}^-$ such that $u_0^j < 0$, we proceed as follows. We construct a new set of indices $\mathcal{T}_0' = \mathcal{T}_0 \dot{-} \{j\}$ and apply a modified algorithm (3.52). The corresponding modification consists in guaranteeing that the particular approximations fulfills (3.50). To this end let us define a number

$$\bar{\alpha}_{k+1} = \min \frac{b_i - (a_i, x_k)}{(a_i, p_{k+1})},$$

with the minimum taken over all i such that $(a_i, p_{k+1}) > 0$. Let α_{k+1} be the quantity introduced in (3.52). If $\alpha_{k+1} < \bar{\alpha}_{k+1}$, then put $x_{k+1} = x_k + \alpha_{k+1}p_{k+1}$ and continue (3.52). If $\alpha_{k+1} \geq \bar{\alpha}_{k+1}$, then put $x_{k+1} = x_k + \bar{\alpha}_{k+1}p_{k+1}$ and stop (3.52). Thus we either find by (3.52) the minimum of $\mathcal{L}$ on the set determined by $\mathcal{T}_0'$, or we stop the calculation (provided $\alpha_{k+1} \geq \bar{\alpha}_{k+1}$). In both cases, the approximation x_{k+1} thus obtained is regarded as the initial approximation of x_0.

2^0 If $(I - P_{\mathcal{T}_0})\mathcal{L}'(x_0) \neq 0$, then we apply algorithm (3.52), calculating again both α_{k+1}, $\bar{\alpha}_{k+1}$ in each step, and stopping as in the previous case. The value x_{k+1} thus obtained is chosen to be the new initial approximation and the entire procedure given above is repeated. Again, it is possible to prove that the above algorithm converges after a finite number of steps, provided the problem of minimization of $\mathcal{L}$ on (3.50) has a unique solution. In applications, we achieve validity of the condition $\mathcal{T}^0 \neq \emptyset$ by neglecting some stable boundary conditions when constructing the stiffness matrix.

2.4 Dual Variational Formulation of the Problem with Bounded Zone of Contact

We shall again study the contact problem with a bounded range of contact, formulated in terms of displacements in section 2.1.1, but we will now introduce the dual variational formulation, that is, the formulation in terms of stresses. We shall restrict our considerations to polygonal boundaries $\partial\Omega'$, $\partial\Omega''$.

Similarly to Chapter 1 (see section 1.1.1), we will use the method of a saddle point to derive the dual variational formulation. We will use the notation from section 2.1.3, using the term primal variational problem when referring to the formulation from definition 1.3.

Let us introduce the space S of new parameters

$$S = \{\mathcal{N} = (\mathcal{N}_{ij}),\ \ i,j = 1,2,\ \ \mathcal{N}_{ij} \in L_2(\Omega),\ \ \mathcal{N}_{ij} = \mathcal{N}_{ji}\}$$

(with $\Omega = \Omega' \cup \Omega''$) and set

$$\mathcal{N}_{ij} = e_{ij}(v). \tag{4.1}$$

Then we can write

$$\mathcal{L}(v) = \frac{1}{2}\int_{\Omega} c_{ijkl}\mathcal{N}_{ij}\mathcal{N}_{kl}\,dx - L(v) = \mathcal{L}_1(\mathcal{N}, v), \tag{4.2}$$

concluding that the primal problem is equivalent to the minimization of the functional $\mathcal{L}_1(\mathcal{N}, v)$ with the constraints (4.1). This problem can be solved by means of Lagrangian multipliers λ and the following Lagrangian:

$$\mathcal{H}([\mathcal{N}, v], \lambda) = \mathcal{L}_1(\mathcal{N}, v) + \int_{\Omega} \lambda_{ij}(e_{ij}(v) - \mathcal{N}_{ij})dx. \tag{4.3}$$

It is easily seen that

$$\sup_{\lambda \in S} \int_{\Omega} \lambda_{ij}(e_{ij}(v) - \mathcal{N}_{ij})dx = \begin{array}{ll} 0 & \text{provided } \mathcal{N} = e(v), \\ +\infty & \text{provided } \mathcal{N} \neq e(v). \end{array}$$

Hence every solution u of the primal problem fulfils the condition

$$\mathcal{L}(u) = \inf_{v \in K} \mathcal{L}(v) = \inf_{\substack{v \in K \\ \mathcal{N} \in S}} \sup_{\lambda \in S} \mathcal{H}([\mathcal{N}, v], \lambda). \tag{4.4}$$

The problem

$$\sup_{\lambda \in S} \inf_{[\mathcal{N}, v] \in S \times K} \mathcal{H}([\mathcal{N}, v], \lambda) \tag{4.5}$$

will be called the dual variational problem.

In order to find the relations between (4.4) and (4.5), we use the theory of a saddle point (see Ekeland and Temam (1974), or Céa (1971).

First, we present

Lemma 4.1. *Let $\{u, \lambda\}$ be a saddle point of the functional $\mathcal{H}$ on the Cartesian product $\mathcal{A} \times \mathcal{B}$. Then,*

$$\mathcal{H}(u, \lambda) = \inf_{v \in \mathcal{A}} \sup_{\mu \in \mathcal{B}} \mathcal{H}(v, \mu) = \sup_{\mu \in \mathcal{B}} \inf_{v \in \mathcal{A}} \mathcal{H}(v, \mu).$$

Proof. See, for example, Céa (1971), Chapter 5. □

Lemma 4.2. *Let $\{[\tilde{\mathcal{N}}, \tilde{v}], \tilde{\lambda}\}$ be a saddle point of the functional $\mathcal{H}$ (defined in (4.3), (4.2)) on $\mathcal{W} \times S$, where $\mathcal{W} = S \times K$.*

Then there exists a solution u of the primal problem and

$$\tilde{\mathcal{N}} = e(u), \quad \tilde{u} = u, \quad \tilde{\lambda} = \tau(u),$$

where $e(u)$ and $\tau(u)$ are the corresponding tensors of strain and stress.

Proof. The definition of the saddle point implies

$$\delta_\lambda \mathcal{H}([\tilde{\mathcal{N}}, \tilde{v}], \tilde{\lambda}) = 0 \Longleftrightarrow \tilde{\mathcal{N}}_{ij} = e_{ij}(\tilde{v}), \tag{4.6}$$

$$\delta_{\mathcal{N}} \mathcal{H}([\tilde{\mathcal{N}}, \tilde{v}], \tilde{\lambda}) = 0 \Longleftrightarrow \tilde{\lambda}_{ij} = c_{ijkl} \tilde{\mathcal{N}}_{kl}, \tag{4.7}$$

$$\delta_v \mathcal{H}([\tilde{\mathcal{N}}, \tilde{v}], \tilde{\lambda}) \geq 0 \quad \forall v \in K. \tag{4.8}$$

(Here $\delta_\lambda \mathcal{H}$, etc., stand for the partial Gâteaux derivative of $\mathcal{H}$ with respect to the variable λ, etc.) Inequality (4.8) together with (4.6), (4.7) implies (cf. (1.22))

$$A(\tilde{v}, v - \tilde{v}) \geq L(v - \tilde{v}) \quad \forall v \in K,$$

hence, $\tilde{v}$ is a solution of the primal problem, $\tilde{v} = u$. □

Lemma 4.3. *Let u be a solution of the primal problem. Then $\{[e(u), u], \tau(u)\}$ is a saddle point of the functional $\mathcal{H}$ on $\mathcal{W} \times S$.*

Proof. We have to verify that all $\mu \in S$ and $[\mathcal{N}, v] \in \mathcal{W}$ satisfy the inequalities

$$\mathcal{H}([e(u), u], \mu) \leq \mathcal{H}([e(u), u], \tau(u)) \leq \mathcal{H}([\mathcal{N}, v], \tau(u)).$$

However, the left inequality becomes an identity as a consequence of the definition of $\mathcal{H}$. The right inequality can be written in the form

$$\frac{1}{2} \int_\Omega c_{ijkl} e_{ij}(u) e_{kl}(u) dx - L(u) \leq \frac{1}{2} \int_\Omega c_{ijkl} \mathcal{N}_{ij} \mathcal{N}_{kl} dx - L(v)$$

$$+\int_\Omega c_{ijkl}e_{kl}(u)(e_{ij}(v)-\mathcal{N}_{ij})dx,$$

which is equivalent to

$$\frac{1}{2}\int_\Omega c_{ijkl}(\mathcal{N}_{ij}-e_{ij}(u))(\mathcal{N}_{kl}-e_{kl}(u))dx$$

$$+\int_\Omega c_{ijkl}e_{kl}(u)e_{ij}(v-u)dx\geq L(v-u)\quad\forall v\in K,\quad\forall\mathcal{N}\in S.$$

However, this inequality easily follows from the positive definiteness of the coefficients c_{ijkl} (see (1.4)) and from the definition of u. □

Lemma 4.4. *Let there exist a solution u of the primal problem. Then*

$$\mathcal{L}(u)=inf_{\mathcal{N}\in S,v\in K}\sup_{\lambda\in S}\mathcal{H}([\mathcal{N},v],\lambda)=\sup_{\lambda\in S}\inf_{\mathcal{N}\in S,v\in K}\mathcal{H}([\mathcal{N},v],\lambda).\qquad(4.9)$$

Proof. Follows from (4.4) and lemmas 4.3, 4.1. □

We will now simplify the dual problem (4.5) by eliminating the variables $[\mathcal{N},v]$. Let us consider the "inner problem"

$$\inf_{[\mathcal{N},v]\in\mathcal{W}}\mathcal{H}([\mathcal{N},v],\lambda),$$

where λ is a fixed element. Evidently,

$$\inf_{\mathcal{W}}\mathcal{H}=\inf_{\mathcal{N}\in S}\mathcal{H}_1(\mathcal{N},\lambda)+\inf_{v\in K}\mathcal{H}_2(v,\lambda),\qquad(4.10)$$

where

$$\mathcal{H}_1(\mathcal{N},\lambda)=\frac{1}{2}\int_\Omega c_{ijkl}\mathcal{N}_{ij}\mathcal{N}_{kl}dx-\int_\Omega\lambda_{ij}\mathcal{N}_{ij}dx,$$

$$\mathcal{H}_2(v,\lambda)=\int_\Omega\lambda_{ij}e_{ij}(v)dx-L(v).\qquad(4.11)$$

We easily deduce that

$$\inf_{\mathcal{N}\in S}\mathcal{H}_1(\mathcal{N},\lambda)=\mathcal{H}_1(\tilde{\mathcal{N}},\lambda)=-\frac{1}{2}\int_\Omega a_{ijkl}\lambda_{ij}\lambda_{kl}dx,\qquad(4.12)$$

where a_{ijkl} are the coefficients of the inverse generalized Hooke's Law and $\tilde{\mathcal{N}}_{ij}=a_{ijkl}\lambda_{kl}$.

Let there exist $v_0\in K$ such that $\mathcal{H}_2(v_0,\lambda)<0$. As K is a convex cone, $tv_0\in K$ and

$$\mathcal{H}_2(tv_0,\lambda)\to-\infty\quad\text{for } t\to+\infty.$$

If $\mathcal{H}_2(v,\lambda) \geq 0$ for all $v \in K$, then

$$\inf_{v \in K} \mathcal{H}_2(v,\lambda) = \mathcal{H}_2(0,\lambda) = 0.$$

Denoting

$$K^+_{F,P} = \{\lambda \in S \mid \mathcal{H}_2(v,\lambda) \geq 0 \quad \forall v \in K\}, \tag{4.13}$$

we thus obtain

$$\inf_{v \in K} \mathcal{H}_2(v,\lambda) = \begin{array}{ll} 0 & \text{provided } \lambda \in K^+_{F,P}, \\ -\infty & \text{provided } \lambda \notin K^+_{F,P}. \end{array} \tag{4.14}$$

Lemma 4.5. *Let there exist a solution u of the primal problem. Then $K^+_{F,P}$ is a nonempty closed and convex subset of S.*

Proof. We will show that the stress tensor $\tau(u) \in K^+_{F,P}$. Indeed, we have

$$\int_\Omega \tau_{ij}(u)e_{ij}(v-u)dx \geq L(v-u) \quad \forall v \in K.$$

Substituting here $v = u + w$, where w is an arbitrary element of K, we obtain

$$\mathcal{H}_2(w,\tau(u)) = \int_\Omega \tau_{ij}(u)e_{ij}(w)dx - L(w) \geq 0 \quad \forall w \in K.$$

The fact that $K^+_{F,P}$ is both closed and convex is evident. □

The relations (4.10), (4.12), and (4.14) imply that

$$\inf_{[\mathcal{N},v] \in \mathcal{W}} \mathcal{H}([\mathcal{N},v],\lambda) = \begin{array}{ll} -\mathcal{S}(\lambda) & \text{provided } \lambda \in K^+_{F,P}, \\ -\infty & \text{provided } \lambda \notin K^+_{F,P}, \end{array}$$

where

$$\mathcal{S}(\lambda) = \frac{1}{2}\int_\Omega a_{ijkl}\lambda_{ij}\lambda_{kl}\,dx.$$

Hence, we have

$$\sup_{\lambda \in S} \inf_{[\mathcal{N},v] \in \mathcal{W}} \mathcal{H}([\mathcal{N},v],\lambda) = \sup_{\lambda \in K^+_{F,P}} [-\mathcal{S}(\lambda)] = -\inf_{\lambda \in K^+_{F,P}} \mathcal{S}(\lambda).$$

If there exists a saddle point of $\mathcal{H}$ on $\mathcal{W} \times S$, then

$$-\inf_{\lambda \in K^+_{F,P}} \mathcal{S}(\lambda) = -\mathcal{S}(\tau(u)) = \mathcal{L}(u), \tag{4.15}$$

where u is the solution of the primal problem. Indeed, this follows from lemmas 4.2, 4.1, and 4.4.

On the other hand, lemma 4.3 guarantees the existence of a saddle point provided there exists a solution of the primal problem. In section 2.2.1 we dealt with the problem of existence and uniqueness of a solution of the primal problem. According to theorem 2.2, sufficient conditions for the existence of a solution are

$$L(y) \leq 0 \quad \forall y \in K \cap \mathbf{R},$$

$$L(y) < 0 \quad \forall y \in K \cap \mathbf{R} \dot{-} \mathbf{R}^*, \tag{4.16}$$

and the difference of two arbitrary solutions belongs to the subspace $\mathbf{R}$ of displacements of rigid bodies. Hence, both the strain tensor $e(u)$ and the stress tensor $\tau(u)$ are uniquely determined. Thus, we can assert that if (4.16) holds, then there is a saddle point of $\mathcal{H}$ on $\mathcal{W} \times S$, its second component is uniquely determined and (4.15) holds.

In other words, if (4.16) holds, then the *dual problem* to find $\lambda \in K^+_{F,P}$ such that

$$\mathcal{S}(\lambda) \leq \mathcal{S}(\mu) \quad \forall \mu \in K^+_{F,P} \tag{4.17}$$

has a unique solution $\lambda = \tau(u)$, where u is an arbitrary solution of the primal problem. Thus, we have obtained a *uniquely* solvable formulation of a certain class of contact problems, in contrast to the primal variational formulation (see section 2.2.1, where we eventually restricted our considerations solely to the cases of one-dimensional subspaces of virtual displacements of rigid bodies).

Remark 4.1. The existence and uniqueness of solution of the dual problem (4.17) can be proved directly by using lemma 4.5 and the fact that the functional $\mathcal{S}$ is strictly convex.

Remark 4.2. Let us point out that the dual problem possesses a unique solution whenever the primal problem has a solution. However, both the existence and uniqueness of the dual problem are obtained immediately, provided the set $K^+_{F,P}$ is nonempty. This leads to a conjecture that the dual problem may have a solution even if the primal problem has none. We do not intend to study this problem; let us only mention the fact that $K^+_{F,P}$ is nonempty only if condition $(4.16)_1$ is fulfilled. Hence, this inequality is a necessary condition for the existence of solution of both the primal and dual problems (see lemma 2.1).

Interpretation of the Set $K^+_{F,P}$. For the purposes of approximation, it is useful to study in more detail the structure of the set $K^+_{F,P}$ of admissible stress fields.

Lemma 4.6. 1^0 *Let $\lambda \in K^+_{F,P}$ be sufficiently smooth. Then λ fulfills the following conditions:*

$$\frac{\partial \lambda_{ij}}{\partial x_j} + F_i = 0 \quad \text{in } \Omega = \Omega' \cup \Omega'', \quad i = 1,2; \tag{4.18}$$

$$\lambda_{ij} n_j = P_i \quad \text{on } \Gamma_\tau = \Gamma'_\tau \cup \Gamma''_\tau, \quad i = 1,2; \tag{4.19}$$

$$T_t(\lambda) = 0 \quad \text{on } \Gamma_0, \tag{4.20}$$

$$T_t(\lambda') = T_t(\lambda'') = 0 \quad \text{on } \Gamma_K, \tag{4.21}$$

$$T_n(\lambda') = T_n(\lambda'') \leq 0 \quad \text{on } \Gamma_K. \tag{4.22}$$

2^0 Conversely, let $\lambda \in S$ be sufficiently smooth and satisfy (4.18)–(4.22). Then, $\lambda \in K^+_{F,P}$.

Proof. 1^0 Integrating by parts we obtain for all $v \in K$:

$$\int_\Omega \lambda_{ij} e_{ij}(v)\,dx = -\int_\Omega v_i \frac{\partial \lambda_{ij}}{\partial x_j} dx + \int_{\partial\Omega' \cup \partial\Omega''} [T_n(\lambda) v_n + T_t(\lambda) v_t]\,ds$$

$$\geq \int_\Omega F_i v_i\,dx + \int_{\Gamma_\tau} P_i v_i\,ds.$$

Substituting $v_i^M = \pm\varphi_i \in C_0^\infty(\Omega^M)$, $M = {}'\!,{}''$, we find (4.18). Thus, we obtain

$$\int_{\partial\Omega' \cup \partial\Omega''} T_i(\lambda) v_i\,ds \geq \int_{\Gamma_\tau} P_i v_i\,ds \quad \forall v \in K.$$

Choosing $v_i = \pm\psi_i$ such that the traces of ψ have a support in Γ_τ, we obtain (4.19).

Now take $v' = 0$ and v'' such that $v''_n = 0$, $v''_t = \pm\psi$ on Γ_0, where the support of ψ is in Γ_0. Hence, (4.20) follows.

It remains to analyze the inequality

$$\int_{\Gamma_K} [T_n(\lambda') v'_n + T_t(\lambda') v'_t + T_n(\lambda'') v''_n + T_t(\lambda'') v''_t]\,ds \geq 0 \quad \forall v \in K. \tag{4.23}$$

Let $v \in V$ be such that $v'_t = v''_t = 0$ and $v'_n = -v''_n = \pm\varphi$ on Γ_K, $\varphi \in C_0^\infty(\Gamma_K)$. Thus, we find that

$$\int_{\Gamma_K} [T_n(\lambda') - T_n(\lambda'')]\varphi\,ds = 0.$$

Hence,

$$T_n(\lambda') = T_n(\lambda'') \quad \text{on } \Gamma_K.$$

If we choose $v \in V$ such that $v'_n = v''_n = 0$, $v''_t = 0$, $v'_t = \pm\varphi$ on Γ_K, then (4.23) implies that $T_t(\lambda') = 0$ on Γ_K. The identity $T_t(\lambda'') = 0$ can be analogously derived.

Finally, we have

$$\int_{\Gamma_K} T_n(\lambda)(v'_n + v''_n)ds \geq 0 \quad \forall v \in K,$$

hence, $T_n(\lambda) \leq 0$ by virtue of the inequality $v'_n + v''_n \leq 0$ that holds on Γ_K.

2^0 Let us multiply (4.18) by a function v_i, where $v \in K$, and integrate by parts over Ω. Then,

$$0 = -\int_\Omega \lambda_{ij}\frac{\partial v_i}{\partial x_j}dx + \int_\Omega F_i v_i\, dx + \int_{\partial\Omega' \cup \partial\Omega''} \lambda_{ij} n_j v_i\, ds$$

$$= -\int_\Omega \lambda_{ij} e_{ij}(v)dx + L(v) + \int_{\Gamma_K} T_n(\lambda)(v'_n + v''_n)ds.$$

The definition of K, together with (4.22), implies that the last integral is nonnegative, hence $\lambda \in K^+_{F,P}$. □

2.4.1 Approximation of the Dual Problem

To obtain an approximation of the dual problem, it is necessary to construct finite-dimensional approximations of the set $K^+_{F,P}$ of admissible stress fields. To this end we first find a particular solution $\bar{\lambda}$ of the nonhomogeneous equations (4.18), (4.19), and then we will write $\lambda = \bar{\lambda} + \tau$ for $\lambda \in K^+_{F,P}$, where τ is a "self-equilibriated" stress field.

Since the resultant of the system of forces F_i, P_i (which act on the body Ω'') is generally nonzero, we must introduce reactions (normal loads) $T_n(\bar{\lambda})$ on Γ_K, which naturally act on both bodies Ω' and Ω''.

Lemma 4.7. *If* $\bar{\lambda} \in S$ *fulfills conditions* (4.18), (4.19), *then*

$$\int_{\Gamma_K} T_i(\bar{\lambda}'')ds = -\left[\int_{\Omega''} F_i dx + \int_{\Gamma''_\tau} P_i ds + \int_{\Gamma_0} T_i(\bar{\lambda}'')ds\right], \quad i = 1, 2. \tag{4.24}$$

Proof. Immediately follows from (4.18) and (4.19):

$$-\int_{\Omega''} F_i dx = \int_\Omega \frac{\partial \bar{\lambda}''_{ij}}{\partial x_j}dx = \int_{\partial\Omega''} \bar{\lambda}''_{ij} n''_j ds$$

$$= \int_{\Gamma''_\tau} P_i\, ds + \int_{\Gamma_0} T_i(\bar{\lambda}'')ds + \int_{\Gamma_K} T_i(\bar{\lambda}'')ds,$$

and this yields (4.24). □

With regard to lemma 4.7, we can select the simplest distribution of reaction pressure forces $T_n(\bar{\lambda})$ on Γ_K.

Example 4.1. Let Γ_0 consist of line segments parallel to the x_1-axis, while Γ_K is such a segment that $n_1'' > 0$ on Γ_K. We can choose $\bar{\lambda} \in K_{F,P}^+$ such that

$$T_n(\bar{\lambda}') = T_n(\bar{\lambda}'') = g \quad \text{on } \Gamma_K, \tag{4.25}$$

where

$$g = -\left[\int_{\Omega'} F_1 \, dx + \int_{\Gamma_\tau''} P_1 \, ds\right] / \int_{\Gamma_K} dx_2 = \text{const.} \tag{4.26}$$

Indeed, using the identities $T_1(\bar{\lambda}) = T_t(\bar{\lambda}) = 0$ on Γ_0 and $dx_2 = n_1'' ds$, $T_t(\bar{\lambda}) = 0$ on Γ_K, we find that the choice (4.25), (4.26) satisfies condition (4.24). Further, we know that $(4.16)_1$ is a necessary condition for the existence of solution, which in our case means (see example 2.1):

$$V_1'' = \int_{\Omega''} F_1 \, dx + \int_{\Gamma_\tau''} P_1 \, ds \geq 0. \tag{4.27}$$

Hence $g \leq 0$, and condition (4.22) is also fulfilled.

Example 4.2. Let Γ_0 and Γ_K be the same as in the previous example and let $n_2'' > 0$. We can choose $T_n(\bar{\lambda}') = T_n(\bar{\lambda}'') = 0$ on Γ_K, and

$$T_t(\bar{\lambda}') = T_t(\bar{\lambda}'') = -V_1''/\int_{\Gamma_K} dx_1 = \text{const.} \quad \text{on } \Gamma_K. \tag{4.28}$$

Then evidently $\bar{\lambda} \notin K_{F,P}^+$, unless $V_1'' = 0$ holds.

Example 4.3. Let $\Gamma_0 = \emptyset$ and let Γ_K be a segment parallel to the x_1-axis. We can choose $\bar{\lambda}$ such that

$$T_n(\bar{\lambda}'') = -T_2(\bar{\lambda}'') = V_2''/\int_{\Gamma_K} ds = g \quad \text{on } \Gamma_K, \tag{4.29}$$

where

$$V_2'' = \int_{\Omega''} F_2 \, dx + \int_{\Gamma_\tau''} P_2 \, ds.$$

Since $V_2'' \leq 0$ is a necessary condition for the existence of solution (according to $(4.16)_1$), we have $T_n(\bar{\lambda}) \leq 0$ on Γ_K, and hence (4.22) holds as well.

Remark 4.3. It is not necessary for $\bar{\lambda}$ to belong to the set $K_{F,P}^+$. Nonetheless, from the practical point of view it is suitable that $\bar{\lambda}$ satisfy $T_i(\bar{\lambda}) =$

const on Γ_K. Namely, this is the case when we can construct internal approximations of the set $K^+_{F,P}$. This offers some advantages.

An algorithm for the construction of $\bar{\lambda}$ will be presented in section 2.4.13.

In the following, let us consider the situation from example 4.1 and show in detail the approximation of the set $K^+_{F,P}$. Let $\bar{\lambda}$ satisfy condition

$$\int_\Omega \bar{\lambda}_{ij} e_{ij}(v)dx = L(v) + \int_{\Gamma_K} g(v'_n + v''_n)ds \quad \forall v \in V. \tag{4.30}$$

We easily find that

$$\lambda \in K^+_{F,P} \Longleftrightarrow \lambda - \bar{\lambda} \equiv \tau \in \mathcal{U}_0,$$

with

$$\mathcal{U}_0 = \left\{ \tau \in S \mid \int_\Omega \tau_{ij} e_{ij}(v)dx \geq -g \int_{\Gamma_K} (v'_n + v''_n)ds \ \ \forall v \in K \right\}.$$

Lemma 4.8. 1^0 *Let $\tau \in \mathcal{U}_0$ be sufficiently smooth. Then τ fulfills the homogeneous equations* (4.18), (4.19), (4.20), (4.21) *and*

$$T_n(\tau') = T_n(\tau'') \leq -g \quad \text{on } \Gamma_K. \tag{4.31}$$

2^0 *Let Ω be divided into a finite number of closed subdomains K_r,*

$$\bar{\Omega} = \bigcup_r K_r, \quad \mathring{K}_r \cap \mathring{K}_s = \emptyset \quad \text{for } r \neq s$$

($\mathring{K}$ denotes the interior of the set K_r). Let $\tau \in S$ fulfill the homogeneous equations (4.18) *in each subdomain $\mathring{K}_r$, the homogeneous boundary conditions* (4.19), (4.20), (4.21), *and the inequalities* (4.31). *Moreover, let the stress tensor $T(\tau)$ be continuous when crossing an arbitrary common boundary of two adjacent subdomains, that is,*

$$T(\tau)|_{K_r} + T(\tau)|_{K_s} = 0 \quad \text{on } K_r \cap K_s, \quad \forall r \neq s.$$

Then, $\tau \in \mathcal{U}_0$.

Proof. Analogous to that of lemma 4.6.

Remark 4.4. If condition (4.27) is fulfilled, then $\mathcal{U}_0$ is nonempty, since it contains the zero element. Besides, $\mathcal{U}_0$ is a convex and closed subset of S. □

Substituting $\lambda = \bar{\lambda} + \tau$ into definition (4.17) of the dual problem, we obtain an *equivalent dual problem:* find $\tau^0 \in \mathcal{U}_0$ such that

$$\mathcal{J}(\tau^0) \leq \mathcal{J}(\tau) \quad \forall \tau \in \mathcal{U}_0, \tag{4.32}$$

where

$$\mathcal{J}(\tau) = \frac{1}{2} \int_\Omega a_{ijkl} \tau_{ij} (\tau_{kl} + 2\bar{\lambda}_{kl}) dx.$$

2.4.11. Equlibrium Model of Finite Elements. Since the admissible stress fields in problem (4.31) are required to satisfy the homogeneous equations of equilibrium, it is necessary to construct finite-dimensional subspaces of tensor fields with the same property. For this purpose, we can use the equilibrium model of finite elements, proposed by Watwood and Hartz (1968).

This model consists of triangular block elements, which are formed by joining the vertices of a general triangle with its center of gravity. In each subtriangle we define three linear functions—the components of the self-equilibriated stress field. The stress vector is continuous when crossing any boundary between two subtriangles.

In a triangle K let us define the space of self-equilibriated linear stress fields

$$M(K) = \left\{ \begin{array}{c} \tau_{11} = \beta_1 + \beta_2 x_2 + \beta_3 x_2, \\ \tau_{22} = \beta_4 + \beta_5 x_1 + \beta_6 x_2, \\ \tau_{12} = \tau_{21} = \beta_7 - \beta_6 x_1 - \beta_2 x_2 \end{array} \right\},$$

where $\beta \in \mathbf{R}^7$ is an arbitrary vector. Immediately we see that $\partial \tau_{ij} / \partial x_j = 0$ in K for every $\tau \in M(K)$, $i = 1, 2$.

Further, let us consider a block element $K = \cup_{i=1}^3 K_i$ according to Figure 22 and define

$$N(K) = \{\tau = (\tau^1, \tau^2, \tau^3) \,|\, \tau^i = \tau|_{K_i} \in M(K_i),$$

$$T(\tau^i) + T(\tau^{i-1}) = 0 \quad \text{on } \overline{0a_i},\ i = 1, 2, 3\}.$$

(The last condition expresses the continuity of the stress vectors on the sides $\overline{0a_i}$.)

Theorem 4.1. *Given an arbitrary exterior load $\bar{T}$ of the triangle K, such that* (i) *it is linear along each side of the triangle K,* (ii) *it satisfies the conditions of the total equilibrium, then there exists a unique stress field $\tau \in N(K)$ such that*

$$T(\tau) = \bar{T} \quad \text{on } \partial K.$$

Proof. See Hlaváček (1979), theorem 2.1. □

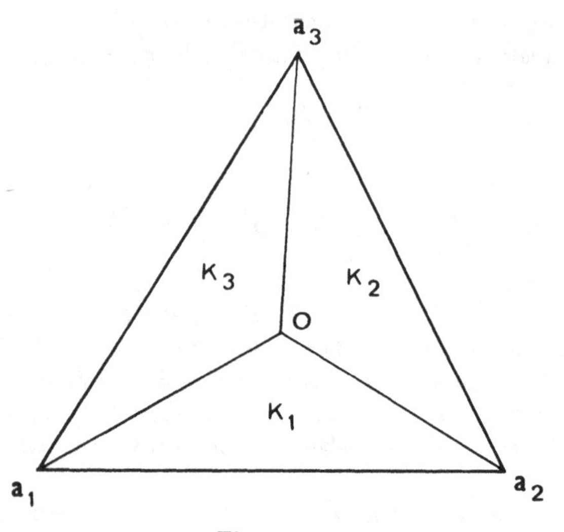

Figure 22

Let G be a bounded polygonal domain, $h \in (0,1]$ a real parameter, and $\mathcal{T}_h$ a triangulation of the domain G. Let us define

$$h = \max_{K \in \mathcal{T}_h} \operatorname{diam} K,$$

$$N_h(G) = \{\tau \in S(G) \mid \tau|_K \in N(K) \quad \forall K \in \mathcal{T}_h,$$
$$T(\tau)|_K + T(\tau)|_{K'} = 0 \quad \forall K \cap K'\},$$
$$E(G) = \{\tau \in [H^1(G)]^4 \cap S(G) \mid \partial\tau_{ij}/\partial x_j = 0 \quad \text{in } G, \quad i = 1,2\}.$$

Theorem 4.2. *There exists a linear continuous mapping r_h: $E(G) \to N_h(G)$ such that for every $\tau \in E(G) \cap [H^2(G)]^4$ the estimate*

$$\|\tau - r_h\tau\|_{[L^2(G)]^4} \leq Ch^2 \|\tau\|_{[H^2(G)]^4} \tag{4.33}$$

holds, where C is independent of h and τ, provided the system of triangulations $\{\mathcal{T}_h\}$ is regular.

Proof. See Hlaváček (1979), theorem 2.5. Estimate (4.33) also follows from the results of Johnson and Mercier (1978). □

Remark 4.5. The mapping r_h from theorem 4.2 is defined locally on every block element $K \in \mathcal{T}_h$ in the following way: the stress vector $T_k(\tau)$ on each side of $S_j \subset \partial K$ is projected in $L^2(S_j)$ to the subspace $P_1(S_j)$ of linear functions. These projections of $\mathcal{T}$ uniquely determine the stress field $r_h\tau|_K \in N(K)$, as follows from theorem 4.1.

Remark 4.6. Every stress field $\tau \in N_h(G)$ satisfies the homogeneous equations of equilibrium $\partial\tau_{ij}/\partial x_j = 0$, $i = 1, 2$, in the domain G in the sense of distributions.

2.4.12. Applications of the Equilibrium Model. Let us assume that both Ω' and Ω'' are bounded polygonal domains. Define the approximations of the set $\mathcal{U}_0$ by

$$\mathcal{U}_{0h} = \mathcal{U}_0 \cap N_h(\Omega),$$

where

$$N_h(\Omega) = \{(\tau', \tau'') \mid \tau^M \in N_h(\Omega^M), \quad M = \text{'},\text{''}\}.$$

We say that $\tau^h \in \mathcal{U}_{0h}$ is an *approximation of the solution of the dual problem, if*

$$J(\tau^h) \leq J(\tau) \quad \forall \tau \in \mathcal{U}_{0h}. \tag{4.34}$$

Lemma 4.9. *If $V_1'' \geq 0$ (see (4.27)), then there exists a unique solution of problem (4.34).*

Proof. The set $N_h(\Omega)$ evidently is a linear finite-dimensional subset of S, hence it is closed and convex. Using remark 4.4, we deduce that $\mathcal{U}_h$ is also closed, convex, and nonempty. As the functional J is differentiable and strictly convex, this easily yields the existence and uniqueness of τ^h. □

An algorithm for finding τ^h will be presented in the next section. Here we will deal with an estimate of the error

$$\|\lambda - \lambda^h\|_{0,\Omega} = \|\tau^0 - \tau^h\|_{0,\Omega},$$

with $\lambda = \bar{\lambda} + \tau^0$, $\lambda^h = \bar{\lambda} + \tau^h$, $\|\cdot\|_{0,\Omega}$ being the norm in $[L^2(\Omega)]^4$.

The main result is included in the following theorem.

Theorem 4.3. *Let Γ_0 consist of line segments parallel to the x_1-axis and let Γ_K be a segment such that $n_1'' > 0$ on Γ_K. Let us assume that $\tau^0|_{\Omega^M} \in [H^2(\Omega^M]^4$, $M = \text{'},\text{''}$, and $T_n(\tau^0) \in H^2(\Gamma_K)$. Let the system of triangulations $\{\mathcal{T}_h\}$ be (α, β)-regular and satisfy the following conditions: between Γ_K and Γ_0 in the domain Ω'' and between Γ_K and Γ_u in Ω', the triangulation $\mathcal{T}_h$ is inscribed into smooth "vaulted strips" with bounded curvature and a slope $|\theta| < \Theta < \frac{\pi}{2}$ (Θ independent of h), which are perpendicular to Γ_0 and Γ_K (see Figure 23). Then the estimate*

$$\|\tau^0 - \tau^h\|_{0,\Omega} \leq C(\tau^0)h^{3/2}, \tag{4.35}$$

with C independent of h holds.

Proof. Based on the idea of unilateral approximations like the proof of theorem 6.6 in section 1.1.62. A detailed proof may be found in Haslinger and Hlaváček (1981 b). □

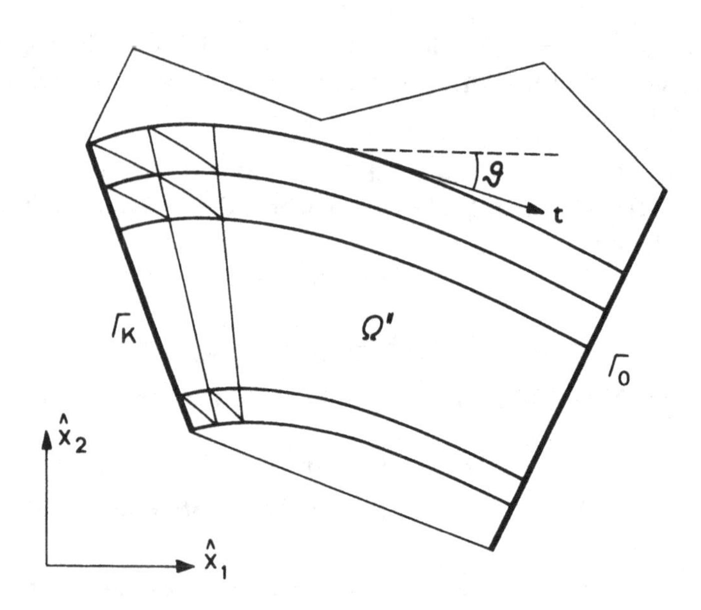

Figure 23

Remark 4.7. Let us briefly consider the situation from example 3, that is, let $\Gamma_0 = \emptyset$ and Γ_K be a segment parallel to the x_1-axis. Let us construct a particular stress field $\bar{\lambda}$ (which satisfies condition (4.30)), define $\mathcal{U}_0$ (the equivalent dual problem (4.32)), $\mathcal{U}_{0h}$, and the approximations of the solution of (4.34). If $V_2'' \leq 0$, then $\mathcal{U}_0$ contains the zero element and there is a unique approximation τ^h. Then an analog of theorem 4.3 holds, by only replacing the condition concerning the vaulted strips by the following condition: The triangulation $\mathcal{T}_h$ in Ω'' includes a fixed rectangle $AUBC$ (see Figure 24) independent of h, with $\overline{AU} = \Gamma_K$, which is divided into rectangular elements. The triangulation $\mathcal{T}_h$ in Ω' fulfills the same conditions as in theorem 4.3.

2.4.13. Algorithm for Approximations of the Dual Problem. Let us first present a survey of the results by Watwood and Hartz (1968), which can be immediately used in the algorithm for the solution of problem (4.34), that is, for finding the stress field $\tau^h \in \mathcal{U}_{0h}$.

The behavior of the stress field can be expressed on each subtriangle K_i in the following way:

$$\tau = \begin{bmatrix} \tau_{11} \\ \tau_{22} \\ \tau_{12} \end{bmatrix}$$

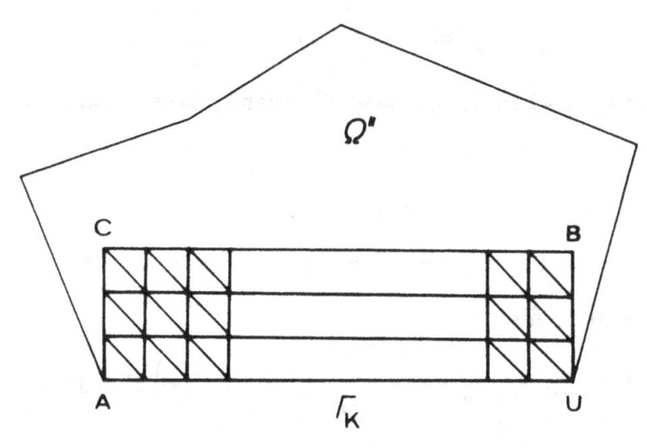

Figure 24

$$= \frac{E}{\sqrt{AE_0}} \begin{bmatrix} \sqrt{A} & 0 & 0 & 0 & x_1 & 0 & x_2 \\ 0 & \sqrt{A} & 0 & x_1 & 0 & x_2 & 0 \\ 0 & 0 & \sqrt{A} & 0 & -x_2 & -x_1 & 0 \end{bmatrix} S = MS, \quad (4.36)$$

where $S \in \mathbf{R}^7$ is the vector of coefficients, E is the Young module, E_0 a dimensionless quantity equal to the reference module, and A the area of the triangle K_i.

Let the origin of the Cartesian coordinates (x_1, x_2) coincide with the center of gravity of K_i. Let us consider a material, homogeneous and isotropic in K_i, with the Poisson constant σ. Then

$$\int_{K_i} a_{mjkl}\tau_{mj}\tau_{kl}\,dx = S^T f S,$$

where the matrix f has the form

$$f = \frac{tE}{E_0} \begin{bmatrix} \alpha A, & & & & & & \\ \beta A, & \alpha A & & & & & \\ 0 & 0 & \gamma A, & & & & \\ 0 & 0 & 0 & \alpha\delta_1, & & \text{symmetry} & \\ 0 & 0 & 0 & \beta\delta_1, & \alpha\delta_1 + \gamma\delta_2, & & \\ 0 & 0 & 0 & \alpha\delta_{12}, & (\beta+\gamma)\delta_{12}, & \alpha\delta_2 + \gamma\delta_1, & \\ 0 & 0 & 0 & \beta\delta_{12}, & \alpha\delta_{12}, & \beta\delta_2, & \alpha\delta_2, \end{bmatrix},$$

t being the thickness of the element K_i,

$$\delta_j = A^{-1} \int_{K_i} (x_j)^2 dx_1 dx_2, \quad j = 1, 2,$$

$$\delta_{12} = A^{-1} \int_{K_i} x_1 x_2 dx_1 dx_2,$$

α, β, γ constants, which in the case of a plane stress assume the values

$$\alpha = 1, \quad \beta = -\sigma, \quad \gamma = 2(1+\sigma),$$

while in the case of a plane strain,

$$\alpha = 1 - \sigma^2, \quad \beta = -\sigma(1+\sigma), \quad \gamma = 2(1+\sigma).$$

Similarly, we could find

$$\int_{K_i} a_{mjkl} \tau_{mj} \bar{\lambda}_{kl} dx = \left(\int_{K_i} \bar{\lambda}^T B^{-1} M \, dx \right) S = B_0^T S,$$

with $\bar{\lambda}^T = (\bar{\lambda}_{11}, \bar{\lambda}_{22}, \bar{\lambda}_{12})$, M the matrix from formula (4.36), and B^{-1} the (3×3)-matrix of the inverse Hooke's Law ($e = B^{-1}\tau$).

Each component of the stress vector on the side $\overline{a_i a_{i+1}}$ of the triangle K can be expressed in terms of the exterior parameters $S^* \in \mathbf{R}^4$ and a continuous parameter $p \in [-1, 1]$ as follows:

$$T_1(p) = S_1^* + S_2^* p, \quad T_2(p) = S_3^* + S_4^* p.$$

For instance, let us consider the side $\overline{a_2 a_3}$ (see Figure 22). Then, at the point a_2 we have $p = -1$, and at a_3 we have $p = 1$ and

$$S^* = \frac{1}{2 l_a \sqrt{A}} \overset{(a)}{C} S,$$

where l_a is the length of the side $\overline{a_2 a_3}$, while $\overset{(a)}{C}$ is the following (4×7)-matrix:

$$\overset{(a)}{C} = \left[\begin{matrix} -2\sqrt{(A)}(Y_2 - Y_3), & 0 & 2\sqrt{(A)}(X_2 - X_3), & 0 \\ 0 & 0 & 0 & 0 \\ 0 & 2\sqrt{(A)}(X_2 - X_3), & -2\sqrt{(A)}(Y_2 - Y_3), & X_2^2 - X_3^2, \\ 0 & 0 & 0 & -(X_2 - X_3)^2, \end{matrix} \right.$$

$$\left. \begin{matrix} -2(X_2 Y_2 - X_3 Y_3), & -(X_2^2 - X_3^2), & -(Y_2^2 - Y_3^2), \\ -2(X_2 - X_3)(Y_2 - Y_3), & (X_2 - X_3)^2, & (Y_2 - Y_3)^2, \\ Y_2^2 - Y_3^2, & 2(X_2 Y_2 - X_3 Y_3), & 0 \\ -(Y_2 - Y_3)^2, & -2(X_2 - X_3)(Y_2 - Y_3), & 0 \end{matrix} \right].$$

By a cyclic permutation of indices we can find the matrices $\overset{(b)}{C}$ and $\overset{(c)}{C}$. Let us denote the exterior parameters on side b by S_5^*, S_6^*, S_7^*, S_8^*, and on side c by S_9^*, S_{10}^*, S_{11}^*, and S_{12}^*. Then, the total vector $S^* \in \mathbf{R}^{12}$ satisfies

$$S^* = CS, \tag{4.37}$$

where the (12 × 7)-matrix C consists of the matrices $\overset{(a)}{C}, \overset{(b)}{C}, \overset{(c)}{C}$:

$$C = \frac{1}{2\sqrt{A}} \begin{bmatrix} l_a^{-1} & \overset{(a)}{C} \\ l_b^{-1} & \overset{(b)}{C} \\ l_c^{-1} & \overset{(c)}{C} \end{bmatrix}.$$

The conditions of continuity of the stress vector on the common sides have the form

$$S_i^* + S_j^* = 0, \tag{4.38}$$

where the indices i, j correspond to the same basis function, but to adjacent triangles.

From the definition of $\mathcal{U}_{0h} = \mathcal{U}_0 \cap N_h(\Omega)$, it is easily seen that $\tau \subset \mathcal{U}_{0h}$ if and only if all the constraints of the form (4.38) hold,

$$S_j^* = 0 \quad \text{on each side } \overline{a_i a_{i+1}} \subset \bar{\Gamma}_\tau, \tag{4.39}$$

$$S_j^* t_1 + S_{j+2}^* t_2 = 0 \quad \text{on } \bar{\Gamma}_0 \cup \Gamma_K \tag{4.40}$$

(where t_k are the components of the tangent vector and (4.40) holds independently of $\tau^M \in N_h(\Omega^M)$, $M = {}',{}''$), and

$$[S_j^* n_1' + S_{j+2}^* n_2' - (S_{j+1}^* n_1' + S_{j+3}^* n_2')]_{\Omega'} \le -g \quad \text{on } \Gamma_K,$$

$$[S_j^* n_1' + S_{j+2}^* n_2' + (S_{j+1}^* n_1' + S_{j+3}^* n_2')]_{\Omega'} \le -g \quad \text{on } \Gamma_K, \tag{4.41}$$

conditions of the form (4.38) hold on the common sides of any two triangles belonging to Γ_K.

Instead of working with all components of the vector S, it is recommended to *reduce* these *parameters* by eliminating the interior degrees of freedom in each triangular block. Let us write the conditions of continuity on the segments $\overline{0a_i}$ (see Figure 22) in the form

$$A_u S = 0, \tag{4.42}$$

where A_u is a (12 × 21)-matrix and S a (21 × 1)-matrix. There exists a regular (21 × 21)-matrix Q such that

$$A_u Q = [I \quad 0], \tag{4.43}$$

where I is the unit matrix. Naturally, the matrix Q is not uniquely determined. It even suffices to replace I in (4.43) by an arbitrary regular (12 × 12)-matrix. Besides, in the following we need only the last nine columns of the matrix Q (which form the matrix Q_1). The identity (4.43) can be obtained, for example, by the Gauss elimination of the matrix A_u^T. Let us carry out the transformation

$$S = Q\hat{S} = [Q_0 \quad Q_1]\begin{bmatrix} \hat{S}^u \\ \hat{S}^1 \end{bmatrix} \tag{4.44}$$

cutting Q between its 12th and 13th column and $\hat{S}$ in a corresponding fashion. After substituting in (4.42), we obtain

$$Q_u Q\hat{S} = [I \quad 0]\hat{S} = 0 \Rightarrow \hat{S}^u = 0,$$

and the transformation (4.44) reduces to

$$S = Q_1\hat{S}^1, \tag{4.45}$$

with Q_1 a (21×9)-matrix and $\hat{S}^1$ a (9×1)-matrix. The parameters $\hat{S}_k^1$, $k = 1, \ldots, 9$, are the degrees of freedom of the triangular block element. In the following, we will write $Z_k = \hat{S}_k^1$, that is, $\hat{S}^1 = Z$.

Now the functional $J(\tau)$ of the equivalent dual problem can be written in terms of Z:

$$\begin{aligned} J(\tau) &= \frac{1}{2}\sum_{K\in\tau_h}\sum_{i=1}^{3}\int_{K_i} a_{mjkl}\tau_{mj}(\tau_{kl} + 2\bar{\lambda}_{kl})dx \\ &= \sum_{K\in\tau_h}\sum_{i=1}^{3}\left(\frac{1}{2}S^T fS + b_0^T S\right) = \sum_{K\in\tau_h}\left(\frac{1}{2}S^T FS + B_0^T S\right) \\ &= \sum_{K\in\tau_h}\left(\frac{1}{2}Z^T Q_1^T FQ_1 Z + B_0^T Q_1 Z\right) = \frac{1}{2}Z^T AZ + b^T Z = \mathcal{Y}(Z); \end{aligned}$$

where the vectors S and Z successively correspond to a subtriangle, to a triangular block, or to the whole triangulation. The $(N \times N)$-matrix A is now positive definite.

Substituting (4.37) and (4.45) into the conditions of the form (4.38)–(4.41), we obtain constraints

$$DZ = 0, \tag{4.46}$$

$$EZ \leq -g, \tag{4.47}$$

where D, E are matrices of the types $(r_1 \times N)$, $(r_2 \times N)$, respectively, and g is a vector whose all components are equal to g.

Let us define a set

$$\mathcal{B} = \{Z \in \mathbf{R}^N \mid Z \text{ fulfills (4.46) and (4.47)}\}.$$

Thus, we arrive at the problem to find $\sigma \in \mathcal{B}$ such that

$$\mathcal{Y}(\sigma) \leq \mathcal{Y}(Z) \quad \forall Z \in \mathcal{B}. \tag{4.48}$$

This problem can be solved, for example, by applying *Uzawa's algorithm* (see Céa (1971), Chapter 4, section 5.1). Denote $r = r_1 + r_2$ and

$$B = \begin{bmatrix} D \\ E \end{bmatrix}, \quad G = \begin{bmatrix} 0 \\ g \end{bmatrix},$$

the matrices being of the types $(r \times N)$, $(r \times 1)$, respectively. Define the set of Lagrangian multipliers

$$\Lambda = \{y \in \mathbf{R}^r \mid y_j \geq 0 \quad \text{for } j = r_1 + 1, \ldots, r\}.$$

We choose $y^0 \in \Lambda$ and calculate $z^0 \in \mathbf{R}^N$ from the system

$$Az^0 = -b - B^T y^0.$$

If we have y^n, z^n, then the values y^{n+1}, z^{n+1} are determined from the conditions

$$y^{n+1} = P_\Lambda[y^n + \rho(Bz^n + G)],$$
$$Az^{n+1} = -b - B^T y^{n+1},$$

where P_Λ denotes the projection to the set Λ (that is, $(P_\Lambda t)_j = t_j$ for $j = 1, \ldots, r_1$, $(P_\Lambda t)_j = \max\{0, t_j\}$ for $j = r_1 + 1, \ldots, r$), and $\rho \in \mathbf{R}$ is a sufficiently small parameter.

It can be proven that $z^n \to \sigma$ in $\mathbf{R}^N$ for $n \to \infty$, where σ is the solution of (4.48) (which is unique according to lemma 4.9), provided the matrix B has the full rank, that is, r.

In conclusion, let us suggest the *construction of a particular stress field* $\bar{\lambda}$. Let us again consider the situation in example 4.1 and choose $T_n(\bar{\lambda}') = T_n(\bar{\lambda}'') = g$ on Γ_K (see (4.25), (4.26)). In accordance with the interpretation of the set $K^+_{F,P}$ and (4.30) (cf. lemmas 4.6 and 4.8), we can proceed as follows:

Choose $\lambda^{(1)} \in S$, which fulfills conditions (4.18) in $\Omega = \Omega' \cup \Omega''$ (by direct integration with respect to x_1 or x_2).

Assume that F is a constant and P a piecewise linear vector field. Then, $\lambda^{(1)}_{ij}$ are linear polynomials and $T_i(\lambda^{(1)})$ is linear on each side of the polygonal boundary $\partial\Omega' \cup \partial\Omega''$.

We put $\bar{\lambda} = \lambda^{(1)} + \lambda^{(2)}$ and seek for $\lambda^{(2)}$ in the space $N_h(\Omega)$, imposing the following boundary conditions:

$$T(\lambda^{(2)}) = P - T(\lambda^{(1)}) \quad \text{on } \Gamma_\tau, \tag{4.49}$$

$$T_t(\lambda^{(2)}) = -T_t(\lambda^{(1)}) \quad \text{on } \Gamma_0, \tag{4.50}$$

$$T_t(\lambda^{(2)\prime}) = T_t(\lambda^{(2)\prime\prime}) = -T_t(\lambda^{(1)}) \quad \text{on } \Gamma_K, \tag{4.51}$$

$$T_n(\lambda^{(2)\prime}) = T_n(\lambda^{(2)\prime\prime}) = g - T_n(\lambda^{(1)}) \quad \text{on } \Gamma_K. \tag{4.52}$$

Since the right-hand sides in (4.49)–(4.52) are piecewise linear functions, there exists $\lambda^{(2)} \in N_h(\Omega)$ which satisfies these conditions. We can construct it using the procedure introduced above. We use the parameters Z and formulae (4.36), (4.45) and (4.37), and write conditions of continuity of form (4.38), as well as the boundary conditions (4.49)–(4.52) in terms of Z via (4.37) and (4.45). The undetermined reactions $T_n(\lambda^{(2)})$ on Γ_0 and $T_i(\lambda^{(2)})$ on Γ_u can be chosen in such a way that the resulting system of linear equations is solvable. The choice of these reactions is in accordance with the conditions of total equilibrium of the bodies Ω' and Ω'', respectively.

2.5 Contact Problems with Friction

In the preceding part of the book we studied the contact problem of two elastic bodies without friction, when the tangent component of the stress vector on the contact zone is $T_t = 0$. It is clear that the assumption of zero friction between Ω', Ω'' does not fully comport with the real situation and consequently, it is desirable to include the influence of friction in our considerations. For the sake of simplicity, we will study the contact between an elastic body Ω and a perfectly rigid foundation. Extension to the contact of two elastic bodies is possible (see Jarušek (1982)). The finite nonzero friction will be expressed heuristically, by means of Coulomb's Law: on the contact surface Γ_K of the elastic body with the perfectly rigid foundation we assume

$$u_n \leq 0, \quad T_n \leq 0, \quad u_n T_n = 0 \tag{5.1}$$

$$T_t = T - nT_n, \quad |T_t| \leq \mathcal{F}|T_n|, \quad (\mathcal{F}|T_n| - |T_t|)u_t = 0, \quad u_t T_t \leq 0 \tag{5.2}$$

where $\mathcal{F}$ is the friction coefficient, $\mathcal{F} \geq 0$ on Γ_K.

Let us first consider the problem with "given friction", when the unknown normal component $T_n(u)$ is replaced by a given slip stress $g_n \geq 0$. In this case, (5.2) is replaced by

$$|T_t| \leq \mathcal{F} g_n, \quad (\mathcal{F} g_n - |T_t|)u_t = 0, \quad u_t T_t \leq 0 \quad \text{on } \Gamma_K. \tag{5.3}$$

Let us pass to the variational formulation of the problem with given friction. Thus, let $\Omega \subset \mathbf{R}_2$ be a domain with a Lipschitzian boundary and let $\partial\Omega = \bar{\Gamma}_u \cup \bar{\Gamma}_P \cup \bar{\Gamma}_K$, where Γ_u, Γ_P, Γ_K are open disjoint subsets of $\partial\Omega$; moreover, Γ_u and Γ_K are nonempty. In the symbols of the spaces involved, we will not explicitly indicate whether or not their elements are vector functions. This will be clear from the context.

Let $u^0 \in H^1(\Omega)$, $F \in L^2(\Omega)$, $g_n \in L^2(\Gamma_K)$, $g_n \geq 0$, $P \in L^2(\Gamma_P)$. Let a closed convex set of admissible displacements K be given by

$$K = \{v \in H^1(\Omega) \mid v = u^0 \text{ on } \Gamma_u, \; v_n \leq 0 \text{ on } \Gamma_K\}. \tag{5.4}$$

A function $u \in K$ is called a weak solution of the Signorini problem with friction for given g_n, if

$$\forall v \in K: \; a(u, v - u) + \int_{\Gamma_K} \mathcal{F} g_n(|v_t| - |u_t|) ds$$

$$\geq \int_\Omega F_i(v_i - u_i) dx + \int_{\Gamma_P} P_i(v_i - u_i) ds, \tag{5.5}$$

where

$$u_t = u - n u_n, \quad a(u, v) = \int_\Omega c_{ijkm} \epsilon_{ij}(u) \epsilon_{km}(v) dx,$$

c_{ijkm} are bounded measurable functions in Ω that fulfill the conditions of symmetry (1.3) and of positive definiteness (1.4) in the domain Ω. Similar to section 2.1.3, we will prove *formal equivalence* of the classical and weak formulations of our problem. Let us show in more detail how the friction conditions (5.3) will be derived. To do so, we will assume that we have already proved the validity of the equilibrium conditions

$$\frac{\partial \tau_{ij}}{\partial x_j} + F_i = 0 \quad \text{in } \Omega, \quad i = 1, 2,$$

the boundary conditions

$$\tau_{ij} n_j = P_i \quad \text{on } \Gamma_P, \quad i = 1, 2,$$

and the unilateral boundary conditions (5.1). Integrating by parts in (5.5) and employing all the above conditions, we obtain

$$\int_{\Gamma_K} T_t(v_t - u_t) ds + \int_{\Gamma_K} \mathcal{F} g_n(|v_t| - |u_t|) ds \geq 0 \quad \forall v \in K. \tag{5.6}$$

Let $v \in K$ have the form $v = u \pm \psi$, where $\psi_n = 0$ on Γ_K, $\psi = 0$ on Γ_u. Then

$$\int_{\Gamma_K} T_t(\pm \psi_t) ds + \int_{\Gamma_K} \mathcal{F} g_n(|u_t \pm \psi_t| - |u_t|) ds \geq 0 \quad \forall \psi, \; \psi_n = 0 \text{ on } \Gamma_K,$$

and hence

$$\pm \int_{\Gamma_K} T_t \psi_t ds \leq \int_{\Gamma_K} \mathcal{F} g_n |\psi_t| ds,$$

that is,

$$\left| \int_{\Gamma_K} T_t \psi_t ds \right| \leq \int_{\Gamma_K} \mathcal{F} g_n |\psi_t| ds \quad \forall \psi, \ \psi_n = 0 \ \text{ on } \Gamma_K.$$

Hence, the first inequality in (5.3) easily follows. Since $u_t \leq |u_t|$ on Γ_K, we obtain from the results just proven that

$$T_t u_t + \mathcal{F} g_n |u_t| \geq 0 \quad \text{on } \Gamma_K. \tag{5.7}$$

Now let ψ be such that $\psi_n = 0$, $\psi_t = -u_t$ on Γ_K. Substituting $v = u + \psi$ into (5.6) we obtain

$$-\int_{\Gamma_K} T_t u_t ds - \int_{\Gamma_K} \mathcal{F} g_n |u_t| ds \geq 0.$$

Hence and from (5.7), we conclude

$$T_t u_t + \mathcal{F} g_n |u_t| = 0 \quad \text{on } \Gamma_K,$$

which is an equivalent expression of the remaining conditions in (5.3).

In the sequel, we will assume that the relation between the stress tensor and the strain tensor is described by Hooke's Law for homogeneous isotropic bodies. In that case, we have

$$\tau_{ij} = \lambda \delta_{ij} e_{kk} + 2\mu e_{ij},$$

where $\lambda, \mu > 0$ are Lamé's constants. Furthermore, we will assume that $\Gamma_P = \emptyset$, Γ_K is a sufficiently differentiable part of $\partial\Omega$, and similarly, that $\mathcal{F} \geq 0$ is a sufficiently smooth function with a compact support in Γ_K. By $H^{1/2}(\partial\Omega)$ let us denote the space of traces of functions from $H^1(\Omega)$ (the meaning of this notation will be seen from the following, see also Nečas (1967), and Fučik, John, and Kufner (1977)). Further, by $H^{-1/2}(\partial\Omega)$ let us denote the space of functionals over $H^{1/2}(\partial\Omega)$. Let $H^{1/2}(\Gamma_K) \subset H^{1/2}(\partial\Omega)$ be the space of such v's that vanish on Γ_u, and denote by $H^{-1/2}(\Gamma_K)$ the dual to $H^{1/2}(\Gamma_K)$.

We will say that $g_n \in H^{-1/2}(\Gamma_K)$ is ≤ 0 if the duality fulfills $(g_n, v) \leq 0$ for all $v \geq 0$, $v \in H^{1/2}(\Gamma_K)$. Defining $(\mathcal{F} g_n, v) \overset{df}{=} (g_n, \mathcal{F} v)$ for $v \in H^{1/2}(\partial\Omega)$, we see that definition (5.5) can be extended to $g_n \in H^{-1/2}(\Gamma_k)$, $g_n \leq 0$, in the form

$$u \in K, \ \forall v \in K, \ a(u, v - u) - (\mathcal{F} g_n, |v_t| - |u_t|) \geq \int_\Omega F_i(v_i - u_i) dx. \tag{5.8}$$

Let us recall that $w \in H^1(\Omega)$ implies $|w| \in H^1(\Omega)$, and

$$\| \, |w| \, \|_{1,\Omega} \leq \|w\|_{1,\Omega}. \tag{5.9}$$

From definition (5.8) it is seen that the function u satisfies in Ω (in the sense of distributions) the system of Lamé's equations

$$(\lambda + \mu)\Delta u_i + \mu \frac{\partial}{\partial x_i}(\operatorname{div} u) = -F_i; \tag{5.10}$$

therefore, it is reasonable to define $T_n(u)$ for a solution u of problem (5.8) (on the basis of Green's theorem) by

$$(T_n(u), v_n) \stackrel{df}{=} a(u, v) - \int_\Omega f_i v_i \, dx, \tag{5.11}$$

for $v = 0$ on Γ_u, $v_t = 0$ on Γ_K. Hence, $T_n(u \in H^{-1/2}(\Gamma_K))$. Thus we define: $u \in K$ is a solution of the Signorini problem with friction, if (5.8) holds and $g_n = T_n(u)$. Consequently, if we define a mapping $\Phi : g_n \mapsto T_n(u(g_n))$, then our task is to find a fixed point of this mapping. Since compactness of the mapping cannot be expected and the authors have not succeeded in finding any kind of monotonicity for it, none of the classical fixed point theorems or methods from the theory of monotone operators can be applied. This is why the theory to be explained is a little more complicated. In this book we will show that the mapping just mentioned is a weakly continuous mapping from $L^2(\Gamma_K)$ into itself. Then the existence of a fixed point follows from the so-called "weak Schauder theorem," which, for the reader's convenience, we will prove for a special case of the Hilbert space, namely, the separable one. Naturally, it is necessary to find a closed convex set which the mapping maps into itself. This will be achieved by the smallness of the friction coefficient. We will also introduce (without proof) estimates of the friction coefficient which comport with our theory. They appear to suit practical requirements. We will present all the main ideas, as well as methods for the proofs. For the sake of brevity we will not repeat analogous proofs, leaving them to the kind reader. For detailed proofs we refer him or her to the paper by Nečas, Jarušek, and Haslinger (1980); however, spaces $H^{-1/2+\alpha}(\Gamma_K)$, $0 < \alpha < 1/4$ are considered there instead of $L^2(\Gamma_K)$ (for the definition, see below).

Schauder Theorem (Weak Version). *Let H be a separable Hilbert space. Let A be a mapping from $K \subset H$ into K, where K is a closed, bounded, convex set. Let A be a weakly continuous operator, that is, let $u_n \rightharpoonup u$ (weak convergence) imply $Au_n \rightharpoonup Au$. Then there is a fixed point of the operator A in K, that is, $u \in K$ such that $Au = u$.*

Proof. Let $y_i \in K$, $i = 1, 2, \ldots$, be such points that $\overline{co\{y_1, y_2, \ldots\}} = K$, where co stands for the convex hull. Let $\{x_i\}_{i=1}^{\infty}$ be an orthonormal basis in H and P_n the projector to the subspace H_n of linear combinations of $x_1, x_2, \ldots, x_n$. Let $\epsilon_k = 1/k$. There exist points $y_1, y_2, \ldots, y_{m(k)}$ such that for $x \in K$,

$$\max_{x \in K} \min_{i=1,2,\ldots,m(k)} \|P_k A(x) - P_k y_i\| < \frac{1}{k}. \tag{5.12}$$

Let $a_i(x) = \max(0, \frac{1}{k} - \|P_k A(x) - P_k y_i\|)$. Set

$$S_k(x) = \sum_{i=1}^{m(k)} a_i(x) y_i / \sum_{i=1}^{m(k)} a_i(x). \tag{5.13}$$

Finally, let $K_k = co\{y_1, y_2, \ldots, y_{m(k)}\}$. Evidently, $S_k(K_k) \subset K_k$. Since the operator A is weakly continuous, the operator S_k is continuous from K_k into K_k, hence by Brouwer's theorem[7] there is a fixed point $z_k \in K_k$, $S_k z_k = z_k$. Now we choose a subsequence $z_{k_\ell} \rightharpoonup z \in K$. However, for $x \in K$ we have

$$\|P_k A(x) - P_k S_k(x)\| = \left\| \sum_{i=1}^{m(k)} a_i(x)(P_k A(x) - P_k y_i) / \sum_{i=1}^{m(k)} a_i(x) \right\| \leq \frac{1}{k}. \tag{5.14}$$

Let $w \in H_t$. We have $(w, A(z_{k_\ell})) = (w, A(z_{k_\ell}) - S_{k_\ell}(z_{k_\ell})) + (w, z_{k_\ell})$, and, for $k_\ell \geq t$, $|(w, A(z_{k_\ell}) - S_{k_\ell}(z_{k_\ell}))| = |(w, P_{k_\ell} A(z_{k_\ell}) - P_{k_\ell} S_{k_\ell}(z_{k_\ell}))| \leq \|w\| \frac{1}{k_\ell}$. Hence, with regard to the continuity of A we have $(w, A(z)) = (w, z)$; since

$$\overline{\bigcup_{n=1}^{\infty} H_n} = H,$$

we conclude that $z = A(z)$. $\square$

For the sake of simplicity of our considerations (which are complex enough even then), we restrict ourselves in this chapter to the case $\Omega = P$, where P is the infinite strip

$$P = \{(x_1, x_2) \in \mathbf{R}^2; \ x_1 \in \mathbf{R}^1; \ x_2 \in (0, 1)\}. \tag{5.15}$$

[7] Brouwer's theorem: let T be a continuous operator from a closed, bounded, convex set $K \subset \mathbf{R}^n$ into itself. Then there is a fixed point. For the proof, see Ljusternik, Sobolev (1965).

2.5.1 The Problem with Prescribed Normal Force

Let $w \in H^1(P)$, where P is the strip (5.15). Hence, we have

$$\int_P \left[\left(\frac{\partial w}{\partial x_1}\right)^2 + \left(\frac{\partial w}{\partial x_2}\right)^2 + w^2\right] dx = \|w\|_{1,P}^2 < \infty. \tag{5.16}$$

Denote by $\hat{u}(\xi, x_2)$ the Fourier transform of the function u in the variable x_1. We easily verify that the functions from $C^1(\bar{P})$ vanishing for large $|x_1|$ are dense in $H^1(P)$. (Let us denote this family by $\mathcal{H}$.) Let v be such a function. Its Fourier transform is defined by

$$\hat{v}(\xi_1, x_2) \stackrel{df}{=} \int_{-\infty}^{\infty} v(x_1, x_2) e^{-ix_1\xi_1} dx_1. \tag{5.17}$$

The functions from $C^1(\mathbf{R}^1)$ vanishing for large $|x_1|$ satisfy the Parsevel identity (see Schwartz (1959))

$$\int_{-\infty}^{\infty} |v|^2 dx_1 = \frac{1}{2\pi} \int_{-\infty}^{\infty} |\hat{v}|^2 d\xi_1, \tag{5.18}$$

which also makes it possible to define $\hat{u}$ for functions $u \in H^1(P)$ by a limiting process. The Parseval identity also implies

$$\int_P \left[\left(\frac{\partial u}{\partial x_1}\right)^2 + \left(\frac{\partial u}{\partial x_2}\right)^2 + u^2\right] dx$$

$$= \frac{1}{2\pi} \int_{-\infty}^{\infty} \int_0^1 \left[\xi_1^2 |\hat{u}|^2 + \left|\frac{\partial \hat{u}}{\partial x_2}\right|^2 + |\hat{u}|^2\right] d\xi_1 dx_2. \tag{5.19}$$

The Fourier transform also offers a possibility of characterizing traces (we know that the trace $u(\cdot, 0)$ of a function $u \in H^1(P)$ belongs to $L^2(\mathbf{R}^1)$).

Lemma 5.1.

$$\int_{-\infty}^{\infty} |\hat{u}(\xi_1, 0)|^2 (1 + |\xi_1|) d\xi_1 \le c\|u\|_{1,P}^2. \tag{5.20}$$

Proof. It suffices to prove (5.20) for a dense subset $\mathcal{H} \subset H^1(P)$ (see above). First, let us extend the function $v \in \mathcal{H}$ to the strip $\mathbf{R}^1 \times (-1, 2)$ by setting

$$v(x_1, x_2) = v(x_1, -x_2) \quad \text{for } -1 < x_2 < 0, \tag{5.21}$$

$$v(x_1, x_2) = v(x_1, 2 - x_2) \quad \text{for } 1 < x_2 < 2. \tag{5.22}$$

Let $\eta \in \mathcal{D}((-1,2))$, $\eta(x) = 1$ for $x \in [0,1]$. Put $w(x_1,x_2) \stackrel{\mathrm{df}}{=} \eta(x_2)v(x_1,x_2)$ for $x \in \mathbf{R}^2$. Then

$$\|w\|^2_{1,R^2} \leq c\|v\|^2_{1,P}. \tag{5.23}$$

Let

$$\hat{w}(\xi_1,\xi_2) = \int_{\mathbf{R}^2} w(x_1,x_2)e^{-i(x,\xi)}\,dx.$$

Then, however, the inverse formula holds:

$$\hat{w}(\xi_1,0) = \hat{v}(\xi_1,0) = \frac{1}{2\pi}\int_{-\infty}^{\infty} \hat{w}(\xi_1,\xi_2)d\xi_2. \tag{5.24}$$

Hence,

$$\int_{-\infty}^{\infty} |\hat{v}(\xi_1,0)|^2(1+\xi_1^2)^{1/2}d\xi_1 \leq \frac{1}{4\pi^2}\int_{-\infty}^{\infty}(1+\xi_1^2)^{1/2}d\xi_1,$$

$$\int_{-\infty}^{\infty} |\hat{w}(\xi_1,\xi_2)|^2(1+|\xi|^2)d\xi_2 \int_{-\infty}^{\infty}(1+|\xi|^2)^{-1}d\xi_2. \tag{5.25}$$

Since

$$\int_{-\infty}^{\infty}(1+|\xi|^2)^{-1}d\xi_2 = \pi(1+\xi_1^2)^{-1/2},$$

we obtain the required result: indeed, it immediately follows from the Parseval identity for two-dimensional Fourier transformations

$$\int_{\mathbf{R}^2}|w|^2dx = \frac{1}{(2\pi)^2}\int_{\mathbf{R}^2}|\hat{w}|^2d\xi, \tag{5.26}$$

$$\|w\|^2_{1,\mathbf{R}^2} = \frac{1}{(2\pi)^2}\int_{\mathbf{R}^2}|\hat{w}|^2(1+|\xi|^2)d\xi. \quad \square \tag{5.27}$$

The analog of relation (5.27) in $\mathbf{R}^1$ is

$$\frac{1}{2\pi}\int_{\mathbf{R}^1}(1+\xi^2)|f|^2d\xi = \int_{\mathbf{R}^1}[f^2+(f')^2]dx; \tag{5.28}$$

hence, it is natural to define the space $H^k(\mathbf{R}^1)$, $0 \leq k < \infty$, as the Hilbert space with the inner product

$$\frac{1}{2\pi}\int_{-\infty}^{\infty}(1+|\xi|)^{2k}\hat{f}(\xi)\overline{\hat{g}(\xi)}d\xi. \tag{5.29}$$

Thus, lemma 5.1 asserts that the mapping $v \mapsto v(\cdot,0)$ is a linear, bounded mapping from $H^1(P)$ into $H^{1/2}(\mathbf{R}^1)$. Evidently, $H^{1/2}(\mathbf{R}^1) \hookrightarrow L^2(\mathbf{R}^1)$.

Let us now denote by $H^{-k}(\mathbf{R}^1)$, $\infty > k \geq 0$, the dual space to $H^k(\mathbf{R}^1)$; if $f \in H^{-k}(\mathbf{R}^1)$, we will write the duality in the form (f, w).

Let us consider the strip P and let $\Gamma_K = \{x \in \mathbf{R}^2; x_2 = 0\}$, $\Gamma_u = \{x \in \mathbf{R}^2; x_2 = 1\}$.

Further, let $u^0 \in H^1(P)$, let $u^0 = 0$ on Γ_K, $F \in L^2(P)$, $g_n \in H^{-1/2}(\mathbf{R}^1)$, $\mathcal{F} \in \mathcal{D}(\mathbf{R}^1)$, $\mathcal{F} \geq 0$. We will write $g_n \leq 0$ if $(g_n, w) \leq 0$ for $w \geq 0$, $w \in H^{1/2}(\mathbf{R}^1)$. Thus, consider $g_n \leq 0$ and let $K = \{v \in H^1(P); v_n \leq 0$ on Γ_K, $v = u^0$ on $\Gamma_u\}$. A function $u \in H^1(P)$ solves the Signorini contact problem in the strip P with a prescribed normal force, if

$$u \in K, \quad \forall v \in K,$$

$$a(u, v-u) - (\mathcal{F} g_n, |v_1| - |u_1|) \geq \int_P F_i(v_i - u_i)dx \stackrel{df}{=} (F, v-u)_{0,P}. \quad (5.30)$$

Theorem 5.1. *There exists a unique solution of problem* (5.30), *and the estimate*

$$\|u\|_{1,P} \leq c[\|u^0\|_{1,P} + \|F\|_{0,P}] \quad (5.31)$$

holds. The mapping $g_n \mapsto u$ from $H^{-1/2}(\mathbf{R}^1)$ into $H^1(P)$ is 1/2-Hölderian.

Proof. It is easy to verify that problem (5.30) is equivalent to finding the minimum of the functional

$$\mathcal{T}(v) \stackrel{df}{=} \frac{1}{2}a(v,v) - (\mathcal{F} g_n, |v_1|) - \int_P f_i v_i dx \quad (5.32)$$

on the convex set K; obviously, K is closed. However, now Korn's inequality (the proof of which we will return to later)

$$\forall v \in H^1(P), \quad v = 0 \quad \text{on } \Gamma_u \Rightarrow a(v,v) \geq c\|v\|_{1,P}^2 \quad (5.33)$$

implies the coerciveness of the functional on K:

$$\lim_{\|v\| \to \infty, v \in K} \mathcal{T}(v) = \infty. \quad (5.34)$$

However, the functional $\mathcal{T}(v)$ is convex, and thus also weakly lower- semi-continuous (for every $c \in \mathbf{R}^1$, the set of $v \in K$ with $\mathcal{T}(v) \leq c$ is convex, closed, and hence weakly closed, which yields the implication $v_n \rightharpoonup v \Rightarrow \underline{\lim}_{n\to\infty} \mathcal{T}(v_n) \geq \mathcal{T}(v)$). This makes the application of the fundamental theorem 1.5 from Chapter 1 possible. The uniqueness of solution: let w^1, w^2 be two solutions. Then

$$a(w^2, w^1 - w^2) \geq (F, w^1 - w^2)_{0,P} + (\mathcal{F} g_n, |w_1^1| - |w_1^2|),$$

$$a(w^1, w^2 - w^1) \geq (F, w^2 - w^1)_{0,P} + (\mathcal{F} g_n, |w_1^2| - |w_1^1|), \quad (5.35)$$

which implies $a(w^2 - w^1, w^2 - w^1) \leq 0$, consequently $w^1 = w^2$ by (5.33). Now set $v = u^0$ in (5.30). We obtain

$$a(u - u_0, u - u_0) \leq (\mathcal{F} g_n, |u_1|) - (F, u^0 - u)_{0,P} - a(u_0, u - u_0)$$
$$\leq -(F, u^0 - u)_{0,P} - a(u^0, u - u^0), \tag{5.36}$$

which together with Korn's inequality yields

$$\|u - u_0\|_{1,P} \leq c(\|F\|_{0,P} + \|u^0\|_{1,P}).$$

Further, let $g_n^1, g_n^2 \in H^{-1/2}(\mathbf{R}^1)$ and let u^1, u^2 be the corresponding solution. Then, similar to (5.35), we obtain

$$a(u^2 - u^1, u^2 - u^1) \leq (\mathcal{F} g_n^2 - \mathcal{F} g_n^1, |u_1^2| - |u_1^1|)$$
$$\leq c(\|g_n^2 - g_n^1\|_{-1/2,\mathbf{R}^1}(\|u_1^2\|_{1/2,\mathbf{R}^1} + \|u_1^1\|_{1/2,\mathbf{R}^1})). \tag{5.37}$$

Now inequality (5.20), and Korn's inequality (5.33) and (5.37) yield

$$\|u^2 - u^1\|_{1,P}^2 \leq c_1(\|u^2\|_{1,P} + \|u^1\|_{1,P})\|g_n^1 - g_n^2\|_{-1/2,\mathbf{R}^1}. \quad \square \tag{5.38}$$

We now move on to the proof of Korn's inequality.

Lemma 5.2. *Let $w \in H^1(P)$, $w = 0$ on Γ_u. Then*

$$\int_P \frac{\partial w_i}{\partial x_j} \frac{\partial w_i}{\partial x_j} dx \leq c \int_P e_{ij}(w) e_{ij}(w) dx. \tag{5.39}$$

Proof. We may again assume $w \in C^1(P)$, $w = 0$ for large $|x_1|$. Without loss of generality, let us consider the strip $\mathbf{R}^1 \times (0, \pi)$ and $w = 0$ for $x_2 = 0$ (we have interchanged Γ_K and Γ_u). For $-\pi < x_2 < 0$ let us put $w_1(x_1, x_2) = -w_1(x_1, -x_2)$, $w_2(x_1, x_2) = w(x_1, -x_2)$. Let

$$a_k = \frac{1}{\sqrt{\pi}} \int_{-\pi}^{\pi} \hat{w}_1(\xi_1, x_2) \sin kx_2 dx_2, \quad k = 1, 2, \ldots,$$

$$b_k = \frac{1}{\sqrt{\pi}} \int_{-\pi}^{\pi} \hat{w}_2(\xi_1, x_2) \cos kx_2 dx_2,$$

$$b_0 = \frac{1}{2\pi} \int_{-\pi}^{\pi} \hat{w}_2(\xi_1, x_2) dx_2.$$

Then, evidently

$$\int_{-\infty}^{\infty} \int_{-\pi}^{\pi} \left[\left(\frac{\partial w_1}{\partial x_1} \right)^2 + \frac{1}{2} \left(\frac{\partial w_1}{\partial x_2} + \frac{\partial w_2}{\partial x_1} \right)^2 + \left(\frac{\partial w_2}{\partial x_2} \right)^2 \right] dx_1 dx_2$$

$$= 2\int_{-\infty}^{\infty}\int_{0}^{\pi}\left[\left(\frac{\partial w_1}{\partial x_1}\right)^2 + \frac{1}{2}\left(\frac{\partial w_1}{\partial x_2} + \frac{\partial w_2}{\partial x_1}\right)^2 + \left(\frac{\partial w_2}{\partial x_2}\right)^2\right] dx_1 dx_2; \tag{5.40}$$

consequently,

$$\int_{-\infty}^{\infty}\int_{-\pi}^{\pi}\left[\left(\frac{\partial w_1}{\partial x_1}\right)^2 + \left(\frac{\partial w_1}{\partial x_2} + \frac{\partial w_2}{\partial x_1}\right)^2 + \left(\frac{\partial w_2}{\partial x_2}\right)^2\right] dx_1 dx_2$$

$$= \frac{1}{2\pi}\int_{-\infty}^{\infty}\sum_{k=0}^{\infty}[\xi_1^2|a_k|^2 + k^2|b_k|^2 + \frac{1}{2}(k^2|a_k|^2 + \xi_1^2|b_k|^2$$

$$+ 2k\xi_1 Rei a_k \bar{b}_k)]d\xi_1$$

$$\geq \frac{1}{4\pi}\int_{-\infty}^{\infty}\left[\sum_{k=0}^{\infty}(\xi_1^2|a_k|^2 + k^2|a_k|^2 + \xi_1^2|b_k|^2 + k^2|b_k|^2)\right] d\xi_1$$

$$= \int_{-\infty}^{\infty}\int_{-\pi}^{\pi}\left[\left(\frac{\partial w_1}{\partial x_1}\right)^2 + \left(\frac{\partial w_1}{\partial x_2}\right)^2 + \left(\frac{\partial w_2}{\partial x_1}\right)^2 + \left(\frac{\partial w_2}{\partial x_2}\right)^2\right] dx_1 dx_2. \ \square \tag{5.41}$$

Lemma 5.3. *Every w with $w = 0$ on Γ_u satisfies*

$$\int_P (w_1^2 + w_2^2) dx_1 dx_2$$

$$\leq c\int_P\left[\left(\frac{\partial w_1}{\partial x_1}\right)^2 + \left(\frac{\partial w_1}{\partial x_2}\right)^2 + \left(\frac{\partial w_2}{\partial x_1}\right)^2 + \left(\frac{\partial w_2}{\partial x_2}\right)^2\right] dx_1 dx_2. \tag{5.42}$$

Proof (for smooth w). We have

$$w_i(x_1, x_2) = -\int_{x_2}^{1}\frac{\partial w_i}{\partial x_2}(x_1, \eta)d\eta, \quad i = 1, 2,$$

hence,

$$w_i^2(x_1, x_2) \leq \int_0^1\left(\frac{\partial w_i}{\partial x_2}(x_1, \eta)\right)^2 d\eta$$

and (5.42) follows by integrating the last formula. $\square$

In accordance with (5.11) of the introduction we will define for a solution u of problem (5.30):

$$(T_n(u), v_2) = -a(u, v) + (F, v)_{0,P}, \tag{5.43}$$

where $v \in H^1(P)$, $v = 0$ and Γ_u and $v_1 = 0$ on Γ_K. In order to justify definition (5.43), we must establish an "inverse assertion" to lemma 5.1:

Lemma 5.4. *There is a continuous linear operator E from $H^{1/2}(\mathbf{R}^1)$ into $H^1(P)$, such that Ew on Γ_K coincides with w (in the sense of traces) and*

$$\|Ew\|_{1,P} \le c\|w\|_{1/2,\mathbf{R}^1}. \tag{5.44}$$

Proof. Put $\hat{f}(\xi, x_2) \stackrel{df}{=} \hat{w}(\xi_1)e^{-(1+|\xi_1|)x_2}\eta(x_2)$, where $\eta \in \mathcal{D}(\mathbf{R}^1)$, $\eta(0) = 1$. Then obviously

$$\int_{-\infty}^{\infty}\int_0^1 \left[|\hat{f}|^2 + \xi_1^2|\hat{f}|^2 + \left|\frac{\partial \hat{f}}{\partial x_2}\right|^2\right] d\xi_1 dx_2$$

$$\le c\int_{-\infty}^{\infty} |\hat{f}|^2(1+\xi_1^2)^{1/2} d\xi_1, \tag{5.45}$$

and taking into account (5.19), we obtain the assertion. □

Theorem 5.2. *Let u be a solution of problem* (5.30). *Then $T_n(u) \le 0$ on Γ_K.*

Proof. Let $w_1 \equiv 0$, $w_2 = 0$ on Γ_u, $w_2 \ge 0$ on Γ_0. From (5.43) we obtain

$$(T_n(u), w_2) = -a(u, w) + (F, w)_{0,P} \le 0, \tag{5.46}$$

and setting $v = u + w$ in (5.30), we conclude by (5.46) that

$$(T_n(u), w_2) \le (\mathcal{F} g_n, |u_1|) \le 0. \quad \square$$

2.5.2 Some Auxiliary Spaces

The reader will find a more detailed information about many results of this section in Nečas (1967), and Fučik, John, and Kufner (1977).

In the foregoing section, we introduced in the space $H^\alpha(\mathbf{R}^1)$ the form

$$\left(\frac{1}{2\pi}\int_{-\infty}^{\infty} |\hat{w}|^2(1+|\xi|)^{2\alpha} d\xi\right)^{1/2} \stackrel{df}{=} \|w\|_{\alpha,\mathbf{R}^1}. \tag{5.47}$$

Let us define

$$\|w\|^2_{1/2,\mathbf{R}^1} \stackrel{df}{=} \int_{-\infty}^{\infty}\int_{-\infty}^{\infty} \frac{[w(x+h) - w(x)]^2}{h^2} dx dh. \tag{5.48}$$

Now we have

$$c\left(\frac{1}{2}\right) \stackrel{df}{=} \frac{1}{\pi}\int_{-\infty}^{\infty} \frac{\sin^2 t}{t^2} dt = 1. \tag{5.49}$$

Then the relation between (5.47) and (5.48) is expressed by

Lemma 5.5. *Let* $w \in H^{1/2}(\mathbf{R}^1)$. *Then*

$$\frac{1}{2\pi}\int_{-\infty}^{\infty} |\hat{w}|^2(1+|\xi|)d\xi = \|w\|^2_{0,\mathbf{R}^1} + \frac{1}{2\pi}\|w\|^2_{1/2,\mathbf{R}^1}. \tag{5.50}$$

Proof. Done by direct calculation. □

The most important result of this section is the lemma on "reiteration" of the fractional differentiation:

Lemma 5.6. *Set* $w_h(x) = w(x-h)$. *Let* $w \in H^1(\mathbf{R}^1)$. *Then*

$$\int_{-\infty}^{\infty} \frac{\|w_{-h} - w\|^2_{1/2,\mathbf{R}^1}}{h^2} dh = \int_{-\infty}^{\infty} |\hat{w}|^2(1+|\xi|)|\xi|\, d\xi. \tag{5.51}$$

Proof. Proceeds again by direct calculation. □

From the definitions of the spaces $H^k(\mathbf{R}^1)$, $k \geq 0$, and of $H^{-k}(\mathbf{R}^1)$ as their duals, it is seen that $H^{-k}(\mathbf{R}^1)$ coincides with the space of tempered distributions g whose Fourier transforms satisfy

$$\frac{1}{2\pi}\int_{-\infty}^{\infty} |\hat{g}|^2(1+|\xi|)^{-2k} d\xi \stackrel{df}{=} \|g\|^2_{k,\mathbf{R}^1} < \infty. \tag{5.52}$$

(For the definition of tempered distributions, see Schwartz (1950).)

Again by direct calculation, we obtain

Lemma 5.7. *Let* $w \in H^{-1/2}(\mathbf{R}^1)$. *Then*

$$\int_{-\infty}^{\infty} \frac{\|w_{-h} - w\|^2_{-1/2,\mathbf{R}^1}}{h^2} dh = \int_{-\infty}^{\infty} |\hat{w}|^2(1+|\xi|)^{-1}|\xi|\, d\xi. \tag{5.53}$$

If we intend to consider traces from the spaces $H^1(\mathbf{R}^1)$, then it is natural to introduce such Sobolev spaces whose traces are exactly the spaces $H^1(\mathbf{R}^1)$. Therefore, we define the space $H^1_{1/2}(P)$ as the subspace of those functions w from $H^1(P)$ that fulfill

$$\int_{-\infty}^{\infty}\int_0^1\int_{-\infty}^{\infty}\sum_{i=1}^{2}\left[\frac{\partial w}{\partial x_i}(x_1+h,x_2) - \frac{\partial w}{\partial x_i}(x_1,x_2)\right]^2 h^{-2}dx_1dx_2dh < \infty. \tag{5.54}$$

Again we obtain:

Lemma 5.8.

$$\int_{-\infty}^{\infty}\int_0^1\int_{-\infty}^{\infty}\sum_{i=1}^{2}\left[\frac{\partial w}{\partial x_i}(x_1+h,x_2) - \frac{\partial w}{\partial x_i}(x_1,x_2)\right]^2 h^{-2}dx_1dx_2dh$$

$$= \int_{-\infty}^{\infty} \int_{0}^{1} \left[\left| \frac{\partial \hat{w}}{\partial x_2} \right|^2 + |\hat{w}\xi_1|^2 \right] |\xi_1| d\xi_1 dx_2.$$

Therefore, let us equip the space $H^1_{1/2}(P)$ with the norm

$$\|w\|^2_{1,1/2,P}$$

$$\stackrel{\text{df}}{=} \frac{1}{2\pi} \int_{-\infty}^{\infty} \int_{0}^{1} \left[\left| \frac{\partial \hat{w}}{\partial x_2} \right|^2 + |\hat{w}\xi_1|^2 + |\hat{w}|^2 \right] (1 + |\xi_1|) d\xi_1 dx_2. \quad (5.55)$$

Now let us define the space $H^1_{-1/2}(P)$ as the subspace of those functions w from: $L^2(P)$ that fulfill

$$\|w\|^2_{1,-1/2,P}$$

$$\stackrel{\text{df}}{=} \frac{1}{2\pi} \int_{-\infty}^{\infty} \int_{0}^{\infty} \left[\left| \frac{\partial \hat{w}}{\partial x_2} \right|^2 + |\hat{w}\xi_1|^2 + |\hat{w}|^2 \right] (1+|\xi_1|)^{-1} d\xi_1 dx_2 < \infty. \quad (5.56)$$

Lemmas 5.8, 5.6, and 5.1 yield:

Lemma 5.9.

$$\|w(\cdot,0)\|_{1,\mathbf{R}^1} \leq c\|w\|_{1,1/2,P}. \quad (5.57)$$

In the same way as in lemma 5.1 we can prove:

Lemma 5.10.

$$\|w(\cdot,0)\|_{0,\mathbf{R}^1} \leq c\|w\|_{1,-1/2,P}. \quad (5.58)$$

In conclusion let us introduce a lemma which can be proved in the same way as lemma 5.4.

Lemma 5.11. *There exists a continuous linear operator E from $L^2(\mathbf{R}^1)$ into $H^1_{-1/2}(P)$ such that $E\omega$ equals ω on Γ_K, vanishes on Γ_u and*

$$\|E\omega\|_{1,-1/2,P} \leq c\|\omega\|_{0,\mathbf{R}^1}.$$

2.5.3 Existence of Solution of the Problem with Friction

First of all, we prove a fundamental lemma:

Lemma 5.12. *Let $g_n \in L^2(\mathbf{R}^1)$, $u^0 \in H^1_{1/2}(P)$, $F \in H^1(P)$,*[8] *$g_n \leq 0$. Let $\mathcal{F} \in \mathcal{D}(\mathbf{R}^1)$ be fixed, $\mathcal{F} \geq 0$, and let $\mathcal{F}_t = t\mathcal{F}$, $t > 0$. Then the solution u of problem (5.30) with friction coefficient $\mathcal{F}_t$ satisfies*

$$\|u\|_{1,1/2,P} \leq c_1 t\|g_n\|_{0,\mathbf{R}^1} + c_2(\|u^0\|_{1,1/2,P} + \|F\|_{1,P}). \quad (5.59)$$

[8]This assumption can be weakened.

Proof. Set $v = u_{-h} - u^0_{-h} + u^0$ in (5.30). This yields

$$a(u, u_{-h} - u^0_{-h} + u^0 - u) - t(\mathcal{F} g_n, |u_1|_{-h} - |u_1|)$$
$$\geq (F, u_{-h} - u^0_{-h} + u^0 - u)_{0,P}. \tag{5.60}$$

By a translation of coordinates we obtain from (5.30)

$$a(u_{-h}, v_{-h} - u_{-h}) - t(\mathcal{F}_{-h}(g_n)_{-h}, |v_1|_{-h} - |u_1|_{-h})$$
$$\geq (F_{-h}, v_{-h} - u_{-h})_{0,P}. \tag{5.61}$$

Substituting here $v = u_h - u^0_h + u^0$, we obtain from (5.60), (5.61)

$$a(u_{-h} - u, u_{-h} - u) \leq a(u_{-h} - u, u^0_{-h} - u^0)$$
$$+ t(\mathcal{F}_{-h}(g_n)_{-h} - \mathcal{F} g_n, |u_1|_{-h} - |u_1|)$$
$$- (F - F_{-h}, u_{-h} - u^0_{-h} - (u - u^0))_{0,P}. \tag{5.62}$$

First of all, notice that

$$(\mathcal{F}_{-h}(g_n)_{-h} - \mathcal{F} g_n, |u_1|_{-h} - |u_1|) = ((g_n)_{-h} - g_n,$$
$$\mathcal{F}(|u_1|_{-h} - |u_1|)) + ((g_n)_{-h}, \mathcal{F}_h - \mathcal{F})(|u_1|_{-h} - |u_1|)). \tag{5.63}$$

Now $w \in H^{1/2}(\mathbf{R}^1)$ satisfies

$$\|\mathcal{F} w\|_{1/2,\mathbf{R}^1} \leq c\|w\|_{1/2,\mathbf{R}^1}, \tag{5.64}$$

$$\|(\mathcal{F}_{-h} - \mathcal{F})w\|_{1/2,\mathbf{R}^1} \leq c|h|\|w\|_{1/2,\mathbf{R}^1}, \tag{5.65}$$

while for $w \in H^1(P)$ we have

$$|(F - F_{-h}, u_{-h} - u + u^0 - u^0_{-h})| \leq ch^2\|F\|_{1,P}(\|u\|_{1,P} + \|u^0\|_{1,P}). \tag{5.66}$$

Let us multiply inequality (5.62) by h^{-2} and use Korn's inequality (5.39). Taking into account the lemmas of the preceding section, (5.64)–(5.66), inequality (5.31), and the relation $\||v|\|_{1,\mathbf{R}^1} \leq \|v\|_{1,\mathbf{R}^1}$, we arrive at inequality (5.59). □

Lemma 5.13. *Under the assumption of lemma* 5.12 *we have*

$$\|T_n(u)\|_{0,\mathbf{R}^1} \leq c_3 t\|g_n\|_{0,\mathbf{R}^1} + c_4(\|u^0\|_{1,1/2,P} + \|F\|_{1,P}). \tag{5.67}$$

Proof. We employ lemma 5.11, lemma 5.2, and inequality

$$|a(u,v)| \leq c\|u\|_{1,1/2,P}\|v\|_{1,-1/2,P}. \quad \square \tag{5.68}$$

Thus, we obtain:

Theorem 5.3. *Under the assumptions of lemma* 5.12 *there exists a solution of the Signorini problem with friction for small* t.

Proof. For small t we have a ball B in $L^2(\mathbf{R}^1)$ such that the set of $g_n \in B$ with $g_n \leq 0$ is mapped into itself. Let $g_n^k \rightharpoonup g_n$. Then $\mathcal{F} g_n^k \to \mathcal{F} g_n$ in $H^{-1/2}(\mathbf{R}^1)$ (see Fučik, John, and Kufner (1977)), hence, $u^k \to u$ in $H^1(P)$ by theorem 5.1. Consequently, $T_n(u(g_n^k)) \to T_n(u(g_n))$ in $H^{-1/2}(\mathbf{R}^1)$. Since $H^{1/2}(\mathbf{R}^1)$ is dense in $L^2(\mathbf{R}^1)$, we thus obtain $T_n(u(g_n^k)) \rightharpoonup T_n(u(g_n))$ in $L^2(\mathbf{R}^1)$. Thus, we may apply the weak version of the Schauder theorem. □

Remark 5.1. In the above mentioned paper by Nečas, Jarušek and Haslinger (1980), some fine estimates are used to prove that a fixed point, and hence a solution, is obtained for $t < \sqrt{2\mu/\lambda + 3\mu}$ provided $\max_{x \in \mathbf{R}^1} |\mathcal{F}(x)| = 1$.

2.5.4 Algorithms for the Contact Problem with Friction for Elastic Bodies

In this section we will show how to proceed when approximating contact problems with friction. We shall present two iteration methods which have been successfully employed to obtain the numerical solution of the above problem. Nevertheless, the proof of their convergence still remains open.

In the first part, we will discuss the so-called direct iterations and their applicability to the solution of the Signorini problem with friction. (For simplicity, we thus assume that Ω'' is a perfectly rigid obstacle.)

In the second part, we will deal with the method of alternating iterations and their applicability to the solution of contact problems for two elastic bodies. Moreover, in this part we will also consider the semicoercive cases. In both situations, we consider a bounded contact zone.

2.5.41. Direct Iterations. In section 2.5.1, a solution of the Signorini problem with friction given by Coulomb's Law was defined in terms of a fixed point of the operator Φ on the set $H_-^{-1/2}(\Gamma_K)$. As is usual in problems of this type, we will use the method of successive approximations to find the fixed point of Φ on $H_-^{-1/2}(\Gamma_K)$:

$$\text{given } g_0 \in H_-^{-1/2}(\Gamma_K); \quad g_{k+1} = \Phi(g_k). \tag{5.69}$$

Let us point out that the convergence of $\{g_k\}_{k=1}^{\infty}$ to a fixed point of Φ will *not* be proven; nonetheless, the authors' experience shows the applicability of the method to be very good, in particular as concerns the rate of convergence.

Thus, each iteration step is defined as the solution of the Signorini problem with a given friction. In the sequel, we will discuss in more detail how to approximate such a single iteration step.

Let the elastic body be represented by a bounded domain $\Omega \subset \mathbf{R}^2$, whose Lipschitzian boundary $\partial\Omega$ consists of three disjoint and open in $\partial\Omega$ parts Γ_u, Γ_P, Γ_K, that is,

$$\partial\Omega = \bar{\Gamma}_u \cup \bar{\Gamma}_P \cup \bar{\Gamma}_K.$$

Let Γ_u, Γ_K be nonempty. We assume that

$$u = 0 \quad \text{on } \Gamma_u;$$

unilateral boundary conditions and the conditions involving friction on Γ_K are prescribed by (5.1), (5.3). In the following, we will briefly write g instead of the product $\mathcal{F} g_n$.

Let us put

$$\mathbf{V} = \{v \in (H^1(\Omega))^2 \mid v = 0 \quad \text{on } \Gamma_u\},$$
$$K = \{v \in \mathbf{V} \mid v_n \le 0 \quad \text{on } \Gamma_K\}.$$

We will call a function $u \in K$ a *variational solution* of the Signorini problem with the given friction g, if it satisfies

$$a(u, v-u) + j(v) - j(u) \ge L(v-u) \quad \forall v \in K, \tag{5.70}$$

with

$$a(u,v) = \int_\Omega c_{ijkl}\epsilon_{kl}(u)\epsilon_{ij}(v)dx,$$

$$L(v) = \int_\Omega F_i v_i dx + \int_{\Gamma_P} P_i v_i ds, \quad F \in (L^2(\Omega))^2, \quad P \in (L^2(\Gamma_P))^2,$$

$$j(v) = \int_{\Gamma_K} g|v_t|ds, \quad g \in L^2(\Gamma_K), \quad g \ge 0 \quad \text{on } \Gamma_K.$$

Moreover, c_{ijkl} fulfill the usual conditions of symmetry and ellipticity. An equivalent expression of (5.70) is the problem

$$\text{find } u \in K : \mathcal{Y}(u) \le \mathcal{Y}(v) \quad \forall v \in K, \tag{5.70'}$$

where $\mathcal{Y}(v) = \frac{1}{2}a(v,v) + j(v) - L(v)$.

We already know from the results of the preceding section that (5.70) has a unique solution u.

The main difficulty from the standpoint of the choice of a suitable algorithm for solution of problem (5.70) is the presence of the nondifferentiable term $j(v)$. To remove it we will use the same method as in the scalar case in section 1.1.2.

Evidently,

$$j(v) = \sup_{\mu \in \Lambda} (g\mu, v_t)_{0,\Gamma_K},$$

where

$$\Lambda = \{\mu \in L^2(\Gamma_K) \mid |\mu| \leq 1 \quad \text{on supp } g, \\ \mu = 0 \quad \text{on } \Gamma_K \setminus \text{supp } g\},$$

and $(\cdot,\cdot)_{0,\Gamma_K}$ stands for the scalar product of functions in $L^2(\Gamma_K)$. Hence,

$$\inf_{v \in K} \mathcal{Y}(v) = \inf_{v \in K} \sup_{\mu \in \Lambda} \mathcal{L}(v, \mu),$$

where $\mathcal{L} : [H^1(\Omega)]^2 \times \Lambda \to \mathbf{R}^1$ is the Lagrangian function given by

$$\mathcal{L}(v, \mu) = \mathcal{Y}(v) + (g\mu, v_t)_{0,\Gamma_K}.$$

Instead of problem (5.70′), we will consider the problem: find a saddle point (w, λ) of the Lagrangian function $\mathcal{L}$ on $K \times \Lambda$, that is,

$$\mathcal{L}(w, \mu) \leq \mathcal{L}(w, \lambda) \leq \mathcal{L}(v, \lambda) \quad \forall (v, \mu) \in K \times \Lambda,$$

or equivalently,

$$\text{find} \quad (w, \lambda) \in K \times \Lambda \quad \text{such that}$$
$$a(w, v - w) + (g\lambda, v_t - w_t)_{0,\Gamma_K} \geq L(v - u) \quad \forall v \in K$$
$$(g(\mu - \lambda), w_t) \leq 0 \quad \forall \mu \in \Lambda. \tag{5.71}$$

The relation between the formulations (5.70′) and (5.71) is expressed in

Theorem 5.4. *There is a unique solution* (w, λ) *of problem* (5.71), *and it satisfies*

$$w = u, \quad g\lambda = T_t(u), \tag{5.72}$$

where $u \in K$ *is the unique solution of* (5.70′).

Proof. The existence of the solution (w, λ) of problem (5.71) is a consequence of Korn's inequality on the space $\mathbf{V}$, of the boundedness of the convex set Λ, and of lemma 5.4 from section 1.1.51. By applying Green's theorem we prove (5.72) and, since u is unique, the solution (w, λ) of problem (5.71) is unique as well. □

(5.71) will be called the *mixed variational formulation* of the Signorini problem with given friction. In this way the problem of minimizing a non-differentiable functional $\mathcal{Y}$ is replaced by the problem of finding a saddle point of the functional $\mathcal{L}$ on $K \times \Lambda$.

In order to obtain an approximation of (5.71), we will use the method of finite elements. To this end, let us suppose that $\Omega \subset \mathbf{R}^2$ is a *polygonal* domain, $\{\mathcal{T}_h\}$, $h \to 0_+$ a *regular* system of triangulations of $\bar{\Omega}$ which is consistent with the partition of $\partial\Omega$ into Γ_u, Γ_K and Γ_P. Furthermore, let us also assume that Γ_K consists of a *single segment*.[9] We associate each $\mathcal{T}_h$ with a finite-dimensional space $\mathbf{V}_h$ of piecewise linear vector functions:

$$\mathbf{V}_h = \{v_h \in [C(\bar{\Omega})]^2 | v_h|_{T_i} \in [P_1(T_i)]^2$$

$$\forall T_i \in \mathcal{T}_h, \quad v_h = 0, \quad \text{on } \Gamma_u\},$$

and with a convex, closed subset $K_h \subset \mathbf{V}_h$:

$$K_h = \{v_h \in \mathbf{V}_h \,|\, (v_h \cdot n)(a_i) \leq 0 \quad \forall i = 1, 2, \ldots, m\}. \tag{5.73}$$

Here $a_1, a_2, \ldots, a_m$ denote the nodes of $\mathcal{T}_h$ lying on Γ_K. It is immediately seen from the definition of K_h that $K_h \subset K$ for all $h \in (0,1)$; that is, K_h are *interior* approximations of K.

Let $\{\mathcal{T}_H\}$, $H \in (0,1)$, be a partition of Γ_K consistent with the boundary of supp g in Γ_K, whose nodes we denote correspondingly by $b_1, b_2, \ldots, b_M$, $H = \max_i |\overline{b_i b_{i+1}}|$. Generally, these points need not coincide with the nodes $a_1, \ldots, a_m$. In the following, we will write $h = H$ if and only if $m = M$ and $a_i = b_i$ for all $i = 1, \ldots, m$. Let

$$L_H = \{\mu_H \in L^2(\Gamma_K) \,|\, \mu_H|_{b_i b_{i+1}} \in P_0(b_i b_{i+1}), \quad i = 1, \ldots, M\},$$

and

$$\Lambda_H = \{\mu_H \in L_H \, |\mu_H| \leq 1 \text{ on supp } g, \quad \mu_H = 0 \text{ on} \Gamma_K \setminus \text{supp } g\}.$$

By an *approximation* of problem (5.71), we mean the problem of finding a saddle point (w_h, λ_H) of $\mathcal{L}$ on $K_h \times \Lambda_H$:

$$\mathcal{L}(w_h, \mu_H) \leq \mathcal{L}(w_h, \lambda_H) \leq \mathcal{L}(v_h, \lambda_H) \quad \forall (v_h, \mu_H) \in K_h \times \Lambda_H, \tag{5.74}$$

or in an equivalent form

$$a(w_h, v_h - w_h) + (g\lambda_H, v_{ht} - w_{ht}) \geq L(v_h - w_h) \quad \forall v_h \in K_h$$

$$(g(\mu_H - \lambda_H), w_{ht})_{0,\Gamma_K} \leq 0 \quad \forall \mu_h \in \Lambda_H. \tag{5.75}$$

Theorem 5.5. *Problem* (5.74) *has a solution* (w_h, λ_H) *for all* $h, H \in (0,1)$. *Moreover, its first component is uniquely determined.*

[9] This assumption is not necessary.

Proof. The existence of a solution (w_h, λ_H) is proved in the same way as in the continuous case. The uniqueness of the first component is a consequence of the fact that the mapping $v_h \to \mathcal{L}(v_h, \mu_H)$ is strictly convex for all $\mu_H \in \Lambda_H$. □

The other component λ_H need not in general be uniquely determined. Now we will give sufficient conditions which guarantee its uniqueness. Let us put

$$K_h^0 = \{v \in \mathbf{V}_h \mid v_h \cdot n \leq 0 \quad \text{on } \Gamma_K\}. \tag{5.76}$$

Then, as an immediate consequence of remark 5.5 from section 1.1.52 we have

Theorem 5.6. *If*

$$(g\mu_H, v_{ht})_{0,\Gamma_K} = 0 \quad \forall v_h \in K_h^0 \Rightarrow \mu_H = \Theta \quad \text{on } \Gamma_K, \tag{5.77}$$

then the second component λ_H is uniquely determined.

Remark 5.2. Let us assume that g is piecewise constant on Γ_K. Then the product $g\mu_H \in L_H$, while $v_{ht} = v_h \cdot t$ for $v_h \in K_h^0$ is piecewise linear on Γ_K with vertices at the points $a_1, \ldots, a_m$. In this case, condition (5.77) expresses the fact that the set of piecewise linear functions over the above mentioned partition is sufficiently rich or, in other words, that the ratio $h_{\partial\Omega}/H$ is "sufficiently" small. The number $h_{\partial\Omega}$ is defined to be $\max_i |\overline{a_i a_{i+1}}|$.

A natural question occurs, namely, what is the relation between (w, λ) and (w_h, λ_H). To answer it we will use the results of section 1.1.52.

Theorem 5.7. *Let $h \to 0_+$ if and only if $H \to 0_+$ and let the intersection $\bar{\Gamma}_K \cap \bar{\Gamma}_u$ consist of a finite number of points. Then*

$$w_h \to w \quad \text{in } [H^1(\Omega)]^2$$

$$\lambda_H \rightharpoonup \lambda \quad (\text{weakly}) \text{ in } L^2(\Gamma_K).$$

Proof. Let us verify all the assumptions of theorem 5.3 from section 1.1.52. Since in our case both K_h and Λ_H are interior approximations of K and Λ, (5.15) and (5.16) from the above-mentioned section are automatically fulfilled. (5.17) holds as well. (5.13), that is, the density of $\{K_h\}$, $h \in (0,1)$, in K is verified analogously as in theorem 3.2 from section 2.3.311. To this end, the following density result: the set $[C^\infty(\bar{\Omega})]^2 \cap K$ is dense in K in the norm of the space $[H^1(\Omega)]^2$ (see Hlaváček, Lovíšek (1977)). It remains to verify that $\{\Lambda_H\}$, $H \in (0,1)$, is dense in Λ. Let $\mu \in \Lambda$ be arbitrary. The symbol $\pi_H \mu$ will denote the orthogonal L^2-projection of μ into the space L_H. Then,

$$\pi_H \mu \to \mu, \quad H \to 0_+, \quad \text{in } L^2(\Gamma_K).$$

Since $|\mu| \leq 1$ a.e. on Γ_K, we have $|\pi_H \mu| \leq 1$ on Γ_K as well, and consequently, $\pi_H \mu \in \Lambda_H$. □

Remark 5.3. Taking into account (5.72), we see that w_h may be taken for an approximation of the field of displacements u and $g\lambda_H$ for an approximation of $T_t(u)$ on Γ_K, where u is a solution of (5.70) or (5.70′), respectively.

Remark 5.4. The result of theorem 5.4 can be improved provided we add some assumptions on smoothness of u. First, it is possible to replace the weak convergence of λ_H to λ by the strong one. Second, even the rate of convergence of $\{w_h, \lambda_H\}$ to $\{w, \lambda\}$ with respect to the parameters h, H can be estimated. However, in accordance with the results of section 2.5.13, it clearly cannot be expected that the solution of (5.70) will be too smooth. Therefore, any further assumptions on the smoothness of u might not be realistic. This is why we restricted ourselves solely to the proof of convergence without any additional assumptions on the smoothness of u.

Let us now consider the question of realization of the approximate problem (5.74). Since we have here a problem of finding a saddle point in a finite dimension, we will use Uzawa's method, described in section 1.1.53. As we already know, each iteration step consists of two parts:

$$\text{knowing } \lambda_H^n \in \Lambda_H,$$

we calculate $u_h^n \in K_h$ as a solution of the minimization problem

$$\mathcal{L}(u_h^n, \lambda_H^n) \leq \mathcal{L}(v_h, \lambda_h^n) \quad \forall v_h \in K_h, \tag{5.78}$$

or

$$a(u_h^n, v_h - u_h^n) + (g\lambda_H^n, v_{ht} - u_{ht}^n)_{0,\Gamma_K} \geq L(v_h - u_h^n) \quad \forall v_h \in K_h, \tag{5.78′}$$

then replace λ_H^n by λ_H^{n+1} according to the rule

$$\lambda_H^{n+1} = P_{\Lambda_H}(\lambda_H^n + \rho g u_{ht}^n), \quad \rho > 0, \tag{5.79}$$

and return to (5.78). The value of $\lambda_H^0 \in \Lambda_H$ is chosen arbitrarily. The symbol P_{Λ_H} stands for the projection into a convex set Λ_H. Its explicit representation is given in remark 5.10 in section 1.1.53.

The theory presented in section 1.1.53 implies that there exist positive numbers $\rho_2 > \rho_1$ such that for all $\rho \in (\rho_1, \rho_2)$ we have

$$u_h^n \to u_h, \quad n \to \infty.$$

Now let us show that under condition (5.77) we obtain the convergence of λ_H^n to λ_H as well. First of all, the sequence $\{\lambda_H^n\}$ is bounded and therefore we can choose a subsequence $\{\lambda_H^{n'}\} \subset \{\lambda_H^n\}$ such that

$$\lambda_H^{n'} \to \lambda_H^*, \quad n' \to \infty. \tag{5.80}$$

Since $\lambda_H^n \in \Lambda_H$ for all n and Λ_H is closed, we have $\lambda_H^* \in \Lambda_H$. Since K_h is a cone with its vertex at 0, it follows from (5.78′) that

$$a(u_h^n, v_h) + (g\lambda_H^n, v_{ht})_{0,\Gamma_K} \geq L(v_h) \quad \forall v_h \in K_h.$$

However, if we restrict ourselves only to functions $v_h \in \overset{\circ}{K}_h$, (which was defined by (5.76)), we can write the equality sign in the preceding inequality:

$$a(u_h^n, v_h) + (g\lambda_H^n, v_{ht})_{0,\Gamma_K} = L(v_h) \quad \forall v_h \in \overset{\circ}{K}_h .$$

By passing to the limit for $n' \to \infty$ we obtain

$$a(u_h, v_h) + (g\lambda_H^n, v_{ht})_{0,\Gamma_K} = L(v_h) \quad \forall v_h \in \overset{\circ}{K}_h . \tag{5.81}$$

On the other hand, (u_h, λ_H) as a solution of problem (5.75) obviously fulfills

$$a(u_h, v_h) + (g\lambda_H, v_{ht})_{0,\Gamma_K} = L(v_h) \quad \forall v_h \in \overset{\circ}{K}_h . \tag{5.82}$$

Subtracting (5.81) from (5.82) and using (5.77), (5.80) we arrive at

$$\lambda_H^{n'} \to \lambda_H, \quad n' \to \infty.$$

Since λ_H is unique, even the whole sequence λ_H^n converges to λ_H. The result just established will be formulated in the next theorem.

Theorem 5.8. *Consider the iteration process* (5.78), (5.79). *Let* (5.77) *hold. Then there exist* $\rho_2 > \rho_1 > 0$ *such that for all* $\rho \in (\rho_1, \rho_2)$,

$$u_h^n \to u_h,$$

$$\lambda_H^n \to \lambda_H, \quad n \to \infty.$$

The above iteration process leads to a sequence of solutions of quadratic programming problems. Evidently, however, in practice it is hardly realizable in this form. Nevertheless, some specific properties of our problem enable us to modify the above algorithm to the form which offers a very effective method of solution. Namely, we have in mind the following properties: first of all, the stiffness matrix remains the same throughout the whole iteration process. Further, the number of components of the approximated vector of displacements u_h subject to the condition on Γ_K is small as compared to the total number of components of u_h. Finally, the linear term of $\mathcal{L}$ changes during the iteration process only those components, which in the given enumeration correspond to the components of u_h on Γ_K. Thanks to these properties, it is possible to eliminate the free components of the vector u_h (corresponding to the nodes not belonging to Γ_K), and to carry out the iteration process with only the other components of u_h.

Let us assume that the enumeration of nodes in the domain Ω is chosen so that the constrained components are placed as the last ones, that is, $x = (x_1, x_2)$, $x \in \mathbf{R}^m$, $x_1 \in \mathbf{R}^n$, $x_2 \in \mathbf{R}^k$, $m = n + k$. We seek for the minimum of the quadratic function

$$\mathcal{Y}(x) = \frac{1}{2}(x, Cx)_{\mathbf{R}^m} - (\mathcal{F}, x)_{\mathbf{R}^m}$$

on the set

$$K_E = \{x = (x_1, x_2) \in \mathbf{R}^m, \quad Bx_2 \leq 0\}, \tag{5.83}$$

where C is a symmetric, positive definite $m \times m$ stiffness matrix, $\mathcal{F}$ is the vector of the right-hand sides obtained by integrating the linear term $L(\cdot) - (g\lambda_H, \cdot)_{0,\Gamma_K}$, and B is a $p \times k$ matrix. As we know, the problem of finding the minimum $x^* = (x_1^*, x_2^*)$, $x_1^* \in \mathbf{R}^n$, $x_2^* \in \mathbf{R}^k$ of the function $\mathcal{Y}$ on K_E is equivalent to finding $x^* \in K_E$, satisfying

$$(Cx^*, y - x^*)_{\mathbf{R}^m} \geq (\mathcal{F}, y - x^*)_{\mathbf{R}^m} \quad \forall y \in K_E. \tag{5.84}$$

Let us write $y = (y_1, y_2)^T$, $\mathcal{F} = (f_1, f_2)^T$, $y_1, f_1 \in \mathbf{R}^n$, $y_2, f_2 \in \mathbf{R}^k$. Analogously, we divide the matrix C into blocks

$$C = \begin{pmatrix} C_{11} & C_{12} \\ C_{21} & C_{22} \end{pmatrix},$$

where C_{11}, C_{22} are square matrices of orders n, k, respectively, while C_{12}, C_{21} are rectangular of types $n \times k$ and $k \times n$, respectively. Evidently $C_{12} = C_{21}^T$.

Now let us choose y in (5.84) so that $y_1 = x_1^* + z_1$, $y_2 = x_2^*$, $z_1 \in \mathbf{R}^n$ arbitrary. Then $y \in K_E$. After multiplication we obtain

$$z_1^T(C_{11}x_1^* + C_{12}x_2^*) = z_1^T f_1 \quad \forall z_1 \in \mathbf{R}^n,$$

or

$$C_{11}x_1^* + C_{12}x_2^* = f_1. \tag{5.85}$$

Now let us choose $y \in K_E$ in (5.84) so that $y_1 = x_1^*$, $y_2 = z_2$, $Bz_2 \leq 0$ arbitrary. After multiplication we obtain

$$(z_2 - x_2^*)^T(C_{21}x_1^* + C_{22}x_2^*) \geq (z_2 - x_2^*) f_2. \tag{5.86}$$

From (5.85) we can express x_1^* in the form

$$x_1^* = C_{11}^{-1}(f_1 - C_{12}x_2^*),$$

which after substituting into (5.86) yields the following relation for x_2^*:

$$(z_2 - x_2^*)^T \tilde{C} x_2^* \geq (z_2 - x_2^*)^T \tilde{f}, \tag{5.87}$$

where $\tilde{f} = f_2 - C_{21}C_{11}^{-1}f_1$, $\tilde{C} = C_{22} - C_{21}C_{11}^{-1}C_{12}$. Now (5.86) implies that x_2^* is the minimum of the quadratic function $\tilde{\mathcal{Y}}(x_2) = (x_2, \tilde{C}x_2)_{\mathbf{R}^k} - (\tilde{f}_2, x_2)_{\mathbf{R}^k}$ on a convex closed set $\tilde{K}_E$, where

$$\tilde{K}_E = \{x_2 \in \mathbf{R}^k \,|\, Bx_2 \leq 0\}.$$

It is possible to show that the matrix $\tilde{C}$ and the vector $\tilde{f}$ can be obtained by means of partial Gauss elimination of free components of x_1. The implementation of this method is presented in detail in the paper by Haslinger and Tvrdý (1982). Thus, if we seek for the minimum of $\mathcal{Y}$ on the set K_E given by (5.83), we can reduce the problem to the problem of finding the minimum of the function $\tilde{\mathcal{Y}}$ on $\tilde{K}_E$. Taking into account the fact that k usually is much smaller than n in problems of this type, the time saved is considerable. If we know x_2^*, we calculate x_1^* from equation (5.85). Thus, it actually suffices to apply Uzawa's algorithm to the function $\tilde{\mathcal{Y}}$.

Remark 5.5. It is possible to introduce still another Lagrangian multiplier in order to remove the constraint $v_n \leq 0$ on Γ_K. Let us again assume that Γ_K consists of a single segment and $\Gamma_P = \emptyset$. Denote

$$\Lambda_1 = H_+^{-1/2}(\Gamma_K) = \{\mu_1 \in H^{-1/2}(\Gamma_K), \langle \mu_1, v_n \rangle \geq 0$$
$$\forall v \in \mathbf{V}, \quad v_n \geq 0 \quad \text{on } \Gamma_K\},$$
$$\Lambda_2 = \{\mu_2 \in L^2(\Gamma_K) \,|\, |\mu_2| \leq 1 \quad \text{a.e. on supp } g,$$
$$\mu_2 = 0 \quad \text{on } \Gamma_K \setminus \text{supp } g\}.$$

Finally, let $\mathcal{L} : \mathbf{V} \times (\Lambda_1 \times \Lambda_2) \to \mathbf{R}^1$ be the Lagrangian function defined by

$$\mathcal{L}(v; \mu_1, \mu_2) = \frac{1}{2}(\tau_{ij}(v), \epsilon_{ij}(v))_0 + \langle \mu_1, v_n \rangle + \langle \mu_2 g, v_t \rangle - (F_i, v_i)_0.$$

By the *mixed variational* formulation of the Signorini problem with given friction, we mean the problem of finding a saddle point $(w; \lambda_1, \lambda_2)$ of the function $\mathcal{L}$ on $\mathbf{V} \times (\Lambda_1 \times \Lambda_2)$. It is possible to prove the following relation between $(w; \lambda_1, \lambda_2)$ and the solution u of the original Signorini problem with given friction:

$$w = u, \quad \lambda_1 = -T_n(u), \quad g\lambda_2 = -T_t(u).$$

Thus, this formulation enables us to approximate at the same time, and independently of one another, the solution u and the normal and tangent stresses on Γ_K.

Let $\Omega \subset \mathbf{R}^2$ be a polygonal domain, $\{\mathcal{T}_h\}$ and $\{\mathcal{T}_H\}$ *regular* systems of triangulations of $\bar{\Omega}$ and $\bar{\Gamma}_K$, respectively. Set

$$\mathbf{V}_h = \{v_h \in [C(\bar{\Omega})]^2 \mid v_h|_{T_i} \in [P_1(T_i)]^2 \quad \forall T_i \in \mathcal{T}_h, \; v_h = 0 \text{ on } \Gamma_u\},$$

$$\Lambda_{1H} = \{\mu_{1H} \in L^2(\Gamma_K) \mid \mu_{1H}|_{b_ib_{i+1}} \in P_0(b_ib_{i+1}), \quad \mu_{1H} \geq 0 \text{ on } \Gamma_K\},$$

$$\Lambda_{2H} = \{\mu_{2H} \in L^2(\Gamma_K) \mid \mu_{2H}|_{b_ib_{i+1}} \in P_0(b_ib_{i+1}),$$
$$|\mu_{2H}| \leq 1 \text{ on supp } g, \quad \mu_{2H} = 0 \text{ on } \Gamma_K \setminus \text{supp } g\}.$$

By an *approximation* of the mixed formulation of the Signorini problem with given friction, we mean the problem of finding a saddle point $(w_h; \lambda_{1H}, \lambda_{2H})$ of the function $\mathcal{L}(v_h; \mu_{1H}, \mu_{2H})$ on $\mathbf{V}_h \times (\Lambda_{1H} \times \Lambda_{2H})$.

Let us now present without proof the most important results concerning the approximation of the Signorini problem with given friction, which is based on the type of mixed variational formulation just introduced. Detailed proofs of these, as well as some other assertions, are found in the paper by Haslinger, and Hlaváček (1982).

If we make no a priori assumptions on the smoothness of the solution u of problem (5.76), then we have

Theorem 5.9. *Let $h \to 0_+$ if and only if $H \to 0_+$. Then*

$$u_h \to u, \quad h \to 0_+ \quad \text{in } [H^1(\Omega)]^2,$$

and

$$\lambda_{2H} \to \lambda_2, \quad H \to 0_+, \quad \text{weakly in } L^2(\Gamma_K).$$

As concerns the behavior of the other components of λ_{1H}, the situation is a little more complicated. In the paper quoted above, this problem is discussed under the assumption that g is a *piecewise constant* function whose points of discontinuity are (some of) the nodes of the partition $\mathcal{T}_H$. We have:

Theorem 5.10. *Let $u \in H^{1+q}(\Omega)$ for some $q > 0$ and $T_n(u) \in L^2(\Gamma_K)$. Moreover, let there exist a constant $\beta > 0$ independent of h, H and such that*

$$\sup_{\mathbf{V}_h} \left(\frac{(\mu_{1H}, v_{hn})_{0,\Gamma_K}}{\|v_h\|_1} + \frac{(\mu_{2H}, v_{ht})_{0,\Gamma_K}}{\|v_h\|_1} \right) \geq \beta \|\mu_H\|_{-1/2,\Gamma_K} \qquad (S)$$

holds for arbitrary $\mu_H = (\mu_{H1}, \mu_{H2}) \in L_H \times L_H$. Then

$$\|u - u_h\|_{[H^1(\Omega)]^2} = o(H^{\tilde{q}}), \quad H \to 0_+,$$

$$\|\lambda - \lambda_H\|_{-1/2,\Gamma_K} = o(H^{\tilde{q}}), \quad H \to 0_+,$$

where $\tilde{q} = \min(q; 1/4)$.

Naturally, the crucial problem is that of when condition (S) is fulfilled. This problem is also solved in the above mentioned paper.

Uzawa's method can again be used for the realization of the mixed variational formulation. While the partial dualization, consisting merely of removing the nondifferentiable term $j(v)$, meant transforming (5.78) into a quadratic programming problem, now the total dualization yields $K_h = \mathbf{V}_h$ and (5.78) is equivalent to the problem of solving a system of linear algebraic equations, in which some components of the right-hand side are being corrected. The reader can easily see that a partial elimination of free components of the displacement vector u_h again leads to forming an effective algorithm of solution.

Remark 5.6. Let us show another useful variational formulation of problem (5.70), which involves only the quantities defined on Γ_K. Consider a triplet of functions $(F, \mu_1, g\mu_2)$, where $F \in [L^2(\Omega)]^2$, $\mu_1 \in \Lambda_1$, $\mu_2 \in \Lambda_2$ with Λ_1, Λ_2 defined in the preceding remark. This triplet defines a generalized force $\mathcal{F} \in \mathbf{V}'$. The corresponding field of displacements $w \in \mathbf{V}$ is given by the relation

$$w = G(\mathcal{F}) = \hat{G}(\mu_1, \mu_2), \tag{5.88}$$

where $G : \mathbf{V}' \to \mathbf{V}$ is the Green operator of our problem.

The *reciprocal variational formulation* of the Signorini problem with given friction is the problem

$$\text{of finding } (\lambda_1, \lambda_2) \in \Lambda_1 \times \Lambda_2 \text{ such that}$$

$$\mathcal{S}(\lambda_1, \lambda_2) \le \mathcal{S}(\mu_1, \mu_2) \quad \forall (\mu_1, \mu_2) \in \Lambda_1 \times \Lambda_2, \tag{5.89}$$

where

$$\mathcal{S}(\mu_1, \mu_2) = \frac{1}{2}\langle \mu_1, \hat{G}(\mu_1, \mu_2) \cdot n \rangle + \frac{1}{2}\langle \mu_2 g, \hat{G}(\mu_1, \mu_2) \cdot t \rangle.$$

It can be shown that between (5.70) and (5.89) the following relation holds:

$$\lambda_1 = T_n(u), \quad g\lambda_2 = T_t(u).$$

(u solves (5.70).) In this form the reciprocal variational formulation is hardly realizable, since the explicit form of G is known only for several particular cases. For this reason it is necessary to use an approximation of G. For instance, the inverse matrix to the stiffness matrix of our problem may serve as such an approximation. The application of this formulation to the solution of contact problems without friction is thoroughly dealt with in the work of Oden and Kukuchi (1979). It is of interest to note here that the Signorini problem with friction governed by Coulomb's Law, if expressed in terms of the recriprocal variational formulation, leads to the solution of the so-called *quasivariational inequality*, with the convex set depending on the solution itself. Indeed, let $\mathcal{K}_{\mathcal{F}g}$ be the convex, closed subset of $[H^{-1/2}(\Gamma_K)]^2$ defined by

$$\mathcal{K}_{\mathcal{F}g} = \{\mu = (\mu_1, \mu_2),\quad \mu_1 \leq 0,\quad |\mu_2| \leq \mathcal{F}g \Longleftrightarrow \langle \mu_2, v_t\rangle + \langle \mathcal{F}g, |v_t|\rangle \geq 0 \quad \forall v \in \mathbf{V}\},\quad g \in H_+^{-1/2}(\Gamma_K).$$

Let $(\lambda_1(g), \lambda_2(g)) \in \mathcal{K}_{\mathcal{F}g}$ be the minimum of the quadratic functional

$$\mathcal{S}(\mu_1, \mu_2) = \frac{1}{2}\langle \mu_1, \hat{G}(\mu_1, \mu_2)\cdot n\rangle + \frac{1}{2}\langle \mu_2, \hat{G}(\mu_1, \mu_2)\cdot t\rangle$$

on $\mathcal{K}_{\mathcal{F}g}$. Again we can show that $\lambda_1(g) = T_n(u)$, $\lambda_2(g) = T_t(u)$, where u solves the Signorini problem with given friction $\mathcal{F}g \in H_+^{-1/2}(\Gamma_K)$. Define a mapping $\tilde{\Phi} : H_+^{-1/2}(\Gamma_K) \to H_+^{-1/2}(\Gamma_K)$ by

$$\tilde{\Phi}(\mathcal{F}g) = -\mathcal{F}\lambda_1(g).$$

If $g = -\lambda_1(g)$, then the solution of the Signorini problem with given friction equals to $-\mathcal{F}\lambda_1$ is the solution of the Signorini problem with friction governed by Coulomb's Law (5.2).

The detailed analysis of the approximation based on the reciprocal variational formulation of the Signorini problem with given friction is given in the paper by Haslinger and Panagiotopoulos (1982).

2.5.42. Alternating Iterations. The contact problem for two elastic bodies with friction can be solved in yet another way, which was suggested in a paper by Panagiotopoulos (1975) and recommended in a research report by Frederikson, Rydholm, and Sjöbolm (1977). Each step of the algorithm consists of two partial problems.

1st step:

1.1. Unilateral contact with a given shear force. $T_t = F_t^{(0)}$. (Choose, e.g., $F_t^{(0)} = 0$.) We compute $T_n = F_n^{(1)}$ on Γ_K.

1.2. Friction with a given normal force $T_n = F_n^{(1)}$ on Γ_K. We compute $T_t = F_t^{(1)}$ on Γ_K.

In the i-th step ($i \geq 2$) we solve the following partial problems:

i.1. Unilateral contact with a given shear force $T_t = F_t^{(i-1)}$. We compute $T_n = F_n^{(i)}$.

i.2. Friction with a given normal force $T_n = F_n^{(i)}$. We compute $T_t = F_t^{(i)}$.

As stopping test of the iteration process we may choose, for instance, a sufficiently small change of the normal contact forces

$$\|F_n^{(i)} - F_n^{(i-1)}\| \le \epsilon_n \|F_n^{(i)}\|,$$

and the friction forces

$$\|F_t^{(i)} - F_t^{(i-1)}\| \le \epsilon_t \|F_t^{(i)}\|.$$

Thus, the algorithm alternates partial problems of two different types. In the following sections we will study both of them in more detail, considering the case of bounded contact zone.

2.5.421. Unilateral Contact with a Given Shear Force. As before we assume that

$$\partial\Omega' \cap \partial\Omega'' = \Gamma_K, \quad \partial\Omega' = \Gamma_u \cup \Gamma_\tau' \cup \Gamma_K, \quad \partial\Omega'' = \Gamma_0 \cup \Gamma_\tau'' \cup \Gamma_K,$$

and the following inequalities hold on Γ_K:

$$u_n' + u_n'' \le 0, \tag{5.90}$$

$$T_n' = T_n'' \le 0, \tag{5.91}$$

$$T_n(u_n' + u_n'') = 0. \tag{5.92}$$

However, in contrast to the former formulations we now prescribe

$$T_t' = T_t'' = F_t, \tag{5.93}$$

where F_t is given in the space $L_2(\Gamma_K)$.

The *primal variational formulation* is defined for the potential energy functional

$$\mathcal{L}(v) = \frac{1}{2}A(v,v) - L(v) - L_t(v), \tag{5.94}$$

where A and L have the same meaning as in (1.21), (1.22), (1.23), and

$$L_t(v) = \int_{\Gamma_K} F_t(v_t' + v_t'')ds. \tag{5.95}$$

A function $u \in K$ is called a weak (variational) solution of the problem, if

$$\mathcal{L}(u) \le \mathcal{L}(v) \quad \forall v \in K, \tag{5.96}$$

where K is defined as in (1.24).

Lemma 5.14. *A solution of problem* (5.96) *exists only if*

$$L(y) + L_t(y) \leq 0 \quad \forall y \in K \cap \mathbf{R} \tag{5.97}$$

(*where* $\mathbf{R}$ *is the space of displacements of the rigid bodies—see section* 2.2.1).

Proof. Follows from the inequality

$$A(u, v-u) \geq L(v-u) + L_t(v-u) \quad \forall v \in K$$

by substituting $v = u + y$. □

Remark 5.7. In the situation from Figure 8 (cf. example 2.1) we have

$$K \cap \mathbf{R} = \{y \mid y' = (0,0), \quad y'' = (a,0), \quad a \leq 0\}.$$

Then (5.97) is equivalent to the condition

$$V_1^t = \int_{\Omega''} F_1 dx + \int_{\Gamma_\tau''} P_1 ds + \int_{\Gamma_K} F_t t_1'' ds \geq 0, \tag{5.98}$$

which means that the resultant of all given exterior forces must not pull the body Ω'' away from the body Ω'.

Lemma 5.15. *If either* $V \cap \mathbf{R} = \{0\}$ *or*

$$L(y) + L_t(y) \neq 0 \quad \forall y \in V \cap \mathbf{R} \dot{-} \{0\}, \tag{5.99}$$

then there is at most one solution of problem (5.96).

Proof. Follows the same lines as that of theorem 2.1. □

Remark 5.8. In the situation from Figure 8 condition (5.99) is equivalent to the condition

$$V_1^t \neq 0. \tag{5.100}$$

Theorem 5.11. *Let* $\mathbf{R}^*$ *be the space of "bilateral" admissible displacements of rigid bodies, defined in* (2.3). *Assume that*

$$K \cap \mathbf{R} = \{0\}, \tag{5.101}$$

or

$$\mathbf{R}^* = \{0\}, \quad K \cap \mathbf{R} \neq \{0\}, \tag{5.102}$$

$$L(y) + L_t(y) < 0 \quad \forall y \in \mathbf{R} \cap K \dot{-} \{0\}. \tag{5.103}$$

Then there is a solution of problem (5.96).

Proof. Analogous to part of theorem 2.4, and therefore is omitted. □

Remark 5.9. By combining conditions (5.99) and (5.101)–(5.103), we obtain sufficient conditions for the existence and uniqueness of solution. In the situation from Figure 8, conditions (5.99) and (5.103) reduce to a single condition: $V_1^t > 0$. Condition $\mathbf{R}^* = \{0\}$ is fulfilled with regard to the position of the part of boundary Γ_K (with respect to Γ_0; i.e., Γ_K consists not only of segments parallel to Γ_0).

Dual Variational Formulation. Let us define the space S as in section 2.4, and the inner product

$$(\tau, \epsilon) = \int_\Omega \tau_{ij}\epsilon_{ij}\,dx, \quad (\Omega = \Omega' \cup \Omega''),$$

$$\mathcal{H}_2(v, \tau) = (\tau, \epsilon(v)) - L(v) - L_t(v), \tag{5.104}$$

$$K_t^+ = \{\tau \in S \mid \mathcal{H}_2(v, \tau)) \geq 0 \quad \forall v \in K\}, \tag{5.105}$$

$$\mathcal{S}(\tau) = \frac{1}{2}\int_\Omega a_{ijkl}\tau_{ij}\tau_{kl}\,dx.$$

The dual variational problem reads as follows:

$$\text{find} \quad \sigma \in K_t^+ \quad \text{such that}$$

$$\mathcal{S}(\sigma) \leq \mathcal{S}(\tau) \quad \forall \tau \in K_t^+. \tag{5.106}$$

Theorem 5.12. *Let there exist a solution u of the primal problem* (5.96). *Then there is a unique solution σ of the dual problem* (5.106) *and it satisfies*

$$\mathcal{L}(u) + \mathcal{S}(\sigma) = 0, \tag{5.107}$$

$$\sigma_{ij} = c_{ijkl}\epsilon_{kl}(u). \tag{5.108}$$

Proof. Can be done analogously to section 2.4, if we replace the functional L throughout by the sum $L + L_t$ and the set $K_{F,P}^+$ by the set K_t^+. □

Interpretation of the set K_t^+.

Lemma 5.16. *Let $\tau \in K_t^+$ be sufficiently smooth. Then τ satisfies the following conditions:*

$$\frac{\partial \tau_{ij}}{\partial x_j} + F_i = 0 \quad \text{in } \Omega, \tag{5.109}$$

$$T(\tau) = P \quad \text{on } \Gamma_\tau, \tag{5.110}$$

$$T_t = 0 \quad \text{on } \Gamma_0, \tag{5.111}$$

$$T_t' = T_t'' = F_t \quad on\ \Gamma_K, \tag{5.112}$$

$$T_n' = T_n'' \leq 0 \quad on\ \Gamma_K. \tag{5.113}$$

Conversely, let τ be sufficiently smooth and satisfy conditions (5.109)–(5.113). *Then $\tau \in K_t^+$.*

Proof. Similar to that of lemma 4.6. □

Approximation of the Dual Variational Problem. The dual problem (5.106) can be approximated by the method of finite elements for the equilibrium model similar to that given in section 2.4.1. First we introduce a particular solution $\bar{\lambda}$ of the equations (5.109)–(5.113). Let us consider the situation from Figure 8 and let the condition

$$V_1^t > 0 \tag{5.114}$$

be fulfilled, which guarantees the existence and uniqueness of solution of the primal problem. First, we establish the condition of total equilibrium of forces and reactions on the body Ω'':

$$-\int_{\Omega''} F_1 dx = \int_{\Gamma_\tau''} P_1 ds + \int_{\Gamma_K} T_1''(\bar{\lambda}) ds = \int_{\Gamma_\tau''} P_1 ds + \int_{\Gamma_K} (T_n''(\bar{\lambda}) n_1'' + F_t t_1'') ds.$$

This implies

$$\int_{\Gamma_K} T_n''(\bar{\lambda}) dx_2 = -V_1^t. \tag{5.115}$$

Thus, let

$$T_n'(\bar{\lambda}) = T_n''(\bar{\lambda}) = -V_1^t \left(\int_{\Gamma_K} dx_2 \right)^{-1} \equiv g_0. \tag{5.116}$$

The constant g_0 is negative due to condition (5.114). Thus, the tensor field $\bar{\lambda}$ fulfills the condition

$$(\bar{\lambda}, \epsilon(v)) = L(v) + L_t(v) + \int_{\Gamma_K} g_0 (v_n' + v_n'') ds \quad \forall v \in V. \tag{5.117}$$

Remark 5.10. If, for example, $F_1 = F_1^0 = \text{const}$, $F_2 = 0$, and P and F_t are piecewise linear, then $\bar{\lambda}$ can be constructed as follows:

$$\bar{\lambda} = \lambda^F + \lambda^0,$$

where

$$\lambda_{11}^F = -F_1^0 x_1, \quad \lambda_{12}^F = \lambda_{22}^F = 0 \quad \text{in } \Omega,$$

and $\lambda^0 \in N_h(\Omega)$ fulfills the modified boundary conditions

$$T(\lambda^0) = P - T(\lambda^F) \quad \text{on } \Gamma_\tau,$$

$$T_t(\lambda^0) \equiv \lambda^0_{12} = 0 \quad \text{on } \Gamma_0,$$

$$T'_t(\lambda^0) = T''_t(\lambda^0) = F_t - T'_t(\lambda^F) \quad \text{on } \Gamma_K.$$

By (5.117) we easily derive that

$$\lambda \in K_t^+ \Longleftrightarrow \lambda - \bar{\lambda} = \tau \in U_0,$$

where

$$U_0 = \left\{ \tau \in S \mid (\tau, \epsilon(v)) \geq -g_0 \int_{\Gamma_K} (v'_n + v''_n) ds \quad \forall v \in K \right\}.$$

We proceed further in the same way as in section 2.4.1. Analogs to lemmas 4.8, 4.9 are valid, too, if we write g_0 instead of g and V_1^t instead of V_1''. If we use the results of section 2.4.11, we also establish the error estimates from section 2.4.12 and the algorithm from section 2.4.13.

2.5.422. Realizability of the Algorithm of Alternating Iterations. Now we will consider a model of a simpler Signorini problem in order to study some questions connected with the possibility of realization of the algorithm of alternating iterations in the semicoercive case, that is, the situation when there exist nontrivial admissible displacements of the rigid body.

In order to grasp the core of the problem, let us consider only one elastic body of a trapezoidal shape, loaded by uniformly distributed horizontal forces and resting on a perfectly rigid foundation Γ_K, where the friction occurs (see Figure 25). So we have the case that $\Omega'' \equiv \Omega$, Ω' is a perfectly rigid body, that is, we have the so-called Signorini problem with friction. Let us first collect the achieved results on existence and then complete the analysis of uniqueness of solution in the case of the partial problem of friction with given normal force.

A. Unilateral contact with given shear force. We assume that $n_1 > 0$, $n_2 < 0$ holds on Γ_K; further,

$$\partial\Omega = \bar{\Gamma}_\tau \cup \bar{\Gamma}_0 \cup \bar{\Gamma}_K$$

and a shear force $F_t \in L_2(\Gamma_K)$ is given.

We denote the potential energy (cf. (5.94), (5.95)) by

$$\mathcal{L}_a(v) = \frac{1}{2} A(v, v) - L(v) - L_t(v),$$

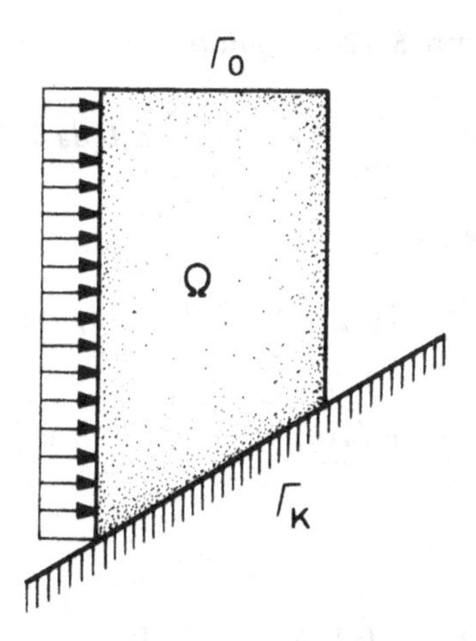

Figure 25

where

$$L(v) = \int_{\Gamma_r} P_1 v_1 \, ds \quad P_1 = \text{const} > 0,$$

$$L_t(v) = \int_{\Gamma_K} F_t v_t \, ds.$$

In our case we have

$$V = \{v \in [H^1(\Omega)]^2 \quad v_n \equiv v_2 = 0 \quad \text{on } \Gamma_0\},$$

$$K = \{v \in V \mid v_n \leq 0 \quad \text{on } \Gamma_K\},$$

and the primal problem reads:

find $u \in K$ such that

$$\mathcal{L}_a(u) \leq \mathcal{L}_a(v) \quad \forall v \in K. \tag{5.118}$$

Theorem 5.13. *A solution u of problem* (5.118) *exists only if*

$$L(y) + L_t(y) \leq 0 \quad \forall y \in K \cap \mathbf{R}.$$

Proof. Follows from lemma 5.14. □

Recall that the set of translations and rotations of a rigid body is

$$\mathcal{R} = \{(y_1, y_2) \mid y_1 = a_1 - bx_2, \quad y_2 = a_2 + bx_1, \quad a_i \in \mathbf{R}^1, \quad b \in \mathbf{R}^1\}.$$

Corollary of Theorem 5.13. *A solution u exists only if*

$$V_1^t = \int_{\Gamma_\tau} P_1 ds + \int_{\Gamma_K} F_t t_1 ds \geq 0.$$

Indeed, in our case we have

$$K \cap \mathcal{R} = \{(y_1, y_2) \mid y_1 = a \leq 0, \quad y_2 = 0\},$$

hence,

$$L(y) + L_t(y) = aV_1^t \leq 0 \quad \forall a \leq 0$$

if and only if $V_1^t \geq 0$.

Theorem 5.14. *If*

$$L(y) + L_t(y) \neq 0 \quad \forall y \in V \cap \mathcal{R} \dot{-} \{0\},$$

then there is at most one solution of problem (5.118).

Proof. Follows from lemma 5.15. □

Corollary of Theorem 5.14. *There is at most one solution u, provided* $V_1^t \neq 0$. *Indeed,*

$$V \cap \mathcal{R} = \{(y_1, y_2) \mid y_1 = a, \quad y_2 = 0\}, \quad a \in \mathbf{R}^1,$$

$$L(y) + L_t(y) = aV_1^t.$$

Theorem 5.15. *If*

$$L(y) + L_t(y) < 0 \quad \forall y \in K \cap \mathcal{R} \dot{-} \{0\},$$

then there is a solution of problem (5.118).

Proof. Follows from Theorem 5.11, (5.102) and (5.103). □

Corollary of Theorems 5.14 and 5.15. *Let*

$$V_1^t > 0. \tag{5.119}$$

Then there is a unique solution of problem (5.118). *Indeed, the uniqueness follows from theorem* 5.14. *The existence is a consequence of theorem* 5.15, *since for* $y \in K \cap \mathcal{R} \dot{-} \{0\}$ *we have* $y = (a, 0)$ *with* $a < 0$, *hence,*

$$L(y) + L_t(y) = aV_1^t < 0.$$

B. Friction with a given normal force. Let a friction coefficient

$$\mathcal{F} \in L^\infty(\Gamma_K), \quad \mathcal{F} > 0 \text{ a.e.}$$

and a normal force $F_n \in L^\infty(\Gamma_K)$ be given.

On Γ_K we have

$$T_n = F_n,$$

$$|T_n| \le g, \quad \text{where} \quad g = \mathcal{F}|F_n|,$$

$$|T_t| < g \Rightarrow u_t = 0,$$

$$|T_t| = g \Rightarrow T_t u_t \le 0.$$

We denote the potential energy by

$$\mathcal{L}_b(v) = \frac{1}{2}A(v,v) - L(v) - L_n(v) + j(v),$$

with

$$L_n(v) = \int_{\Gamma_K} F_n v_n ds, \quad j(v) = \int_{\Gamma_K} g|v_t|ds.$$

The primal problem reads:

$$\text{find} \quad u \in V \quad \text{such that}$$

$$\mathcal{L}_b(u) \le \mathcal{L}_b(v) \quad \forall v \in V. \tag{5.120}$$

Theorem 5.16. *A solution of problem* (5.120) *exists only if*

$$|L(y) + L_n(y)| \le j(y) \quad \forall y \in V \cap \mathcal{R}.$$

Proof. See Duvaut and Lions (1972). □

Corollary of Theorem 5.16. *A solution of problem* (5.120) *exists only if*

$$\left|\int_{\Gamma_\tau} P_1 ds + \int_{\Gamma_K} F_n n_1 ds\right| \le \int_{\Gamma_K} g|t_1|ds.$$

Indeed, in our case we have

$$|L(y) + L_n(y)| = |a|\left|\int_{\Gamma_\tau} P_1 ds + \int_{\Gamma_K} F_n n_1 ds\right|,$$

$$j(y) = |a|\int_{\Gamma_K} g|t_1|ds.$$

Theorem 5.17. *If*

$$|L(y) + L_n(y)| < j(y) \quad \forall y \in \mathcal{R} \cap V \dot{-} \{0\},$$

then there is a solution of problem (5.120).

Proof. See Duvaut and Lions (1972). □

Corollary of Theorem 5.17. *Let*

$$\left|\int_{\Gamma_\tau} P_1 ds + \int_{\Gamma_K} F_n n_1 ds\right| < \int_{\Gamma_K} g|t_1|ds. \tag{5.121}$$

Then there is a solution of problem (5.120)*. Indeed, multiplying* (5.121) *by* $|a|$, $a \neq 0$, *yields the validity of the condition from theorem* 5.17.

Theorem 5.18. *Let condition* (5.121) *be fulfilled. Then there is a unique solution of problem* (5.120).

Proof. Existence follows from the corollary of theorem 5.17. Let us prove the uniqueness.

Let u_1, u_2 be two solutions. Denoting $L_1 = L + L_n$ and taking into account the fact that j is a convex functional, we obtain:

$$A(u_1, u_2 - u_1) - L_1(u_2 - u_1) + j(u_2) - j(u_1) \geq 0,$$

$$A(u_2, u_1 - u_2) - L_1(u_1 - u_2) + j(u_1) - j(u_2) \geq 0.$$

By adding we obtain

$$A(u_1 - u_2, u_2 - u_1) \geq 0,$$

hence,

$$u_1 - u_2 = y \in \mathcal{R} \cap V,$$

$$T_t(u_1) = T_t(u_2) \equiv T_t.$$

The condition of total equilibrium yields

$$L_1(y) + \int_{\Gamma_K} T_t y_t ds = 0. \tag{5.122}$$

Let us assume

$$T_t = g \quad \text{a.e. on } \Gamma_K. \tag{5.123}$$

Then we obtain by (5.122)

$$-L_1(y) = \int_{\Gamma_K} g y_t ds = \pm \int_{\Gamma_K} g|y_t|ds = \pm j(y),$$

and condition (5.121) yields $y = 0$.

Similarly, the assumption

$$T_t = -g \quad \text{a.e. on } \Gamma_K \tag{5.124}$$

leads to the conclusion $y = 0$.

Now let us assume

$$-g < T_t < g \quad \text{on a set } E \subset \Gamma_K, \quad \text{meas } E > 0. \tag{5.125}$$

Denoting $u_2 \equiv u$, $u_1 = u + y$, we have

$$u_t = 0, \quad u_t + y_t = 0 \quad \text{on } E,$$

hence, $y_t = at_1 = 0$ on E, which yields $y = 0$.

The last case to be dealt with is that with

$$T_t = g \quad \text{on } \Gamma_1; \quad \text{meas } \Gamma_1 > 0,$$

$$T_t = -g \quad \text{on } \Gamma_2; \quad \text{meas } \Gamma_2 > 0, \tag{5.126}$$

where $\Gamma_1 \cup \Gamma_2 = \Gamma_K$ (up to a set of zero measure).

Then evidently

$$u_t \leq 0, \quad u_t + y_t \leq 0 \quad \text{on } \Gamma_1,$$

$$u_t \geq 0, \quad u_t + y_t \geq 0 \quad \text{on } \Gamma_2.$$

Let us assume that $y_t = at_1 > 0$ (the case $y_t < 0$ is solved analogously). Hence,

$$u_t \leq -y_t < 0 \quad \text{on } \Gamma_1, \quad u_t \geq 0 \quad \text{on } \Gamma_2. \tag{5.127}$$

However, we have $u_t \in H^{1/2}(\Gamma_K)$, which contradicts (5.127). Hence, again $y = 0$. □

C. Realizability of the algorithm. Let us denote problem (5.18) for $F_t = F_t^{(i-1)}$, $i = 1, 2, \ldots$, as problem (ia T), and problem (5.120) for $F_n = F_n^{(i)}$, $i = 1, 2, \ldots$, as problem (ib T). Recall that the algorithm defined at the beginning of section 2.5.22 consists of the successive solution of the problems

$$(1\text{a T}), \quad (1\text{b T}), \quad (2\text{a T}), \quad (2\text{b T}), \ldots,$$

where $F_t^{(0)}$ is chosen, and $F_n^{(i)} = T_n(u)$, where u is the solution of the problem (ia T), while $F_t^{(j)} = T_t(u)$, where u is the solution of the problem (jb T).

Theorem 5.19. *Let $n_1 > 0$, $n_2 < 0$. Let $\mathcal{F} = \text{const} > 0$.*

$$f_0 = \int_{\Gamma_K} F_t^{(0)} t_1 ds \leq 0, \tag{5.128}$$

$$\int_{\Gamma_K} P_1 ds + f_0 \left(1 + \frac{n_1}{\mathcal{F} t_1}\right) > 0. \tag{5.129}$$

Then the conditions (5.119), (5.121), *which guarantee the existence and uniqueness of solution, are fulfilled for all approximation problems* (ia T), (ib T), *respectively,* $i = 1, 2, \ldots$, *and the identity*

$$\int_{\Gamma_K} F_t^{(i)} t_1 ds = f_0$$

holds for all i.

Proof. Problem (ia T) involves the total equilibrium condition

$$\int_{\Gamma_r} P_1 ds + \int_{\Gamma_K} (F_t^{(i-1)} t_1 + F_n^{(i)} n_1) ds = 0, \tag{E_a}$$

while problem (ib T) involves condition

$$\int_{\Gamma_r} P_1 ds + \int_{\Gamma_K} (F_n^{(i)} n_1 + F_t^{(i)} t_1) ds = 0. \tag{E_b}$$

This immediately implies

$$f_i \equiv \int_{\Gamma_K} F_t^{(i)} t_1 ds = \int_{\Gamma_K} F_t^{(i-1)} t_1 ds = \ldots = f_0.$$

Hence, for problem (ia T) we have

$$V_1^t = \int_{\Gamma_r} P_1 ds + f_{i-1} = \int_{\Gamma_r} P_1 ds + f_0.$$

However, by (5.128) and $t_1 = -n_2 > 0$, we evidently have

$$f_0 \left(1 + \frac{n_1}{\mathcal{F} t_1}\right) \le f_0;$$

thus, we verify by (5.129) that $V_1^t > 0$, that is, condition (5.119) holds.

Consider now problem (ib T) and verify condition (5.121). On the one hand, we obtain from (E_b),

$$\int_{\Gamma_r} P_1 ds + \int_{\Gamma_K} F_n^{(i)} n_1 ds = -f_0; \tag{5.130}$$

on the other hand we can write after substituting from (5.130):

$$\int_{\Gamma_K} g t_1 ds = \int_{\Gamma_K} \mathcal{F} |F_n^{(i)}| t_1 ds = -\mathcal{F} t_1 \int_{\Gamma_K} F_n^{(i)} ds$$

$$= \frac{\mathcal{F}t_1}{n_1}\left(f_0 + \int_\Gamma P_1 ds\right). \tag{5.131}$$

From assumption (5.129) we find

$$f_0 + \int_{\Gamma_r} P_1 ds > -f_0 \frac{n_1}{\mathcal{F}t_1},$$

hence,

$$\frac{\mathcal{F}t_1}{n_1}\left(f_0 + \int_{\Gamma_r} P_1 ds\right) > -f_0 = |f_0|,$$

which is condition (5.121) with respect to equations (5.131), (5.130). □

Remark 5.11. Theorem 5.19 expresses the fact that the mean value of the shear force coincides for all the iterations. Thus, the iteration solution depend on the initial choice of $F_t^{(0)}$. Hence, it also follows that there is no single limit of all the iteration solutions common for all initial choices of $F_t^{(0)}$.

Chapter 3

Problems of the Theory of Plasticity

In this chapter we will deal with variational inequalities of evolution that result from some problems of plasticity. Let us note that we will consider processes that depend on the history of loading, that is, irreversible ones. Elasto-plasticity, which has come into fashion lately, is evidently a mere special case of nonlinear elasticity with generally nonlinear Hooke's Law

$$\sigma_{ij} = \frac{\partial A}{\partial e_{ij}}. \tag{0.1}$$

In addition to nonlinearity of the inequalities (which is of geometrical character as regards contact problems), we have here a physical nonlinearity (0.1), which naturally requires further linearization if approximate methods are to be used. For details we refer the reader to the books by Washizu (1968), Duvaut and Lions (1972), Nečas and Hlaváček (1981), and Kačanov (1974). This theory is justly criticized for its insensitiveness to the history of loading as well as for the inadequacy of the nonlinear relation (0.1) between the Cauchy stress tensor and the tensor of small strains.

In this chapter we will consider the so-called *flow theory of plasticity*. It is rate independent, which limits its validity to shorter time periods, so that it cannot describe such phenomena as creep or fading memory. It is interesting that the method of penalization considered (which will be used to prove the existence of solution) by itself has some features of an approximate method. It approximates the given problem by problems for an elasto-inelastic material with internal state variables (for details, see the book by Nečas and Hlaváček (1981), which includes further references concerning this approach). These models are of independent physical significance, and they are sensitive to creep and fading memory. When studying flow theory we will follow the above quoted book by Nečas and Hlaváček, naturally introducing the problem in a somewhat more general form (which was dealt with as a mathematical model in the fundamental works by Quoc Son Nguyen (1973), Halphen and Quoc Son Nguyen (1975), and theoretically by the above mentioned method in the paper by Nečas and Trávníček (1978), which was preceded by the papers by Nečas and Trávníček (1980), Trávníček (1976)). The same problems were dealt with by a method based on evolution equations with maximally monotone operators in a number of papers by Gröger, (see Gröger (1979), (1978a), (1978b), (1977), (1980)). Gröger and Nečas (1979), and Gröger, Nečas, Trávníček (1979). A partial theoretical solution is also given in the book by Duvaut and Lions (1972), and in C. Johnson (1976a). Gröger's research is continued by Hühnlich (1979).

To introduce the reader to the problem, let us first consider some simple yet typical examples. First of all, let us realize that we are going to follow the loading process together with the corresponding course of solution, which, in other words, means that the quantities observed will be functions of both time and space. Since a physical nonlinearity is involved (passing to the tensor of finite strains is still an unmanaged affair), we will sketch its character in the one-dimensional case of the relation between the stress and strain (that is, by the graph of the stress–strain relation at a fixed point of the one-dimensional bar in question, $e = e(t)$, $\sigma = \sigma(t)$ and the graph $(e(t), \sigma(t))$ belongs to the plotted line). Moreover, we assume that the deformation ϵ is the sum of the elastic and plastic deformations, that is, $\epsilon = e + p$, and that the relation between e, σ is linear; for the sake of simplicity, let $\sigma = e$. The tensor ϵ is assumeed to satisfy $\epsilon = \frac{du}{dx}$, the compatibility condition, where u is the displacement vector, while the stress tensor is assumed to satisfy $\frac{d\sigma}{dx} = 0$, the condition of equilibrium.

Thus, the linear elasticity is represented by the reversible process in Figure 26 and simply means $p \equiv 0$.

The *elasto-perfectly-plastic* case is shown in Figure 27. The solution in the latter case thus fulfills the condition $|\sigma(t)| \leq \sigma_0$.

Let us now assume that by increasing (or decreasing) the stress we would obtain the graph in Figure 28 and, moreover, when reaching the point $\pm\sigma_0$ and then decreasing (increasing) the stress again, we should stay on the line $\sigma = \epsilon$. In other words, this means that in the domain $|\sigma| \leq \sigma_0$ no increment of the plastic deformation occurs. Let us assume that the loading reaches the stress σ_1, and at this point let the stress fall. If we now go back along the line AB, then there is no change of the plastic deformation, and the

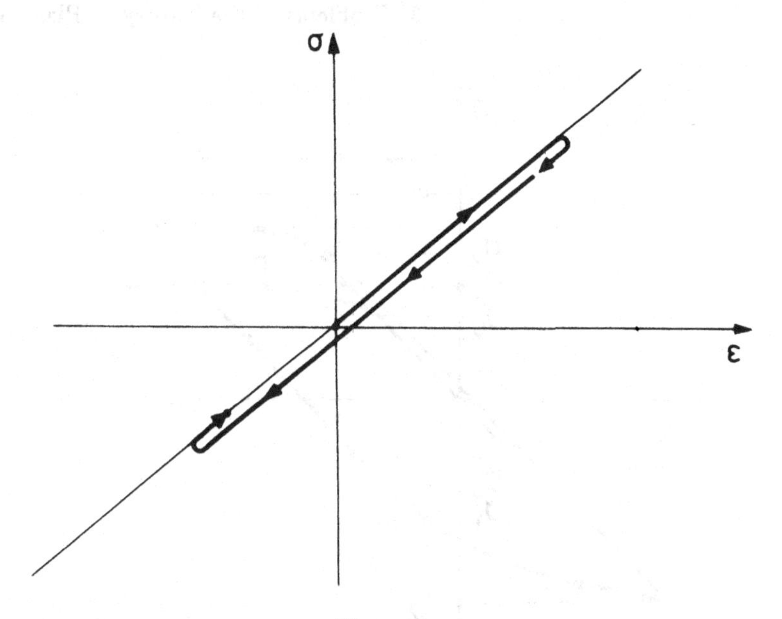

Figure 26

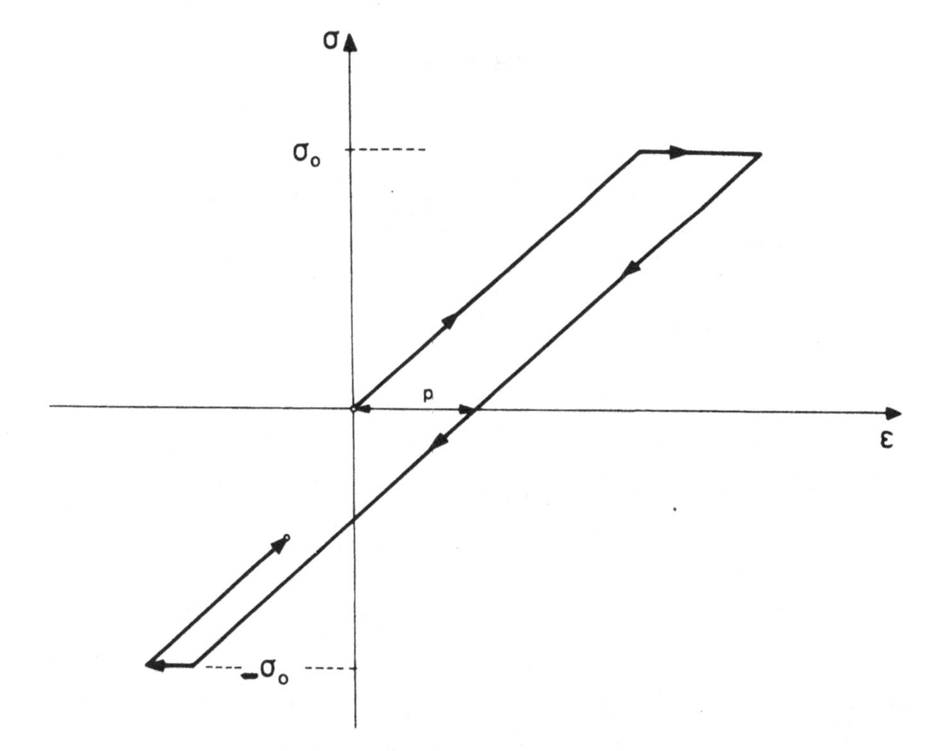

Figure 27

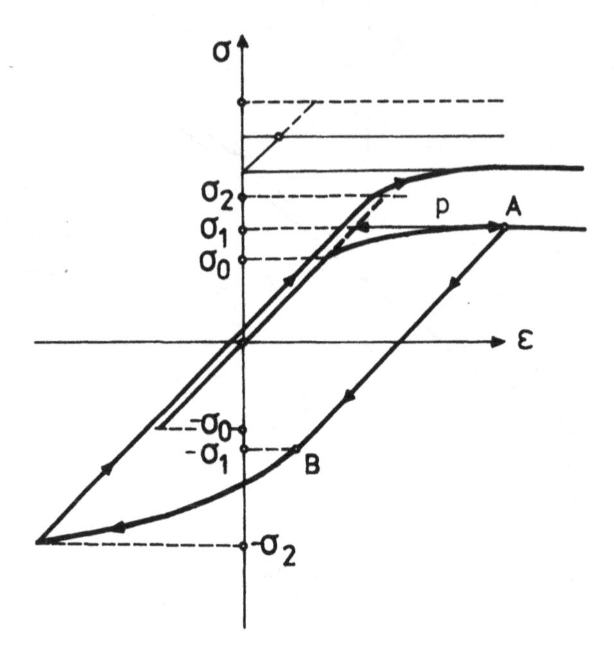

Figure 28

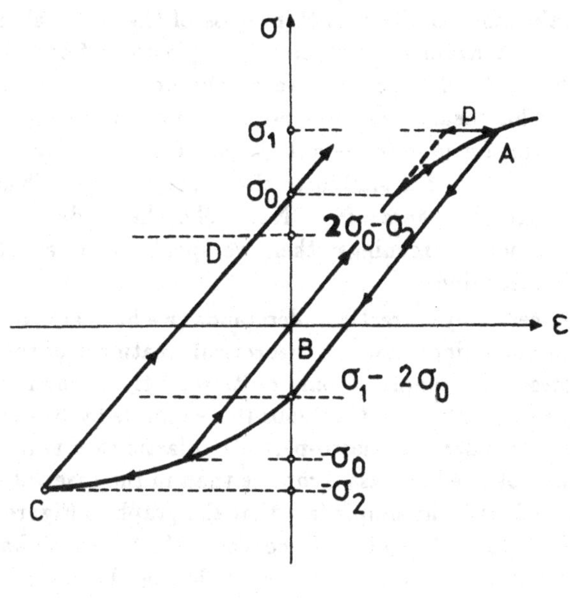

Figure 29

domain of elasticity has grown to the set $|\sigma| \le \sigma_1$. Thus, we obtain the so-called *isotropic strain-hardening.*

Let us again consider the growth process of the stress along the curve in Figure 29. Let again no growth of the plastic deformation occur in the domain $|\sigma| \le \sigma_0$. Let us again reach the point A and go back along the line AB. This means that now no change of the plastic deformation occurs in the domain $|\sigma - (\sigma_1 - \sigma_0)| \le \sigma_0$. Consequently, the center of the convex set $|\sigma| \le \sigma_0$ has shifted to the point $\sigma_1 - \sigma_0$. Thus, we obtain the so-called *kinematic hardening.* Naturally, the reader can imagine a combination of both hardenings; then we speak about an elasto-plastic material with hardening.

The reader certainly correctly understands (we have actually mentioned the fact at the beginning) that the theoretical treatment of these problems has been successfully completed only quite recently, though the problems were formulated as early as in the 1950s (for example by Koiter (1960) and Hodge (1959)). Besides, the elasto-perfectly plastic case is much idealized and the results obtained are less complete than in the case with hardening. This is connected with the simple fact that the graph in Figure 28 does not determine ϵ from the values of σ. As concerns the case with hardening, we will solve it in a little more general setting, following the paper by Nečas and Trávníček (1978). We will also formulate in detail the special cases of both the isotropic and kinematic hardening by introducing the yield surface.

3.1 Prandtl–Reuss Model of Plastic Flow

We will first give the classical formulation of the problem without insisting on precision. Thus, in addition to the domain considered Ω, let a time interval $[0, T]$ be given; it will be seen from the formulation that we can introduce another parameter $t' \in [0, \tau_0]$ such that $t' = t'(t)$ and $dt'/dt > 0$ in $[0, T]$. Let $F(t)$ be the vector function of the body forces and $g(t)$ the vector function of the given stress vector on the boundary. We will deal with the traction boundary value problem; the reader will easily formulate and solve the other boundary value problems. Naturally, for every time moment $t \in [0, T]$ we assume the conditions of total equilibrium:

$$\int_\Omega F(t)dx + \int_{\partial\Omega} g(t)dS = 0, \tag{1.1}$$

$$\int_\Omega (x \times F(t))dx + \int_{\partial\Omega} (x \times g(t))dS = 0. \tag{1.2}$$

The stress tensor $\tau = \tau(t)$ fulfills the condition of equilibrium

$$\frac{\partial \tau_{ij}(t)}{\partial x_j} + F_i(t) = 0 \quad \forall t \in [0,T], \quad i = 1,2,3. \tag{1.3}$$

As concerns the strain tensor ϵ, we assume that it can be written as the sum of two symmetric tensors,

$$\epsilon = e + p, \tag{1.4}$$

where e is the elastic and p the plastic part.

The compatibility of deformations is expressed by the fact that there is a displacement vector u such that

$$\epsilon_{ij} = \frac{1}{2}\left(\frac{\partial u_i}{\partial x_j} + \frac{\partial u_j}{\partial x_i}\right). \tag{1.5}$$

Further, let the function of plasticity $f(\sigma)$ be given, which is assumed to fulfill $f(0) = 0$, $f(\sigma) > 0$ for $\sigma \neq 0$, and to be convex in σ. Of course, we assume that $f(\sigma)$ is invariant when replacing σ_{ij} by σ_{ji}.

Let $\alpha_0 > 0$ and assume that the solution fulfills

$$f(\sigma) \leq \alpha_0. \tag{1.6}$$

For σ and e, let us assume the linear Hooke's Law

$$e_{ij} = A_{ijkl}\sigma_{kl}, \tag{1.7}$$

with A_{ijkl} being functions only of $x \in \Omega$.

Now let us assume that the increment of p can be nonvanishing only for σ with $f(\sigma) = \alpha_0$, and that the condition of normality holds:

$$\dot{p}_{ij} = \lambda \frac{\partial f}{\partial \sigma_{ij}} \quad \text{with} \quad \lambda \geq 0. \tag{1.8}$$

The solution σ is supposed to fulfill

$$\sigma_{ij}\nu_j = g_i \quad \text{on } \partial\Omega. \tag{1.9}$$

Let us assume that σ is the solution of our problem, and that τ is another tensor which satisfies (1.3), (1.6), and (1.9). Then we can formally prove that

$$\int_\Omega A_{ijkl}\dot{\sigma}_{kl}(\tau_{ij} - \sigma_{ij})dx \geq 0. \tag{1.10}$$

Indeed, the compatibility condition implies

$$0 = \int_\Omega \dot{\epsilon}_{ij}(\tau_{ij} - \sigma_{ij})dx, \tag{1.11}$$

but then (1.8) yields

$$\dot{p}_{ij}(\tau_{ij} - \sigma_{ij}) \geq 0, \tag{1.12}$$

which together with (1.11) and (1.7) gives (1.10).

For simplicity we will assume in this section that the loading process started with zero values $F(0) = 0$ and $g(0) = 0$. Evidently, this leads to the initial condition $\sigma(0) = 0$.

In order to be able to define the weak solution, let us first introduce some auxiliary spaces. If as usual $S \in [L^2(\Omega)]^9$, $\sigma \in S \Rightarrow \sigma_{ij} = \sigma_{ji}$, let $C_0^1([0,T], S)$ be the space of continuously differentiable functions vanishing for $t = 0$ and with values in S. Similarly we define $C_0^1([0,T], [L^2(\Omega)]^3)$, $C_0^1([0,T], [L^2(\partial\Omega)]^3)$.

In $C_0^1([0,T], S)$ let us introduce the inner product

$$\int_0^T (\dot{\tau}(t), \dot{\sigma}(t))dt \tag{1.13}$$

(with $(\tau, \sigma) = \int_\Omega \tau_{ij}\sigma_{ij}dx$), and hence the norm

$$\|\tau\| = \left[\int_0^T (\dot{\tau}(t), \dot{\tau}(t))dt\right]^{1/2}, \tag{1.14}$$

and let us find the completion of C_0^1 in this norm. Thus, we obtain the space $H_0^1([0,T], S)$. In the same way, we introduce $H_0^1([0,T], [L^2(\Omega)]^3)$ and $H_0^1([0,T], [L^2(\partial\Omega)]^3)$.

In the same way as in the space of numerical functions, we immediately conclude

$$\|\tau(t_1) - \tau(t_2)\|_S \leq |t_1 - t_2|^{1/2}\|\tau\|_{H_0^1}, \tag{1.15}$$

which implies that the functions τ from H_0^1 are continuous in $[0,T]$ (with values in S); similarly for $[L^2(\Omega)]^3$, $[L^2(\partial\Omega)]^3$.

Definition 1.1. Let $F \in C_0^1([0,T], [L^2(\Omega)]^3)$, $g \in C_0^1([0,T], [L^2(\partial\Omega)]^3)$ and let conditions (1.1), (1.2) be fulfilled for all $t \in [0,T]$. We say that $\tau \in S$ fulfills (1.3) and (1.9) if

$$\int_\Omega \tau_{ij}\epsilon_{ij}(v)dx = \int_{\partial\Omega} g_i v_i dS + \int_\Omega F_i v_i dx \tag{1.16}$$

for all $v \in [W^{1,2}(\Omega)]^3$. A function $\sigma \in H_0^1([0,T],S)$ is a weak solution of an elasto-inelastic body with a perfectly plastic domain, if (1.16) is fulfilled for all $t \in [0,T]$, if

$$f(\sigma(t)) \le \alpha_0 \text{ almost everywhere in } \Omega \tag{1.17}$$

for all $t \in [0,T]$, and if for every $\tau \in H_0^1([0,T],S)$ satisfying both (1.16) and (1.17) and for every $t \in [0,T]$ the inequality

$$\int_0^t dt \int_\Omega A_{ijkl}\sigma_{kl}(\tau_{ij} - \sigma_{ij})dx \ge 0 \tag{1.17'}$$

holds.

In the following, we will assume

$$A_{ijkl} \in L^\infty(\Omega), \quad A_{ijkl} = A_{jikl} = A_{klij}, \tag{1.18}$$

$$A_{ijkl}\eta_{ij}\eta_{kl} \ge c|\eta|^2, \; c > 0, \quad \forall \eta_{ij} = \eta_{ji}, \tag{1.19}$$

$f \in C^2(\mathbf{R}^9)$ and

$$\left|\frac{\partial f}{\partial \sigma}\right| + \left|\frac{\partial^2 f}{\partial \tau \partial \sigma}\right| \le c < \infty \tag{1.20}$$

(here $\frac{\partial f}{\partial \sigma}$ is the tensor with components $\frac{\partial f}{\partial \sigma_{ij}}$ and $\frac{\partial^2 f}{\partial \sigma \partial \tau}$ is the tensor with components $\frac{\partial^2 f}{\partial \sigma_{ij} \partial \tau_{ij}}$).

3.1.1 Existence and Uniqueness of Solution

In S let us introduce the inner product

$$[\sigma, \tau] = \int_\Omega A_{ijkl}\sigma_{ij}\tau_{kl}\,dx. \tag{1.21}$$

It is easy to prove:

Theorem 1.1 (Uniqueness Theorem). *There exists at most one weak solution (according to definition* 1.1).

Proof. Let σ^1, σ^2 be two solutions. Then

$$\int_0^t [\dot\sigma^2, \sigma^1 - \sigma^2]dt \ge 0, \quad \int_0^t [\dot\sigma^1, \sigma^2 - \sigma^1]dt \ge 0,$$

hence

$$0 \ge \int_0^t [\dot\sigma^1 - \dot\sigma^2, \sigma^1 - \sigma^2]dt = \frac{1}{2}[\sigma^1(t) - \sigma^2(t), \sigma^1(t) - \sigma^2(t)]. \qquad \square$$

Theorem 1.2. *Let the assumptions from definition* 1.1 *and* (1.18)–(1.20) *be fulfilled. Let there exist* $\sigma^0 \in C_0^1([0,T],S)$ *satisfying* (1.16) *and* $f(\sigma^0(t)) + \gamma\dot{\sigma}^0(t) \le \alpha_0$ *for* $t \in [0,T]$, *where* $\gamma > 0$. *Then there exists precisely one weak solution of the first boundary value problem of the elasto-inelastic body with a perfectly plastic domain.*

Before proceeding to the proof of theorem 1.2, we will explain the idea leading to the introduction of the penalization functional. As we have already mentioned in the introduction, this idea leads to abstract differential equations which describe the regularized plasticity or, in other words, the plasticity with a yield surface, which is not "infinitely thin." This model has a physical meaning by itself and is not rate independent.

Naturally, we will go back to the formal considerations. Let us assume that if the process reaches the situation $f(\sigma(t)) = \alpha_0$, there is an increase of both the plastic deformation and the level of plasticity $f(\sigma(t))$, hence generally $f(\sigma(t)) > \alpha_0$, but of course $f(\sigma(t))$ does not go too far from the surface $f(\sigma) = \alpha_0$. This results in replacing equation (1.8) by the equation

$$\dot{p} = \frac{1}{\epsilon}[f - \alpha_0]^+ \frac{\partial f}{\partial \sigma}, \quad \epsilon > 0. \tag{1.22}$$

For purely mathematical reasons, let us replace (1.8) by a more suitable equation of the form

$$\dot{p} = \frac{1}{\epsilon}[f - \alpha_0]^+ \frac{\partial f}{\partial \sigma}(1 + ([f(\sigma) - \alpha_0]^+)^2)^{-1/2}. \tag{1.23}$$

Now let us seek σ^ϵ, p^ϵ in the way described at the beginning of this section; that is, by satisfying (1.1)–(1.5), (1.7), (1.9), and (1.23). Our mathematical optimism makes us believe that in a certain sense there exist limits for $\epsilon \to 0_+$. Since $\dot{p}^\epsilon(t) = 0$ for $f(\sigma^\epsilon(t)) < \alpha_0$, the same identity holds for the limit. Since $\dot{p}^\epsilon$ fulfills the condition of normality, the same holds for the limit. However, now the $\dot{p}^\epsilon$'s are in a sense bounded. Since $\partial f/\partial\sigma$ is by assumption a bounded tensor as well, we have

$$\frac{1}{\epsilon}\frac{[f(\sigma^\epsilon(t)) - \alpha_0]^+}{[1 + ([f(\sigma) - \alpha_0]^+)^2]^{1/2}} \le c < \infty,$$

hence, for $\epsilon \to 0$ we have $f(\sigma(t)) \le \alpha_0$. It turns out that our optimism has not deceived us. In the paper by Nečas and Trávníček (1980) a slightly simpler proof of existence of solution is given, which coincides with our course of proof provided the matrix A in relation (1.7) is an identity matrix.

Proof of Theorem 1.2. Let us introduce the penalization functional

$$g(\sigma) = \int_\Omega [\{([f(\sigma) - \alpha_0]^+)^2 + 1\}^{1/2} - 1]dx. \tag{1.24}$$

We find that $g(\sigma)$ is Gâteaux differentiable; that is,

$$\frac{d}{d\lambda}g(\sigma+\lambda\tau)|_{\lambda=0} = Dg(\sigma,\tau) = \int_\Omega \frac{[f(\sigma)-\alpha_0]^+\partial f/\partial\sigma_{ij}}{[1+([f(\sigma)-\alpha_0]^+)^2]^{1/2}}\tau_{ij}dx. \quad (1.24')$$

Hence, we obtain monotonicity of $Dg(\sigma,\cdot)$:

$$Dg(\sigma,\sigma-\tau) - Dg(\tau,\sigma-\tau) \geq 0, \quad (1.25)$$

and

$$|Dg(\sigma,\tau)| \leq c\|\sigma\|_S\|\tau\|_S. \quad (1.26)$$

Let $\epsilon > 0$ and let us look for $\sigma^\epsilon, p^\epsilon \in C_0^1([0,T],S)$ such that

$$e^\epsilon = A\sigma^\epsilon \quad (\text{componentwise: } e^\epsilon_{ij} = A_{ijkl}\sigma^\epsilon_{kl}), \quad (1.27)$$

$$\dot{p}^\epsilon = \frac{1}{\epsilon}\frac{[f(\sigma)-\alpha_0]^+\partial f(\sigma)/\partial\sigma}{[1+([f(\sigma)-\alpha_0]^+)^2]^{1/2}} \overset{\text{df}}{=} \lambda^\epsilon(\sigma)\frac{\partial f(\sigma)}{\partial\sigma}, \quad (1.28)$$

$$\epsilon^\epsilon = e^\epsilon + p^\epsilon \quad \text{is the compatible strain tensor,} \quad (1.29)$$

$$\sigma^\epsilon \quad \text{fulfills (1.16).} \quad (1.30)$$

Problem (1.27)–(1.30) will be transformed to the initial problem for an abstract differential equation. To this end, let $S_0 \subset S$ be the orthogonal complement to the space

$$E = \{\epsilon \in S;\ \epsilon_{ij} = \frac{1}{2}\left(\frac{\partial u_i}{\partial x_j} + \frac{\partial u_j}{\partial x_i}\right),\ u \in [W^{1,2}(\Omega)]^3\},$$

Korn's inequality (see Chapter 2.2) implies that E is closed. Hence, the condition of compatibility of the tensor ϵ is equivalent to condition

$$(\epsilon,\omega) = 0 \quad \forall\omega \in S_0. \quad (1.31)$$

Now (1.16) allows S_0 to be interpreted as the subspace of those tensors from S which satisfy the condition of equilibrium (1.3) with $\mathcal{F} \equiv 0$ and conditions (1.9) with $g \equiv 0$.

Further, let us denote by P the orthogonal projector of S to S_0. Conditions (1.27), (1.29), (1.31) yield

$$P(A\dot{\sigma}^\epsilon + \dot{p}^\epsilon) = 0, \quad (1.32)$$

hence, (1.28) implies

$$PA\dot{\sigma}^\epsilon = -P\left[\lambda^\epsilon(\sigma)\frac{\partial f}{\partial\sigma}\right]. \quad (1.33)$$

Now let us look for $\sigma^\epsilon = \sigma^0 + \tilde{\sigma}^\epsilon$, $\tilde{\sigma}^\epsilon \in S_0$, and put $PA\tilde{\sigma}^\epsilon = a^\epsilon$, $PA \overset{df}{=} B$. However, the operator B has an inverse in S_0, since for $\tilde{\tau} \in S_0$ we have

$$(B\tilde{\tau}, \tilde{\tau}) = (PA\tilde{\tau}, \tilde{\tau}) = (A\tilde{\tau}, \tilde{\tau}) \geq c\|\tilde{\tau}\|_S^2. \tag{1.34}$$

Thus, we may apply the Lax–Milgram Theorem[1] (see Rektorys (1974)).

Hence, we can transcribe (1.33) in the form

$$\dot{a}^\epsilon = -B\dot{\sigma}^0 - P\left[\lambda^\epsilon(\sigma^0 + B^{-1}a^\epsilon)\frac{\partial f}{\partial \sigma}(\sigma^0 + B^{-1}a^\epsilon)\right]. \tag{1.35}$$

Thus, our problem is transformed to that of solving the equation (1.35) in S_0 with the initial condition $a^\epsilon(0) = 0$. From (1.20) we find that the right-hand side of (1.35) is Lipschitzian in the variable a^ϵ, hence, there is a unique solution of (1.35) and by substituting into (1.27), (1.28), and (1.29) we determine the tensors e^ϵ, p^ϵ, ϵ^ϵ.

Now let $\tilde{\tau} \in C([0,T], S_0)$. From (1.27), (1.28), (1.24′), (1.31) we obtain

$$\int_0^t [\dot{\sigma}^\epsilon, \tilde{\tau}]dt + \frac{1}{\epsilon}\int_0^t Dg(\sigma^\epsilon, \tilde{\tau})dt = 0. \tag{1.36}$$

Put $\tilde{\tau} = \gamma(\dot{\sigma}^\epsilon - \dot{\sigma}^0) + \sigma^\epsilon - \sigma^0$. We have

$$\int_0^t Dg(\sigma^\epsilon, \gamma(\dot{\epsilon} - \dot{\sigma}^0) + \sigma^\epsilon - \sigma^0)dt = \gamma g(\sigma^\epsilon(t))$$

$$+ \int_0^t [Dg(\sigma^\epsilon, \sigma^\epsilon - (\gamma\dot{\sigma}^0 + \sigma^0)) - Dg(\gamma\dot{\sigma}^0 + \sigma^0, \sigma^\epsilon - (\gamma\dot{\sigma}^0 + \sigma^0))]dt; \tag{1.37}$$

we have used the condition $f(\sigma^0(t) + \gamma\dot{\sigma}^0(t)) \leq \alpha_0$, which implies $Dg(\gamma\dot{\sigma}^0 + \sigma^0, \sigma^\epsilon - (\gamma\dot{\sigma}^0 + \sigma^0)) = 0$. Hence, the monotonicity condition (1.25) yields

$$\int_0^t Dg(\sigma^\epsilon, \gamma(\dot{\sigma}^\epsilon - \dot{\sigma}^0) + \sigma^\epsilon - \sigma^0)dt \geq \gamma g(\sigma^\epsilon(t)). \tag{1.38}$$

From (1.36) and (1.38) we obtain

$$\gamma\int_0^t [\dot{\sigma}^\epsilon, \dot{\sigma}^\epsilon]dt + \frac{\gamma}{\epsilon}g(\sigma^\epsilon(t)) \leq c\int_0^t ([\dot{\sigma}^0, \dot{\sigma}^0] + [\sigma^0, \sigma^0])dt. \tag{1.39}$$

Let $\epsilon_n \to 0$ be chosen so that $\sigma^{\epsilon_n} \rightharpoonup \sigma$ (weakly) in $H_0^1([0,T], S)$. Since

$$(\sigma^{\epsilon_n}(t), \tau) = \int_0^t (\dot{\sigma}^{\epsilon_n}, \tau)dt$$

[1] If $B = B^*$, then the well-known Riesz Theorem for the inner product $(B\tilde{\sigma}, \tilde{\tau})$ in S_0 applies; the Lax–Milgram theorem is its immediate generalization to nonsymmetric operators.

for all $\tau \in S$, we have $\sigma^{\epsilon_n}(t) \rightharpoonup \sigma(t)$ for all $t \in [0,T]$. Now (1.39) yields

$$g(\sigma^{\epsilon_n}(t)) \le c_1 \epsilon_n, \tag{1.40}$$

and since the functional $g(\sigma)$ is weakly lower semicontinuous by virtue of condition (1.25), we conclude

$$g(\sigma(t)) \le \liminf_{n\to\infty} g(\sigma^{\epsilon_n}(t)) = 0. \tag{1.41}$$

Hence, $f(\sigma(t)) \le \alpha_0$.

Now let $\tau \in H_0^1([0,T], S)$ fulfill (1.16) and (1.17). For $\tilde{\tau} = \sigma^{\epsilon_n} - \tau$ we obtain from (1.36):

$$0 = \int_0^t [\dot{\sigma}^{\epsilon_n}, \sigma^{\epsilon_n} - \tau] dt + \frac{1}{\epsilon} \int_0^t [Dg(\sigma^{\epsilon_n}, \sigma^{\epsilon_n} - \tau)$$

$$- Dg(\tau, \sigma^{\epsilon_n} - \tau)] dt \ge \int_0^t [\dot{\sigma}^{\epsilon_n}, \sigma^{\epsilon_n} - \tau] dt. \tag{1.42}$$

However, then

$$0 \le \limsup_{n\to\infty} \int_0^t [\dot{\sigma}^{\epsilon_n}, \tau - \sigma^{\epsilon_n}] dt = \int_0^t [\dot{\sigma}, \tau] dt$$

$$- \frac{1}{2} \liminf_{n\to\infty} [\sigma^{\epsilon_n}(t), \sigma^{\epsilon_n}(t)] \le \int_0^t [\dot{\sigma}, \tau - \sigma] dt. \quad \square \tag{1.43}$$

Let us note that the yield surface

$$f_\epsilon(\sigma) \overset{\text{df}}{=} [\epsilon^2 + \sigma_{ij}^D \sigma_{ij}^D]^{1/2} - \epsilon, \quad \epsilon > 0,$$

$$\sigma_{ij}^D = \sigma_{ij} - \frac{1}{3} \delta_{ij} \sigma_{kk},$$

fulfills the conditions of theorem 1.2. For $f(\sigma) = [\sigma_{ij}^D \sigma_{ij}^D]^{1/2}$ we additionally have to prove the possibility of the limiting process $\epsilon \to 0$.

3.1.2 Solution by Finite Elements

We shall show how to solve variational evolutionary inequalities of the type (1.10). However, we will restrict ourselves to the case when the body occupies a polyhedral domain $\Omega \subset \mathbf{R}^n$, $n = 2, 3$, and to the displacement boundary value problem ($\partial\Omega = \Gamma_u$). We will follow C. Johnson (1976b), who also proved existence and uniqueness of the solution of this problem.

Let $\mathbf{R}_\sigma$ stand for the space of symmetric matrices of the type $(n \times n)$ (the stress tensors). Let us assume that we are given the function of plasticity

$f : \mathbf{R}_\sigma \to \mathbf{R}$, which is convex and continuous in $\mathbf{R}_\sigma$, and a constant $\alpha_0 > 0$. Denote

$$B = \{\tau \in \mathbf{R}_\sigma, \quad f(\tau) \le \alpha_0\}.$$

Let $f(0) = 0$. We introduce the set of plastically admissible stress fields

$$P = \{\tau \in S \mid \tau(x) \in B \quad \text{a.e. in } \Omega\}.$$

The body forces let be given in the form

$$F(t, x) = \gamma(t) F^0(x),$$

where $F^0 \in [C(\bar{\Omega})]^n$ and $\gamma \in C^2(I)$ is a nonnegative function on the interval $I = [0, T]$, $\gamma(0) = 0$.

We introduce the set of statically admissible stress fields

$$E(t) = \left\{\tau \in S \mid \int_\Omega \tau_{ij}\epsilon_{ij}(v)dx = \int_\Omega F_i(t)v_i dx \quad \forall \in [W_0^{1,2}(\Omega)]^n\right\}$$

and put

$$K(t) = E(t) \cap P.$$

Analogously to definition 1.1, we say that $\sigma \in H_0^1(I, S)$ is a weak solution of the elasto-inelastic body with a perfectly plastic domain if for almost all $t \in I$ we have

$$\sigma(t) \in K(t),$$

$$[\dot{\sigma}(t), \tau - \sigma(t)] \ge 0 \quad \forall \tau \in K(t). \tag{1.44}$$

Let us recall that this definition differs from definition 1.1 in that (1.17′) results only by integrating (1.44) with respect to time. Inequality (1.44) immediately corresponds to (1.10). Zero displacement is given on the boundary $\partial\Omega$.

C. Johnson proved existence and uniqueness of the weak solution of problem (1.44) (see C. Johnson (1976a)), such that $\sigma \in L^\infty(I, S)$ and $\dot{\sigma} \in L^2(I, S)$, under the following assumption:

$$\text{there is } \bar{\chi} \in K(\bar{t}), \text{ where } |\gamma(\bar{t})| = \max_{t \in I} |\gamma(t)|, \tag{1.45}$$

$$\text{and positive constants } C, \delta \text{ such that } |\bar{\chi}(x)| \le C$$

$$\text{for almost all } x \in \Omega \text{ and } \pm(1 + \delta)\bar{\chi} \in P.$$

With the aim of defining approximate solutions of problem (1.44), we introduce finite-dimensional internal approximations of the set $E(t)$,

$$E_h(t) \subset E(t) \quad \forall t \subset I, \quad 0 < h < h_0.$$

The sets $E_h(t)$ can be constructed in $\mathbf{R}^2$ as sums of a particular solution $\chi(t)$ of the equations of equilibrium and the subsets $E_h^0 = N_h(\Omega)$, which were introduced in section 2.4.11. Then the parameter h is the maximal side of the used triangulation $\mathcal{T}_h$. Here we deal with block triangular finite elements and a piecewise linear stress field.

The closed convex set B can be approximated as well, for instance, by the sets

$$B_h = \{\tau \in S \mid f_{i,h}(\tau) \leq \alpha_0, \quad i = 1, \ldots, M_h\},$$

with $f_{i,h}$ are certain linear functions such that $B_h \subset B$. Then

$$P_h = \{\tau \in S \mid \tau(x) \in B_h \ \text{ a.e. in } \Omega\}$$

as an approximation of the set P.

If we now define

$$K_h(t) = E_h(t) \cap P_h,$$

then $K_h \subset K(t)$ for all $t \in I$.

We introduce the discretization of the time interval I. Let N be a positive integer, $k = T/N$, $t_m = mk$, $m = 0, 1, \ldots, N$, $I_m = [t_{m-1}, t_m]$, $\tau^m = \tau(t_m)$, $\partial\tau^m = (\tau^m - \tau^{m-1})/k$.

Instead of the variational inequality (1.44) we shall solve the discrete problem: find $\sigma_{hk}^m \in K_h(t_m)$, $m = 1, \ldots, N$, such that

$$[\partial\sigma_{hk}^m, \tau - \sigma_{hk}^m] \geq 0 \quad \forall \tau \in K_h(t_m), \quad m = 1, \ldots, N, \quad \sigma_{hk}^0 = 0. \tag{1.46}$$

Let us assume that the following analog of assumption (1.45) holds:

There exists $\chi_h \in K_h(\bar{t})$ and positive constants C, δ independent of

h and such that $|\chi_h(x)| \leq C$ for almost all

$$x \in \Omega \text{ and } \pm(1+\delta)\chi_h \in P_h. \tag{1.47}$$

Problem (1.46) is uniquely solvable. This follows from the fact that for all m, σ_{hk}^m is the element which minimizes the strictly convex quadratic functional

$$\frac{1}{2}[\tau, \tau] - [\sigma_{hk}^m, \tau]$$

on the closed convex set $K_h(t_m)$.

Thus, on each time level t_m we have to solve the quadratic programming problem.

3.1.21. A Priori Error Estimates. Let us assume that we have obtained the exact solution σ_{hk}^m of problem (1.46) and let us estimate the error $\sigma^m -$

σ_{hk}^m, where $\sigma^m = \sigma(t_m)$ is the exact solution of the original problem (1.44). To this end it is useful to assume that the partition of the interval I is ordered (possibly nonuniformly) in such a way that $\gamma(t)$ is a monotonic function in each subinterval I_m. However, since the proof of the a priori error estimate changes only inessentially, we will, in the sequel, consider an equidistant partition for the sake of simplicity.

First, for $q = (q^1, \ldots, q^N)$, $q^m \in S$, let us define

$$\|q\|_{l^2(S)} = \left(\sum_{m=1}^{N} k \|q^m\|_S^2 \right)^{1/2} .$$

Lemma 1.1. *If* (1.47) *is valid, then there exist positive constants* C *and* k_0 *such that*

$$\|\partial \sigma_{hk}\|_{l^2(S)} \leq C \tag{1.48}$$

for $h < h_0$, $k < k_0$.

Proof. Analogous to that of lemma 2 in the paper by C. Johnson (1976a) (see also Hlaváček (1980) or 3.2.21). Here we omit the details. □

Let us define

$$\epsilon(h,k) = \inf_{\tau \in \mathcal{K}} \|\sigma - \tau\|_{l^2(S)},$$

where

$$\mathcal{K} = \{\tau = (\tau^1, \ldots, \tau^N) \mid \tau^m \in K_h(t_m), \quad m = 1, \ldots, N\}.$$

The quantity $\epsilon(h,k)$ is actually determined by the approximation properties of the sets $E_h(t)$, P_h and by the regularity of the solution σ, provided we have any information at all about the latter.

Theorem 1.3. *Let assumptions* (1.45), (1.47) *be fulfilled and let* σ, σ_{hk} *be solutions of problems* (1.44), (1.46), *respectively. Then for* k *sufficiently small and* $h < h_0$ *we have*

$$\max_m \|\sigma^m - \sigma_{hk}^m\|_S \leq C(\epsilon^{1/2}(h,k) + k^{1/2}). \tag{1.49}$$

Proof. First of all, we have

$$[\dot{\sigma}, \tau - \sigma] \geq 0 \quad \forall \tau \in K(t), \quad \text{a.e. } t \in I, \tag{1.50}$$

$$[\partial \sigma_{hk}^m, \tau - \sigma_{hk}^m] \geq 0 \quad \forall \tau \in K_h(t_m), \quad m = 1, \ldots, N. \tag{1.51}$$

We extend σ_{hk} to the whole interval I as follows:

$$\sigma_{hk}(t) = \lambda(t)\sigma_{hk}^{m-1} + (1 - \lambda(t))\sigma_{hk}^m,$$

where

$$\lambda(t) = \frac{\gamma(t) - \gamma(t_m)}{\gamma(t_{m-1}) - \gamma(t_m)}, \quad t \in I_m \text{ provided } \gamma(t_m) \neq \gamma(t_{m-1}),$$

$$\lambda(t) = (t_m - t)/k, \quad t \in I_m \text{ provided } \gamma(t_m) = \gamma(t_{m-1}).$$

Then we easily check that $\sigma_{hk}(t) \in E(t)$ for all $t \in I$, taking into account the monotonicity of the function γ in each subinterval I_m. The last property also implies that $0 \leq \lambda(t) \leq 1$. Since P is convex, we also have $\sigma_{hk}(t) \in P$ for all $t \in I$. Summarizing, we conclude that $\sigma_{hk}(t) \in K(t)$ for all $t \in I$.

Now, putting $\tau = \sigma_{hk}$ in (1.50), we obtain

$$[\dot{\sigma}, \sigma_{hk} - \sigma] \geq 0 \quad \text{a.e. in } I.$$

Integrating over I_m we arrive at the inequality

$$[\partial\sigma^m, \sigma_{hk}^m - \sigma^m] \geq \frac{1}{k} \int_{I_m} [\dot{\sigma}, \sigma_{hk}^m - \sigma^m + \sigma(t) - \sigma_{hk}(t)] dt. \tag{1.52}$$

Let us consider $\tau_h \in K$ such that

$$\|\sigma - \tau_h\|_{l^2(S)} \leq 2\epsilon(h, k)$$

and substitute $\tau = \tau_h^m$ in (1.51). Thus, we obtain

$$[\partial\sigma_{hk}^m, \tau_h^m - \sigma_{hk}^m] \geq 0. \tag{1.53}$$

By virtue of (1.52), (1.53) we come to the following inequality for the error $e = \sigma - \sigma_{hk}$:

$$[\partial e^m, e^m] \leq [\partial\sigma_{hk}^m, \tau_h^m - \sigma^m] + |r_m|,$$

where r_m is the right-hand side of (1.52). Multiplying this inequality by k and summing over m yields:

$$\max_m \|e^m\|_S^2 \leq C\|\partial\sigma_{hk}\|_{l^2(S)} \|\tau_h - \sigma\|_{l^2(S)} + 2k \sum_{m=1}^{N} |r_m|. \tag{1.54}$$

For r_m we may write

$$|r_m| \leq \frac{C}{k} \int_{I_m} \|\dot{\sigma}(t)\|_S \left(k\|\partial\sigma_{hk}^m\|_S + k^{1/2} \left(\int_{I_m} \|\dot{\sigma}\|_S^2 ds \right)^{1/2} \right) dt$$

$$\leq Ck^{1/2}\|\partial\sigma_{hk}^m\|_S \left(\int_{I_m} \|\dot{\sigma}\|_S^2 dt \right)^{1/2} + C \int_{I_m} \|\dot{\sigma}(t)\|_S^2 dt$$

$$\leq C \int_{I_m} \|\dot{\sigma}\|_S^2 dt + Ck\|\partial\sigma_{hk}^m\|_S^2.$$

Let us substitute into (1.54). Thus, we obtain the estimate

$$\max_m \|e^m\|_S^2 \leq 2C\|\partial\sigma_{hk}\|_{l^2(S)}\epsilon(h,k)$$

$$+ Ck\left(\sum_{m=1}^{N} k\|\partial\sigma_{hk}^m\|_S^2 + \int_I \|\dot{\sigma}(t)\|_S^2 dt\right) \leq C_1(\epsilon(h,k)+k),$$

by simply taking into account the inequality

$$\int_I \|\dot{\sigma}(t)\|_S^2 dt < \infty$$

and applying lemma 1. This completes the proof of theorem 1.3. □

Remark 1.1. If $n = 2$ and the solution σ is sufficiently smooth, we can prove that $\epsilon(h,k) \leq Ch^2$ with C independent of h, k by using the approximations

$$E_h(t) = \chi(t) \oplus N_h(\Omega), \quad B_h = B.$$

Here $\chi(t)$ is a particular solution of the equations of equilibrium (1.3) and $N_h(\Omega)$ are the spaces of piecewise linear finite elements for the equilibrium model (see section 2.4.11). (In the case $n = 3$ we can use composite tetrahedrons and piecewise linear stress fields—see Křížek (1981).) The proof is a special case of that of theorem 2.4 in the next section, and hence is omitted.

3.2 Plastic Flow with Isotropic or Kinematic Hardening

We start with formal considerations as in the preceding section. Let us again assume that $\dot{p} = 0$ provided $f(\sigma) < \alpha_0$. Now we do not exclude the hardening, that is, the level of the yield surface in the course of the process increases to $f(\sigma) = \alpha > \alpha_0$, when $\dot{p}$ again fulfills the normality condition, but after another fall of stress to the level $f(\sigma) < \alpha$ we have $\dot{p} = 0$. Let

$$\alpha(t) = \max\{\alpha_0, \max_{0\leq t'\leq t} f(\sigma(t'))\} \tag{2.1}$$

be the internal state variable describing the level of the yield surface. Let us assume

$$\dot{p} = \frac{\partial f}{\partial \sigma}\dot{\alpha}. \tag{2.2}$$

This relation is general in the following sense. Let us assume that $\dot{p} = h(f(\sigma))\frac{\partial f}{\partial \sigma}\dot{\alpha}$, where $h(\lambda) > 0$ for $\lambda > 0$. If $f^*(\sigma) = H(f(\sigma))$, where

$$H(\gamma) = \int_0^\gamma h^{1/2}(\varphi)d\varphi,$$

then with α^* for $f^*(\sigma)$ and $\alpha_0^* = H(\alpha_0)$

$$\dot{p} = h(f(\sigma))\frac{\partial f}{\partial \sigma}\dot{\alpha} = \frac{\partial f^*}{\partial \sigma}\dot{\alpha}^*.$$

Putting $\mathcal{F}(\sigma, \alpha) \overset{df}{=} f(\sigma) - \alpha$ we have

$$\dot{p} = \frac{\partial \mathcal{F}}{\partial \sigma}\dot{\alpha}, \quad \dot{\alpha} = -\frac{\partial \mathcal{F}}{\partial \alpha}\dot{\alpha}, \quad \mathcal{F}(\sigma, \alpha) \leq 0. \tag{2.3}$$

The situation just described is often called the *isotropic hardening*, though for justifying such a term we have to assume that $f(\sigma)$ is a function only of the invariants of σ.

Let us now consider a simple case of the so-called *kinematic hardening*. Let β be the tensor of $n \times n$ internal state variables, which are connected with the tensor p of the plastic deformation by the relation

$$\beta = Bp, \tag{2.4}$$

where B is a regular matrix again satisfying $B_{ijkl} \in L^\infty(\Omega)$, $B_{ijkl} = B_{jikl} = B_{klij}$ and $B_{ijkl}\xi_{ij}\xi_{kl} \geq c|\xi|^2$, for $\xi_{ij} = \xi_{ji}$.

Let us seek $\sigma(t)$ such that

$$f(\sigma - \beta) \leq \alpha_0, \tag{2.5}$$

and

$$\dot{p} = \lambda\frac{\partial f}{\partial \sigma}, \quad \text{where } \lambda \geq 0. \tag{2.6}$$

Put $\gamma = B^{-1/2}\beta$ and $\mathcal{F}(\sigma, \gamma) \overset{df}{=} f(\sigma - B^{1/2}\gamma) - \alpha_0$. Again we have

$$\mathcal{F}(\sigma, \gamma) \leq 0, \quad \dot{p} = \lambda\frac{\partial \mathcal{F}}{\partial \sigma}, \quad \dot{\gamma} = -\lambda\frac{\partial \mathcal{F}}{\partial \gamma}, \quad \lambda \geq 0. \tag{2.7}$$

We can also consider the combination of both the isotropic and kinematic hardenings: in the case we additionally set

$$\alpha = \max\left\{\alpha_0, \max_{0 \leq t' \leq t} f(\sigma(t') - B^{1/2}\gamma(t'))\right\}, \tag{2.8}$$

$$\mathcal{F}(\sigma,\gamma,\alpha) \overset{\mathrm{df}}{=} f(\sigma - B^{1/2}\gamma) - \alpha$$

and

$$\dot{p} = \dot{\alpha}\frac{\partial f}{\partial \sigma}. \tag{2.9}$$

Again we have

$$\mathcal{F}(\sigma,\gamma,\alpha) \le 0, \tag{2.9'}$$

$$\dot{p} = \dot{\alpha}\frac{\partial \mathcal{F}}{\partial \sigma}, \quad \dot{\gamma} = -\dot{\alpha}\frac{\partial \mathcal{F}}{\partial \gamma}, \quad \dot{\alpha} = -\dot{\alpha}\frac{\partial \mathcal{F}}{\partial \alpha}. \tag{2.10}$$

Thus, we can see that in all cases we were given a generalized yield function $\mathcal{F}(\sigma,\gamma,\alpha)$ and that the solution fulfilled, in addition to the conditions (2.3), (2.4), (2.5), (2.7), (2.9), also (2.9′) and

$$\dot{p} = \lambda\frac{\partial \mathcal{F}}{\partial \sigma}, \quad \dot{\gamma} = -\lambda\frac{\partial \mathcal{F}}{\partial \gamma}, \quad \dot{\alpha} = -\lambda\frac{\partial \mathcal{F}}{\partial \alpha}, \tag{2.11}$$

where $\lambda \ge 0$ and $\lambda = 0$ provided $\mathcal{F}(\sigma,\gamma,\alpha) < 0$. Further, the hardening has been characterized by

$$\frac{\partial \mathcal{F}}{\partial \gamma_{ij}}\frac{\partial \mathcal{F}}{\partial \gamma_{ij}} + \frac{\partial \mathcal{F}}{\partial \alpha}\frac{\partial \mathcal{F}}{\partial \alpha} \ge c > 0. \tag{2.12}$$

We could suggest further generalizations following the ideas of Nečas, Trávníček (1978), based on the paper by Halphen and Quoc Son Nguyen (1975). However, in this book we will keep the level of generality reached above.

Let us notice another excellent feature of the equations (2.11): the direction of the vector $(\dot{p}, -\dot{\gamma}, -\dot{\alpha})$ coincides with that of the outward normal to the yield surface $\mathcal{F}(\sigma,\gamma,\sigma) = 0$. Consequently, if $\mathcal{F}(\tilde{\sigma},\tilde{\gamma},\tilde{\alpha}) \le 0$, we obtain

$$\dot{p}_{ij}(\tilde{\sigma}_{ij} - \sigma_{ij}) - \dot{\gamma}_{ij}(\tilde{\gamma}_{ij} - \gamma_{ij}) - \dot{\alpha}(\tilde{\alpha} - \alpha) \le 0, \tag{2.13}$$

which after integration over Ω yields

$$(\dot{p}, \tilde{\sigma} - \sigma) - (\dot{\gamma}, \tilde{\gamma} - \gamma) - (\dot{\alpha}, \tilde{\alpha} - \alpha) \le 0, \tag{2.14}$$

where the inner products in both S and L^2 are denoted by the same symbol. If $\tilde{\sigma}$ satisfies (1.16), then in the same way as (1.10) we obtain the fundamental inequality of plasticity

$$[\dot{\sigma}, \tilde{\sigma} - \sigma] + (\dot{\gamma}, \tilde{\gamma} - \gamma) + (\dot{\alpha}, \tilde{\alpha} - \alpha) \ge 0. \tag{2.15}$$

3.2.1 Existence and Uniqueness of Solution of the Plastic Flow Problem with Hardening

First of all, we call the reader's attention to the fact that the method described in this section actually coincides with that presented in section 3.1, and hence also the Prandtl–Reuss model of plastic flow can be solved in the general setting. However, the hardening condition, represented by inequality (2.12), makes it moreover possible to reversely determine λ from (2.11), thus going back to the classical formulation, and also to find the tensor of plastic deformation p (that of elastic deformation e is given by the relation (1.7)), hence $\epsilon = e + p$. Consequently, in this case we determine the strain tensor as well. Moreover, in the case of isotropic hardening we obtain the relation

$$\alpha = \max\left\{\alpha_0, \max_{0\le t'\le t} f(\sigma(t'))\right\}. \tag{2.16}$$

Since we will introduce another easy generalization in this section, which will consist of the fact that generally $\sigma(0) \neq 0$, we will consider the spaces $H^1([0,T],S)$, the closure of $C^1([0,T],S)$ in the norm

$$\left(\int_0^T (\|\tau\|_S^2 + \|\dot{\tau}\|_S^2)dt\right)^{1/2}. \tag{2.17}$$

The space $H^1([0,T],S)$ is obviously a Hilbert space with the inner product

$$\int_0^T [(\tau,\sigma) + (\dot{\tau},\dot{\sigma})]dt. \tag{2.18}$$

Definition 2.1. Let $F \in C^1([0,T],[L^2(\Omega)]^3)$, $g \in C^1([0,T],[L^2(\partial\Omega)]^3)$ and let conditions (1.1), (1.2) be fulfilled for all $t \in [0,T]$. We call $\sigma \in H^1((0,T),S)$, $\gamma \in H^1((0,T),S)$, $\alpha \in H^1((0,T),L^2(\Omega))$, a weak solution of the first problem for an elasto-inelastic body with internal state variables, if $\sigma(t)$ satisfies (1.16) for all $t \in [0,T]$, $\sigma(0) = \sigma_0$ (σ_0 satisfies (1.16)), $\gamma(0) = \gamma_0$, $\alpha(0) = \alpha_0$, and

$$\mathcal{F}(\sigma_0,\gamma_0,\alpha_0) \le 0, \tag{2.19}$$

where $\mathcal{F}(\sigma,\gamma,\alpha)$ is the *generalized yield function,* and for all $t \in [0,T]$ we have

$$\mathcal{F}(\sigma(t),\gamma(t),\alpha(t)) \le 0. \tag{2.20}$$

Let $\tau \in H^1((0,T),S)$, $\delta \in H^1((0,T),S)$, $\beta \in H^1((0,T),L^2(\Omega))$, let $\tau(t)$ satisfy the condition (1.16) for all $t \in [0,T]$. Further, let

$$\mathcal{F}(\tau(t),\delta(t),\beta(t)) \le 0 \tag{2.21}$$

holds for all $t \in [0,T]$. Then the *variational inequality of evolution of the flow theory*

$$\int_0^t [\dot{\sigma}, \tau - \sigma] dt + \int_0^t (\dot{\gamma}, \delta - \gamma) dt + \int_0^t (\dot{\alpha}, \beta - \alpha) dt \geq 0 \tag{2.22}$$

must be fulfilled for all $t \in [0,T]$.

We will assume that $\mathcal{F}$ is twice continuously differentiable with respect to its arguments, that it is convex and all its second derivatives are bounded. Further, we will asume that if $\tau \in C([0,T],S)$, $\gamma \in C([0,T],S)$, $\alpha \in C([0,T],L^2(\Omega))$, then there exist $\delta \in C([0,T],S)$, $\beta \in C([0,T],L^2(\Omega))$ such that $\mathcal{F}(\tau(t),\gamma(t)+\delta(t),\alpha(t)+\beta(t)) \leq 0$ for all $t \in [0,T]$.

Theorem 2.1. *Let us consider the problem of an elasto-inelastic body with interval state variables according to definition 2.1. Let the function $\mathcal{F}(\sigma,\gamma,\alpha)$ satisfy the above formulated assumptions. Then there exists a unique solution of the problem.*

Proof. The uniqueness if proved as easily as in theorem 1.1. The existence will be proved again by the penalization method introduced in section 3.1. Let us seek continuously differentiable functions with values in S (or in $L^2(\Omega)$, respectively): σ^ϵ, γ^ϵ, α^ϵ, e^ϵ, p^ϵ, fulfilling for all $t \in [0,T]$ the following conditions:

$$\sigma^\epsilon \text{ fulfills (1.16)}, \tag{2.23}$$

$$e^\epsilon = A\sigma^\epsilon. \tag{2.24}$$

Let us further introduce the penalization functional

$$g(\sigma,\gamma,\alpha) = \int_\Omega (\{[(\mathcal{F}(\sigma,\gamma,\alpha))^+]^2 + 1\}^{1/2} - 1) dx, \tag{2.25}$$

and let us use the symbols $D_\sigma g$, $D_\gamma g$, $D_\alpha g$ to denote its partial derivatives. Let us also assume that

$$\dot{p}^\epsilon = \frac{1}{\epsilon} D_\sigma g, \quad p^\epsilon(0) = p_0 \in S, \tag{2.26}$$

$$\dot{\gamma}^\epsilon = -\frac{1}{\epsilon} D_\gamma g, \quad \gamma^\epsilon(0) = \gamma_0, \tag{2.27}$$

$$\dot{\alpha}^\epsilon = -\frac{1}{\epsilon} D_\alpha g, \quad \alpha^\epsilon(0) = \alpha_0, \tag{2.28}$$

$$\sigma^\epsilon(0) = \sigma_0, \tag{2.29}$$

$$\text{the tensor } \epsilon = e + p \text{ is compatible.} \tag{2.30}$$

Let P be the projector from the proof of theorem 1.2. Then we have

$$P(A\dot{\sigma}^\epsilon + \dot{p}^\epsilon) = 0. \tag{2.31}$$

Now let $\sigma^0 \in C^1([0,T],S)$ be a tensor such that $\sigma^0(t)$ satisfies (1.16), $\sigma^0(0) = \sigma_0$, and let us look for $\sigma^\epsilon(t) = \sigma^0(t) + \tilde{\sigma}^\epsilon(t)$, $\tilde{\sigma}^\epsilon(t) \in S_0$ for all $t \in [0,T]$. As in the proof of theorem 1.2, put

$$PA\sigma^\epsilon = a^\epsilon. \tag{2.32}$$

Hence, (2.26)–(2.30) imply the conditions

$$\dot{a}^\epsilon = -PA\dot{\sigma}^0 - \frac{1}{\epsilon}D_\sigma\gamma, \quad a^\epsilon(0) = 0, \tag{2.33}$$

where $D_\sigma g = D_\sigma g(\sigma^0 + (PA)^{-1}a^\epsilon, \gamma, \alpha)$,

$$\dot{\gamma}^\epsilon = -\frac{1}{\epsilon}D_\gamma g, \quad \gamma^\epsilon(0) = \gamma_0, \tag{2.34}$$

$$\dot{\alpha}^\epsilon = -\frac{1}{\epsilon}D_\alpha g, \quad \alpha^\epsilon(0) = \alpha_0. \tag{2.35}$$

The assumptions on the function $\mathcal{F}(\sigma,\gamma,\alpha)$ imply that $D_\sigma g$, $D_\gamma g$, $D_\alpha g$ are Lipschitzian, and hence there exists a unique solution in the space $C^1([0,T],S) \times C^1([0,T],S) \times C^1([0,T],L^2(\Omega))$. Now put $\tau = \dot{\sigma}^\epsilon + \sigma^\epsilon - \dot{\sigma}^0 - \sigma^0$, $\gamma = \dot{\gamma}^\epsilon + \gamma^\epsilon - \delta$, $\alpha = \dot{\alpha}^\epsilon + \alpha^\epsilon - \beta$ with $\delta \in C([0,T],S)$, $\beta \in C([0,T],L^2(\Omega))$ and $\mathcal{F}(\dot{\sigma}^0(t) + \sigma^0(t), \delta(t), \beta(t)) \leq 0$ for all $t \in [0,T]$. Thus we can determine $p \in C^1([0,T],S)$ from (2.26). Now (2.27)–(2.28) and (2.31) imply

$$\int_0^t [\dot{\sigma}^\epsilon, \tau]dt + \int_0^t (\dot{\gamma}^\epsilon, \gamma)dt + \int_0^t (\dot{\alpha}^\epsilon, \alpha)dt$$
$$+ \frac{1}{\epsilon}\int_0^t [(D_\sigma g, \tau) + (D_\gamma g, \gamma) + (D_\alpha g, \alpha)]dt = 0. \tag{2.36}$$

Again we have

$$\int_0^t [(D_\sigma g, \tau) + (D_\gamma g, \gamma) + (D_\alpha g, \alpha)]dt$$
$$= g(\sigma^\epsilon(t), \gamma^\epsilon(t), \alpha^\epsilon(t)) - g(\sigma_0, \gamma_0, \alpha_0)$$
$$+ \int_0^t [(D_\sigma g, \sigma^\epsilon - \dot{\sigma}^0 - \sigma^0) + (D_\gamma g, \gamma^\epsilon - \delta) + (D_\alpha g, \alpha^\epsilon - \beta)]dt$$
$$= g(\sigma^\epsilon(t), \gamma^\epsilon(t), \alpha^\epsilon(t)) + \int_0^t [(D_\sigma g(\sigma^\epsilon, \gamma^\epsilon, \alpha^\epsilon), \sigma^\epsilon - \dot{\sigma}^0 - \sigma^0)$$

$$+ (D_\gamma g(\sigma^\epsilon, \gamma^\epsilon, \alpha^\epsilon), \gamma^\epsilon - \delta) + (D_\alpha g(\sigma^\epsilon, \gamma^\epsilon, \alpha^\epsilon), \alpha^\epsilon - \beta)$$
$$- (D_\sigma g(\dot\sigma^0 + \sigma^0, \delta, \beta), \sigma^\epsilon - \dot\sigma^0 - \sigma^0) - (D_\gamma g(\dot\sigma^0, \delta, \beta), \gamma^\epsilon - \delta)$$
$$- (D_\alpha g(\dot\sigma^0 + \sigma^0, \delta, \beta), \alpha^\epsilon - \beta)]dt \geq g(\sigma^\epsilon(t), \gamma^\epsilon(t), \alpha^\epsilon(t)). \tag{2.37}$$

Here we have used the conditions $g(\sigma_0, \gamma_0, \alpha_0) = 0$ and $F(\dot\sigma^0 + \sigma^0, \delta, \beta) \leq 0$, which yield $Dg(\dot\sigma^0 + \sigma^0, \delta, \beta) = 0$. Hence, we obtain from (2.36) and (2.37)

$$\int_0^t [\dot\sigma^\epsilon, \dot\sigma^\epsilon]dt + [\sigma^\epsilon(t), \sigma^\epsilon(t)] + \int_0^t (\dot\gamma^\epsilon, \dot\gamma^\epsilon)dt$$
$$+ (\gamma^\epsilon(t), \gamma^\epsilon(t)) + \int_0^t (\dot\alpha^\epsilon, \dot\alpha^\epsilon)dt + (\alpha^\epsilon(t), \alpha^\epsilon(t))$$
$$+ \frac{1}{\epsilon} g(\sigma^\epsilon(t), \gamma^\epsilon(t), \alpha^\epsilon(t)) \leq C_1. \tag{2.38}$$

Let $\epsilon_n \to 0$, $\epsilon_n > 0$ be a sequence chosen so that $\epsilon^{\epsilon_n} \rightharpoonup \sigma$ in $H^1((0,T), S)$, $\gamma^{\epsilon_n} \rightharpoonup \gamma$ in $H^1((0,T), S)$, $\alpha^{\epsilon_n} \rightharpoonup \alpha$ in $H^1((0,T), L^2(\Omega))$; we again have $\sigma^{\epsilon_n}(t) \rightharpoonup \sigma(t)$, $\gamma^{\epsilon_n}(t) \rightharpoonup \gamma(t)$, $\alpha^{\epsilon_n}(t) \rightharpoonup \alpha(t)$ for all $t \in [0, T]$.

Now (2.38) implies

$$g(\sigma^{\epsilon_n}(t), \gamma^{\epsilon_n}(t), \alpha^{\epsilon_n}(t)) \leq c_1 \epsilon_n, \tag{2.38'}$$

hence again,

$$g(\sigma(t), \gamma(t), \alpha(t)) \leq 0 \Rightarrow F(\sigma(t), \gamma(t), \alpha(t)) \leq 0. \tag{2.39}$$

Let τ, δ, β be a triplet of elements from definition 2.1. Then similar to (1.42), (2.36) yields

$$0 \geq \int_0^t [\dot\sigma^{\epsilon_n}, \sigma^{\epsilon_n} - \tau]dt + \int_0^t (\dot\gamma^{\epsilon_n}, \gamma^{\epsilon_n} - \delta)dt + \int_0^t (\dot\alpha^{\epsilon_n}, \alpha^{\epsilon_n} - \beta)dt, \tag{2.40}$$

and the proof can be completed as previously. □

Theorem 2.2. *Let the assumptions of theorem 2.1 be fulfilled together with condition (2.12). Let σ, γ, α be the solution of the problem. Put*

$$\lambda = \lambda(t, x) = 0 \text{ provided } \mathcal{F}(\sigma(t,x), \gamma(t,x), \alpha(t,x)) < 0 \tag{2.41}$$

(for almost all (t, x) from $(0,T) \times \Omega$),

$$\lambda = \frac{\frac{\partial F}{\partial \sigma}\dot\sigma}{\frac{\partial F}{\partial \gamma}\frac{\partial F}{\partial \gamma} + \frac{\partial F}{\partial \alpha}\frac{\partial F}{\partial \alpha}} \text{ provided}$$

$$\mathcal{F}(\sigma(t,x), \gamma(t,x), \alpha(t,x)) = 0 \tag{2.42}$$

(for almost all (t,x) from $(0,T) \times \Omega$) and solve the equation

$$\dot{p} = \lambda \frac{\partial F}{\partial \sigma}, \quad p(0) = p_0 \tag{2.43}$$

in $H^1((0,T),S)$. Then

$$\dot{\gamma} = -\lambda \frac{\partial F}{\partial \gamma}, \quad \gamma(0) = \gamma_0 \tag{2.44}$$

(in the sense of $H^1((0,T),S)$),

$$\dot{\alpha} = -\lambda \frac{\partial F}{\partial \alpha}, \quad \alpha(0) = \alpha_0 \tag{2.45}$$

(in the sense of $H^1((0,T),L^2(\Omega))$,

$$e(t) + p(t) \text{ is a compatible tensor.} \tag{2.46}$$

Moreover, if $\epsilon_n \to 0$, $\epsilon_n > 0$, then $\sigma^{\epsilon_n} \rightharpoonup \sigma$, $\gamma^{\epsilon_n} \rightharpoonup \gamma$, $\alpha^{\epsilon_n} \rightharpoonup \alpha$, $p^{\epsilon_n} \rightharpoonup p$ (by the proof of theorem (2.1); that is, they converge weakly in the spaces $H^1((0,T),S)$, $H^1((0,T),S)$, $H^1((0,T),L^2(\Omega))$, $H^1((0,T),S)$, respectively. Denote

$$\lambda_n = \frac{1}{\epsilon_n} F(\sigma^{\epsilon_n}, \gamma^{\epsilon_n}, \alpha^{\epsilon_n}) + [1 + (F(\sigma^{\epsilon_n}, \gamma^{\epsilon_n}, \alpha^{\epsilon_n})^+)^2]^{-1/2}. \tag{2.47}$$

Then $\lambda_n \rightharpoonup \lambda$ in $L^2((0,T),L^2(\Omega))$ $(= L^2((0,T) \times \Omega))$, and hence, $\lambda \geq 0$. Further, $\sigma^{\epsilon_n} \to \sigma$ in $C([0,T],S)$, $\gamma^{\epsilon_n} \to \gamma$ in $C([0,T],S)$, $\alpha^{\epsilon_n} \to \alpha$ in $C([0,T],S)$.

Proof. Let us consider the sequences σ^{ϵ_n}, γ^{ϵ_n}, α^{ϵ_n}, p^{ϵ_n}. By virtue of the uniqueness of solution we evidently have (in the respective spaces) $\sigma^{\epsilon_n} \rightharpoonup \sigma$, $\gamma^{\epsilon_n} \rightharpoonup \gamma$, $\alpha^{\epsilon_n} \rightharpoonup \alpha$. From (2.26)–(2.30) we obtain

$$\int_0^t [\dot{\sigma}^{\epsilon_n} \sigma - \sigma^{\epsilon_n}] dt + \int_0^t (\dot{\gamma}^{\epsilon_n}, \gamma - \gamma^{\epsilon_n}) dt$$
$$+ \int_0^t (\dot{\alpha}^{\epsilon_n}, \alpha - \alpha^{\epsilon_n}) dt \geq 0, \tag{2.48}$$

hence,

$$\|\sigma^{\epsilon_n}(t) - \sigma(t)\|_S^2 + \|\gamma^{\epsilon_n}(t) - \gamma(t)\|_S^2$$
$$+ \|\alpha^{\epsilon_n}(t) - \alpha(t)\|_{L^2(\Omega)}^2 \leq \delta_n(t) \to 0. \tag{2.49}$$

However, the functions $\sigma^{\epsilon_n}(t)$ are $\frac{1}{2}$-equi-Hölderian in S (and similarly $\gamma^{\epsilon_n}(t)$, $\alpha^{\epsilon_n}(t)$). Hence, (2.49) implies uniform convergence of σ^{ϵ_n}, γ^{ϵ_n}, α^{ϵ_n} in $C([0,T],S)$, $C([0,T],S)$, $C([0,T],L^2(\Omega))$, respectively. From (2.27), (2.28) we obtain

$$\lambda_n = -\frac{\dot{\gamma}^{\epsilon_n}\frac{\partial F}{\partial \gamma} + \dot{\alpha}^{\epsilon_n}\frac{\partial F}{\partial \alpha}}{\frac{\partial F}{\partial \gamma}\frac{\partial F}{\partial \gamma} + \frac{\partial F}{\partial \alpha}\frac{\partial F}{\partial \alpha}}, \quad \dot{\gamma}^{\epsilon} = -\lambda^{\epsilon}\frac{\partial F}{\partial \gamma}, \tag{2.50}$$

$$\dot{\alpha}^{\epsilon} = -\lambda^{\epsilon}\frac{\partial F}{\partial \alpha}, \quad \dot{p}^{\epsilon} = \lambda^{\epsilon}\frac{\partial F}{\partial \sigma},$$

where

$$\frac{\partial F}{\partial \gamma} = \frac{\partial F}{\partial \gamma}(\sigma^{\epsilon_n}, \gamma^{\epsilon_n}, \alpha^{\epsilon_n})$$

and similarly for $\frac{\partial F}{\partial \alpha}$. Thus, (2.50) implies $\lambda_n \rightharpoonup \lambda$ in $L^2((0,T), L^2(\Omega))$. Hence, $p^{\epsilon_n} \rightharpoonup p$ in $H^1((0,T),S)$ and equations (2.41)–(2.45); (2.46) is a consequence of compatibility of $e^{\epsilon_n} + p^{\epsilon_n}$.

Now let us choose a sequence $\epsilon_n \to 0$, such that $\sigma^{\epsilon_n} \to \sigma$ almost everywhere in $(0,T)\times\Omega$ and similarly $\gamma^{\epsilon_n} \to \gamma$, $\alpha^{\epsilon_n} \to \alpha$. Let $(t,x) \in (0,T)\times\Omega$ be such a point of convergence and let

$$F(\sigma(t,x), \gamma(t,x), \alpha(t,x)) < 0. \tag{2.51}$$

Then for large n we have

$$F(\sigma^{\epsilon_n}(t,x), \gamma^{\epsilon_n}(t,x), \alpha^{\epsilon_n}(t,x)) < 0,$$

which yields $\lambda^{\epsilon_n}(t,x) = 0$.

Let $M_{n_0} \subset (0,T)\times\Omega$ be such a set that

$$F(\sigma^{\epsilon_n}(t,x), \gamma^{\epsilon_n}(t,x), \alpha^{\epsilon_n}(t,x)) < 0$$

for $n \geq n_0$, $(t,x) \in M_{n_0}$. On M_{n_0} we have

$$0 = \int_{M_{n_0}} \lambda^{\epsilon_n}\psi\, dtdx \to \int_{M_{n_0}} \lambda\psi\, dtdx \quad \forall \psi \in L^2(M_{n_0}),$$

hence, $\lambda(t,x) = 0$ almost everywhere in M_{n_0}. If

$$M = \{(t,x) \in (0,T)\times\Omega; (2.51) \text{ is valid}\},$$

then $M = \cup_{n_0=1}^{\infty} M_{n_0}$. Further, $\sigma(t)$, $\gamma(t)$, $\alpha(t)$ have classical derivatives for almost all $(t,x) \in (0,T)\times\Omega$. (This follows from the properties of the spaces H^1; see Nečas (1967).) If at such a point

$$F(\sigma(t,x), \gamma(t,x), \alpha(t,x)) = 0,$$

then $\frac{d}{dt}[F] = 0$ as well, and (2.44), (2.45) imply (2.42).

Remark 2.1. If $F(\sigma, \gamma, \alpha) = f(\sigma) - \alpha$ and $\alpha(0) = \alpha_0 > 0$, then according to theorem 2.2 we have $\dot{\alpha} = \lambda \geq 0$, $\dot{\alpha} = 0$ for $f(\sigma) < \alpha$ and $\dot{\alpha} = \frac{\partial f}{\partial \sigma}\dot{\sigma}$ for $f(\sigma) = \alpha$. It is seen that

$$\alpha(t) \geq \max\{\alpha_0, f(\sigma(t))\} \tag{2.52}$$

for almost all points $x \in \Omega$, since $f(\sigma(t)) \leq \alpha(t)$. On the other hand, $\dot{\alpha} = \frac{\partial f}{\partial \sigma}\dot{\sigma}$ holds only for $f(\sigma) = \alpha$ (otherwise $\dot{\alpha} = 0$). Hence,

$$\alpha(t) \leq \max\{\alpha_0, \max_{0 \leq t' \leq t} f(\sigma(t'))\};$$

but since $\alpha(t)$ is nondecreasing, (2.52) implies

$$\alpha(t) \geq \max\{\alpha_0, \max_{0 \leq t' \leq t} f(\sigma(t'))\}.$$

Consequently, according to theorem 2.2 we conclude (cf. (2.1))

$$\alpha(t) = \max\{\alpha_0, \max_{0 \leq t' \leq t} f(\sigma(t'))\}.$$

3.2.2 Solution of Isotropic Hardening by Finite Elements

In this section we will solve boundary value problems of the theory of plastic flow with isotropic hardening. As we already know, these problems are described by relations (2.1)–(2.3) and lead to the time-dependent variational inequality (2.15), where of course the second term is dropped. We will proceed in the same way as in the paper by Hlaváček (1980), which is based on some results of C. Johnson ((1976b), (1978), (1977)).

We will consider a body in $\mathbf{R}^n$, $n = 2, 3$, occupying a polyhedral domain Ω. Let us denote $I = [0, T]$. The symbol C will, as above, stand for a positive constant that may assume different values at different places. $\mathbf{R}_\sigma$ is the space of symmetric $(n \times n)$-matrices.

Let us assume that a yield function $f : \mathbf{R}_\sigma \rightarrow \mathbf{R}$ is given, which is convex and bounded in $\mathbf{R}_\sigma$, continuously differentiable in $\mathbf{R}_\sigma \dot{-} Q$, where Q is a one-dimensional subspace of $\mathbf{R}_\sigma$ and satisfying

$$f(\lambda\sigma) = |\lambda| f(\sigma) \quad \forall \lambda \in \mathbf{R}, \ \forall \sigma \in \mathbf{R}_\sigma. \tag{3.1}$$

Let us assume that this function also fulfills condition

$$\exists C > 0, \ |\frac{\partial f}{\partial \sigma_{ij}}| < C \quad \forall i, j, \ \forall \sigma \in \mathbf{R}_\sigma \dot{-} Q. \tag{3.2}$$

As an example of a function satisfying the above assumptions let us mention the von Mises yield function (for $n = 3$)

$$f(\sigma) = (\sigma_{ij}^D \sigma_{ij}^D)^{1/2},$$

where $\sigma_{ij}^D = \sigma_{ij} - \frac{1}{3}\delta_{ij}\sigma_{kk}$ is the stress deviator. Let

$$\partial\Omega = \Gamma_u \cup \Gamma_\sigma, \quad \Gamma_u \cap \Gamma_\sigma = \emptyset,$$

where each of the sets Γ_u and Γ_σ is either empty or an open set in $\partial\Omega$.

Let us assume that we are given the (reference) vector of the body forces $F^0 \in [C(\bar{\Omega})]^n$ and of the surface loads $g^0 \in [L_2(\Gamma_\sigma)]^n$. If $\Gamma_u = \emptyset$, let the condition of the total equilibrium be fulfilled, that is, (1.1) and (1.2) for $n = 3$ respectively, and

$$\int_\Omega (x_1 F_2^0 - x_2 F_1^0)\,dx + \int_{\partial\Omega} (x_1 g_2^0 - x_2 g_1^0)\,ds = 0$$

(instead of (1.2)) for $n = 2$.

Let the actual body forces and surface loads be

$$F(t,x) = \gamma(t) F^0(x) \quad \text{in } I \times \Omega,$$

$$g(t,x) = \gamma(t) g^0(x) \quad \text{on } I \times \Gamma_\sigma,$$

with $\gamma : I \to \mathbf{R}$ a nonnegative function from $C^2(I)$, such that

$$\exists t_1 > 0, \ \gamma(t) = 0 \quad \forall t \in [0, t_1]. \tag{3.3}$$

Again, we introduce the set of statically admissible stress fields:

$$E(t) = E(F(t), g(t)) = \{\sigma \in S \mid \int_\Omega \sigma_{ij} \epsilon_{ij}(v)\,dx$$

$$= \int_\Omega F_i(t) v_i\,dx + \int_{\Gamma_\sigma} g_i(t) v_i\,ds \quad \forall v \in V\},$$

where

$$V = \{v \in [W^{1,2}(\Omega)]^n \mid v = 0 \text{ on } \Gamma_u\}.$$

Finally, let us recall the definition (see (2.3))

$$\mathcal{F}(\tau, \alpha) = f(\tau) - \alpha.$$

We denote

$$H = S \times L^2(\Omega),$$

$$B = \{(\tau, \alpha) \in \mathbf{R}_\sigma \times \mathbf{R} \mid \mathcal{F}(\tau, \alpha) \le 0\},$$
$$P = \{(\tau, \alpha) \in H \mid (\tau(x), \alpha(x)) \in B \text{ a.e. in } \Omega\},$$
$$K(t) = (E(t) \times L^2(\Omega)) \cap P, \quad t \in I. \tag{3.4}$$

Let the coefficients of the generalized Hooke's Law satisfy conditions (1.18), (1.19). Further, let a positive constant α_0 be given. For couples (σ, α) we introduce a new symbol $\hat{\sigma}$, for example,

$$\hat{\sigma} = (\sigma, \alpha), \quad \hat{\tau} = (\tau, \beta),$$

and we define the inner products with the corresponding norms:

$$\langle \hat{\sigma}, \hat{\tau} \rangle = \sigma_{ij}\tau_{ij} + \alpha\beta, \quad |\hat{\tau}| = \langle \hat{\tau}, \hat{\tau} \rangle^{1/2},$$

$$(\hat{\sigma}, \hat{\tau})_0 = \int_\Omega \langle \hat{\sigma}, \hat{\tau} \rangle dx, \quad \|\hat{\sigma}\| = (\hat{\sigma}, \hat{\sigma})_0^{1/2},$$

$$\{\hat{\sigma}, \hat{\tau}\} = [\sigma, \tau] + (\alpha, \beta)_{0,\Omega}, \quad |||\hat{\sigma}||| = \{\hat{\sigma}, \hat{\sigma}\}^{1/2}.$$

Similarly to definition 2.1, a weak solution will be a couple $\hat{\sigma} \equiv (\sigma, \alpha) \in H_0^1(I, S) \times H^1(I, L^2(\Omega))$ such that $\alpha(0) = \alpha_0$, $\hat{\sigma}(t) \in K(t)$, and

$$\left\{ \frac{d\hat{\sigma}(t)}{dt}, \hat{\tau} - \hat{\sigma}(t) \right\} \ge 0 \quad \forall \hat{\tau} = (\tau, \alpha) \in K(t) \tag{3.5}$$

holds almost everywhere in I.

Let us point out that our definitions corresponds to $\sigma(0) = 0$ and to the zero displacement on Γ_u. The time-dependent variational inequality (2.22) from definition 2.1 results by integrating (3.5) with respect to the time variable provided $\gamma \equiv 0$ (that is, without the kinematic hardening), and it is equivalent to (3.5). The existence and uniqueness of solution of (3.5) was studied for instance by C. Johnson (1978) for $\partial\Omega = \Gamma_u$, and the case $\partial\Omega = \Gamma_\sigma$ is dealt with in the book by Nečas and Hlaváček (1981).

With the aim of approximately solving problem (3.5), we first introduce finite-dimensional (inner) approximations of the set $E(t)$:

$$E_h(t) = \chi(t) + E_h^0, \quad 0 < h \le h_0 < \infty,$$

where $\chi \in H_0^1(I, S)$ is a fixed stress field such that $\chi(t) \in E(t)$ a.e. in I and $E_h^0 \subset E(0, 0)$ is the finite dimensional subspace of the self-equilibriated stress fields. Then, evidently $E_h(t) \subset E(t)$. The existence of the function χ will be established later in lemma 2.2.

Let $V_h \subset L^2(\Omega)$ be a finite- dimensional subspace—an approximation of $L^2(\Omega)$. Assume that V_h includes constant functions.

Define

$$K_h(t) = (E_h(t) \times V_h) \cap P,$$

so that $K_h(t) \subset K(t)$.

We again introduce a discretization of the time interval I, that is, $k = T/N$, $t_m = mk$, $m = 0, 1, \ldots, N$, $I_m = [t_{m-1}, t_m]$, $\hat{\tau}^m = \hat{\tau}(t_m)$, $\partial\hat{\tau}^m = (\hat{\tau}^m - \hat{\tau}^{m-1})/k$.

Instead of (3.5) we will solve the following approximate problem: find $\hat{\sigma}^m_{hk} \in K_h(t_m)$, such that for $m = 1, \ldots, N$ we have

$$\{\partial\hat{\sigma}^m_{hk}, \hat{\tau} - \hat{\sigma}^m_{hk}\} \geq 0 \quad \forall \hat{\tau} \in K_h(t_m). \tag{3.6}$$

Since $\hat{\sigma}^m_{hk}$ minimizes the strictly convex functional

$$\frac{1}{2}|||\hat{\sigma}|||^2 - \{\hat{\sigma}, \hat{\sigma}^{m-1}\} \tag{3.6'}$$

on the closed convex set $K_h(t_m)$, there exists a unique solution provided $K_h(t_m) \neq \emptyset$. (Lemma 2.2 below yields a sufficient condition for $K_h(t_m) \neq \emptyset$, since $\xi(t) \in K_h(t)$.)

3.2.21. A Priori Error Estimates. First, we prove an important auxiliary result. Define

$$\|q\|_{l^2(H)} = \left(\sum_{m=1}^{N} k\|q^m\|^2\right)^{1/2}$$

for $q = (q^1, \ldots, q^N)$, $q^m \in H$.

Lemma 2.1. *Assume if $\Gamma_\sigma \neq \emptyset$, then there exists a function*

$$\chi^0 \in [L^\infty(\Omega)]^{n^2} \cap E(F^0, g^0). \tag{3.7}$$

Then there exist positive constants C, k_0 such that

$$\|\partial\hat{\sigma}_{hk}\|_{l^2(H)} \leq C \tag{3.8}$$

for $k \leq k_0$ and $0 < h \leq h_0$.

Remark 2.2. Let F^0 be continuous on $\bar{\Omega}$. Then there is $\chi^1 \in S \cap [L^\infty(\Omega)]^{n^2}$ such that

$$\operatorname{div} \chi^1 = -F^0 \quad \text{in } \Omega$$

(χ^1 can be obtained by a mere integration). Let the vector function $g^0 - \chi^1 \cdot \nu$, where ν denotes the unit outward normal, be piecewise linear on Γ_σ with respect to a simplicial partition of Γ_σ. Then condition (3.7) is fulfilled. Indeed, there is a simplicial partition of Ω and a $\chi^2 \in E_h^0$ (see section 2.4.11) such that

$$\chi^2 \cdot \nu = g^0 - \chi^1 \cdot \nu \quad \text{on } \Gamma_\sigma.$$

Putting $\chi^0 = \chi^1 + \chi^2$ we obtain

$$\chi^0 \in [L^\infty(\Omega)]^{n^2}, \ \text{div}\, \chi^0 = -F^0 \text{ in } \Omega, \quad \chi^0 \cdot \nu = g^0 \text{ on } \Gamma_\sigma,$$

which implies $\chi^0 \in E(F^0, g^0)$.

To prove lemma 2.1 we need some auxiliary results:

Lemma 2.2. *Let* (3.7) *be fulfilled. Then there is*

$$\xi(t) = (\chi(t), \varsigma(t)) \in K(t) \quad \forall t \in I, \quad \xi(0) = (0, \alpha_0),$$

and positive constants C, δ_1 *such that*

$$\sup_{t\in I} \|d^j \xi/dt^j\| \le C, \quad j = 0, 1, 2, \tag{3.9}$$

$$\text{dist}(\xi(x,t), \partial B) \ge \delta_1 \quad \forall t \in I, \ a.e. \ in \ \Omega. \tag{3.10}$$

Proof. If $\Gamma_\sigma \neq \emptyset$ we use χ^0; if $\Gamma_\sigma = \emptyset$ then we use

$$\chi^0 \in [L^\infty(\Omega)]^{n^2} \cap E(F^0),$$

obtained by integrating the equations of equilibrium div $\chi^0 = -F^0$ in Ω.

Put

$$\chi(t) = \gamma(t)\chi^0, \quad \varsigma(t) = \gamma(t)C_1 + \alpha_0,$$

where C_1 is a suitably chosen constant. Then evidently $\chi(t) \in E(t)$ for $t \in I$ and

$$f(\chi^0(x)) \le C_1 \quad \text{a.e. in } \Omega$$

for a certain constant $C_1 > 0$, due to the boundedness of χ^0. Thus, we have

$$\mathcal{F}(\chi, \varsigma) = f(\chi) - \varsigma = \gamma(t)[f(\chi^0) - C_1] - \alpha_0 \le -\alpha_0 < 0 \tag{3.11}$$

for all $t \in I$ and almost all $x \in \Omega$. Hence,

$$\xi = (\chi, \varsigma) \in P \ \ \forall t \in I, \ \ \xi(t) \in K(t), \ \ \xi(0) = (0, \alpha_0).$$

Since χ_0 is bounded and $\gamma \in C^2(I)$, we easily find that (3.9) holds.

In order to prove (3.10), let us observe that (3.11), (3.2) imply that there is $\delta_1 > 0$ such that

$$\mathcal{F}(\xi(t) + \rho) \le 0 \quad \forall \rho \in \mathbf{R}_\sigma \times \mathbf{R}, \quad |\rho| \le \delta_1.$$

Hence, $\xi(t) + \rho \in B$, which yields (3.10). $\square$

Let us prove lemma 2.1 by the penalty method. Let π be the projection operator to a closed convex set B in the space $\mathbf{R}_\sigma \times \mathbf{R}$. Let us introduce the penalty functional

$$J_\mu(\hat{\tau}) = \frac{1}{2\mu}\|\hat{\tau} - \pi\hat{\tau}\|^2, \quad \mu > 0, \quad \hat{\tau} \in H. \tag{3.12}$$

Define new approximations

$$\hat{\sigma}^m_{hk\mu} \in E_h(t_m) \times V_h, \quad m = 0, 1, \ldots, N$$

by the identities (omitting for brevity the indices $hk\mu$ in the following:

$$\{\partial\hat{\sigma}^m, \hat{\tau}\} + (J'_\mu(\hat{\sigma}^m), \hat{\tau})_0 = 0 \quad \forall \tau \in E^0_h \times V_h, \tag{3.13}$$

$$\hat{\sigma}^0 = (0, \alpha_0), \quad m = 1, \ldots, N.$$

Notice that the Gateaux derivative of J_μ is

$$J'_\mu(\hat{\sigma}) = \frac{1}{\mu}(\hat{\sigma} - \pi\hat{\sigma}).$$

Problem (3.13) has a unique solution for every m, since $\hat{\sigma}^m$ minimizes the coercive, strictly convex and continuous functional

$$\frac{1}{2}|||\hat{\sigma}|||^2 + kJ_\mu(\hat{\sigma}) - \{\hat{\sigma}^{m-1}, \hat{\sigma}\}$$

on the set $E_h(t_m) \times V_h$, which is closed and convex in H.

Lemma 2.3. *Let condition (3.7) be fulfilled. Then there are positive constants C, k_0 such that for $k \le k_0$, $0 < h \le h_0$ and $\mu > 0$ we have estimates*

(i) $\max_{1 \le m \le N} |||\hat{\sigma}^m_{hk\mu}||| \le C$,

(ii) $\sum_{m=1}^N kJ_\mu(\hat{\sigma}^m_{hk\mu}) \le C$,

(iii) $\sum_{m=1}^N k\|J'_\mu(\hat{\sigma}^m_{hk\mu})\|_{L^1(\Omega)} \le C$,

where $\|z\|_{L^1(\Omega)} = \int_\Omega |z|\,dx$ for $z \in H$.

Proof. (i) Consider $\xi = (\chi, \varsigma)$ from lemma 2.2 and put $\hat{\sigma}^m = \xi^m + \rho^m$, $m = 0, 1, \ldots, N$. Then we have

$$\rho^m = (\bar{\sigma}^m, \beta^m), \quad \bar{\sigma}^m \in E^0_h, \quad \rho^0 = (0, 0).$$

Substitute $\hat{\tau} = \rho^m$ and $\partial\hat{\sigma}^m = \partial\varsigma^m + \partial\rho^m$ in (3.13). We obtain

$$\{\partial\rho^m, \rho^m\} + (J'_\mu(\hat{\sigma}^m), \rho^m)_0 = -\{\partial\xi^m, \rho^m), \quad m = 1, \ldots, N. \tag{3.14}$$

Since J'_μ is monotonic, we have

$$(J'_\mu(\hat{\sigma}^m), \rho^m)_0 = (J'_\mu(\hat{\sigma}^m) - J'_\mu(\varsigma^m), \rho^m)_0 \geq 0. \tag{3.15}$$

Thus, we can write

$$\sum_{m=1}^{M} k\{\partial\rho^m, \rho^m\} \leq -\sum_{m=1}^{M} k\{\partial\xi^m, \rho^m\}, \quad M = 1, \ldots, N. \tag{3.16}$$

On the other hand, we have

$$\sum_{m=1}^{M} k\{\partial\rho^m, \rho^m\} = \sum_{m=1}^{M} \frac{1}{2}(|||\rho^m|||^2 - |||\rho^{m-1}|||^2 + |||\rho^m - \rho^{m-1}|||^2)$$

$$\geq \frac{1}{2}|||\rho^M|||^2, \tag{3.17}$$

and hence by virtue of (3.9) we can estimate

$$\sum_{m=1}^{M} k|\{\partial\xi^m, \rho^m\}| \leq Ck\sum_{m=1}^{M} |||\rho^m||| \leq \frac{1}{2}Ck\sum_{m=1}^{N}(1 + |||\rho^m|||^2)$$

$$\leq C + Ck\sum_{m=1}^{M} |||\rho^m|||^2. \tag{3.18}$$

We will use the discrete analogue of the Gronwall Lemma (see Babuška, Prágر, Vitásek (1966), Chapter 3, lemma 3.3):

Let

$$\varphi(M) \leq \psi(M) + \sum_{r=0}^{M-1} \chi(r)\varphi(r), \quad M = 1, \ldots, m \leq N, \quad \chi(r) \geq 0 \ \forall r.$$

Then

$$\varphi(m) \leq \psi(m) + \sum_{r=0}^{m-1} \chi(r)\psi(r) \prod_{s=r+1}^{m-1} (1 + \chi(s)).$$

The estimates (3.16), (3.17), and (3.18) yield

$$|||\rho^M|||^2 \leq C + C\sum_{r=0}^{M-1} K|||\rho^r|||^2, \quad M = 1, \ldots, N.$$

Setting $\varphi(r) = |||\rho^r|||^2$, $\psi(r) = C$, $\chi(r) = Ck$, we obtain

$$|||\rho^m|||^2 \leq C + \sum_{r=0}^{m-1} Ck(1 + Ck)^{m-1-r} \leq C_1, \tag{3.19}$$

since $(1+Ck)^N \leq C_2$ for $k \leq k_0$.

Finally, by virtue of (3.9) we obtain

$$|||\hat{\sigma}^m||| \leq |||\rho^m||| + |||\xi^m||| \leq C, \quad m = 1, \dots, N.$$

(ii) The convexity of J_μ together with lemma 2.2 implies

$$J_\mu(\hat{\sigma}^m) + (J'_\mu(\hat{\sigma}^m), \xi^m - \hat{\sigma}^m)_0 \leq J_\mu(\xi^m) = 0.$$

Now we conclude from (3.14), (3.15), (3.17) that

$$\sum_{m=1}^N k(J'_\mu(\hat{\sigma}^m), \rho^m)_0 \leq -\sum_{m=1}^N k\{\partial \xi^m, \rho^m\}$$

$$\leq \sum_{m=1}^N k|||\partial \xi^m|||\ |||\rho^m||| \leq CT, \tag{3.20}$$

also making use of (3.19) and (3.9). Hence,

$$\sum_{m=1}^N kJ_\mu(\hat{\sigma}^m) \leq \sum_{m=1}^N k(J'_\mu(\hat{\sigma}^m), \hat{\sigma}^m - \xi^m)_0 \leq C.$$

(iii) Let us consider

$$\hat{\sigma}^m(x) = \xi^m + \rho^m(x) \notin B,$$

and define

$$\jmath(x) = \frac{\hat{\sigma}^m(x) - \pi\hat{\sigma}^m(x)}{|\hat{\sigma}^m(x) - \pi\hat{\sigma}^m(x)|}.$$

($\jmath(x)$ is the unit normal to the superplane L, which separates B and $\hat{\sigma}^m(x)$ at the point $\pi\hat{\sigma}^m(x)$.) Since $\hat{\sigma}^m(x) \notin B$ and $\xi^m(x) + \delta_1\jmath(x) \in B$ by virtue of (3.10), we have

$$\langle \jmath(x), \xi^m(x) + \rho^m(x)\rangle \geq d,$$

$$\langle \jmath(x), \xi^m(x) + \delta_1\jmath(x)\rangle \leq d,$$

where $d = \langle \jmath(x), \pi\hat{\sigma}^m(x)\rangle$. Hence,

$$\langle \jmath, \rho^m\rangle \geq d - \langle \jmath, \xi^m\rangle \geq \delta_1\langle \jmath, \jmath\rangle = \delta_1.$$

Substituting from the definition of $\jmath$, we obtain

$$|\hat{\sigma}^m(x) - \pi\hat{\sigma}^m(x)| \leq \delta_1^{-1}\langle \hat{\sigma}^m(x) - \pi\hat{\sigma}^m(x), \rho^m(x)\rangle.$$

The same inequality evidently holds for $\hat{\sigma}^m(x) \in B$.

Integration over $x \in \Omega$ yields the inequality

$$\|J'_\mu(\hat{\sigma}^m)\|_{L^1(\Omega)} = \frac{1}{\mu}\int_\Omega |\hat{\sigma}^m - \pi\hat{\sigma}^m| dx$$

$$\leq \frac{1}{\delta_1 \mu}\int_\Omega \langle \hat{\sigma}^m - \pi\hat{\sigma}^m, \rho^m\rangle dx = \frac{1}{\delta_1}(J'_\mu(\hat{\sigma}^m), \rho^m)_0.$$

The estimate (iii) is now an easy consequence of the inequality (3.20). □

Lemma 2.4. *Let condition* (3.7) *be fulfilled. Then there are positive constants* C, k_0 *such that*

$$\|\partial\hat{\sigma}_{hk\mu}\|_{l^2(H)} \leq C \tag{3.21}$$

holds for all $k \leq k_0$, $h \leq h_0$, $\mu > 0$.

Proof. The identities (3.13) imply for $m = 2, \ldots, N$

$$\{\partial^2\hat{\sigma}^m, \hat{\tau}\} + (\partial(J'_\mu(\hat{\sigma}^m)), \hat{\tau})_0 = 0 \quad \forall \hat{\tau} \in E_h^0 \times V_h.$$

Choosing $\hat{\tau} = \partial\rho^m$ and substituting $\hat{\sigma}^m = \xi^m + \rho^m$ we obtain

$$\{\partial^2\rho^m, \partial\rho^m\} + (\partial(J'_\mu(\hat{\sigma}^m)), \partial\sigma^m)_0$$

$$= -\{\partial^2\xi^m, \partial\rho^m\} + (\partial(J'_\mu(\hat{\sigma}^m)), \partial\xi^m)_0.$$

By virtue of the monotonicity of J'_μ, the second term on the left-hand side is nonnegative. Summing "by parts" in the last term, we can write

$$\sum_{m=2}^{M} k\{\partial^2\rho^m, \partial\rho^m\} \leq -\sum_{m=2}^{M} k\{\partial^2\xi^m, \partial\rho^m\}$$

$$-\sum_{m=2}^{M-1} k(J'_\mu(\hat{\sigma}^m), \partial^2\xi^{m+1})_0 + (J'_\mu(\hat{\sigma}^M), \partial\xi^M)_0.$$

Here we have used the identity $\partial\xi^2 = (\xi^2 - \xi^1)/k = 0$, which holds for $k < t_1/2$ by virtue of the definition of $\xi(t)$ and (3.3). By (3.9) and (iii) from lemma 2.3, we conclude

$$\left|\sum_{m=2}^{M-1} k(J'_\mu(\hat{\sigma}^m), \partial^2\xi^{m+1})_0\right| \leq C\sum_{m=2}^{M-1} k\|J'_\mu(\hat{\sigma}^m\|_{L^1(\Omega)} \leq C_1.$$

Further,

$$\sum_{m=2}^{M} k\{\partial^2\rho^m, \partial\rho^m\} \geq \frac{1}{2}(|||\partial\rho^M|||^2 - |||\partial\rho^1|||^2),$$

hence, for $M = 2, \dots, N$ we have

$$|||\partial\rho^M|||^2 \le C + C(J'_\mu(\hat{\sigma}^M), \partial\xi^M)_0 + |||\partial\rho^1|||^2 + \sum_{m=2}^{M} Ck|||\partial\rho^m|||^2.$$

By the discrete analogue of the Gronwall Lemma we find for $m = 2, \dots, N$

$$|||\partial\rho^m|||^2 \le C + C(J'_\mu(\hat{\sigma}^m), \partial\xi^m)_0 + C\sum_{r=1}^{m-1} k(J'_\mu(\hat{\sigma}^r), \partial\xi^r)_0.$$

Applying again (iii) from lemma 2.3 to the last term, multiplying by k, and summing, we arrive at the estimate

$$\sum_{m=2}^{N} k|||\partial\rho^m|||^2 \le CT + C\sum_{m=2}^{N} k(J'_\mu(\hat{\sigma}^m), \partial\xi^m)_0 \le C_1.$$

Consequently,

$$|||\partial\hat{\sigma}^m|||^2 \le 2|||\partial\xi^m|||^2 + 2|||\partial\rho^m|||^2 \le C + 2|||\partial\rho^m|||^2,$$

$$\sum_{m=2}^{N} k|||\partial\hat{\sigma}^m|||^2 \le CT + 2\sum_{m=2}^{N} k|||\partial\rho^m|||^2 \le C_2. \tag{3.22}$$

It remains to show that

$$k|||\partial\hat{\sigma}^1|||^2 \le C. \tag{3.23}$$

Substituting $\hat{\sigma}^1 = \xi^1$, $\hat{\tau} = \partial\rho^1$ into (3.13) we obtain

$$|||\partial\rho^1|||^2 + \{\partial\xi^1, \partial\rho^1\} + (J'_\mu(\hat{\sigma}^1), \partial\hat{\sigma}^1)_0 - (J'_\mu(\hat{\sigma}^1), \partial\xi^1)_0 = 0.$$

The third term is nonnegative since $J'_\mu(\hat{\sigma}^0) = 0$ by virtue of $\hat{\sigma}^0 \in B$. By using (iii) from lemma 2.3 we find

$$k|||\partial\rho^1|||^2 \le C,$$

hence, (3.23) holds.

Estimate (3.21) is a consequence of (3.22), (3.23), and of the equivalence of norms $|||\cdot|||$ and $\|\cdot\|$. □

Proof of Lemma 2.1. Let $\mu \to 0$ for a sequence of positive numbers μ. Point (i) of lemma 2.3 implies the existence of such a $C > 0$ that for all $\mu > 0$,

$$\|\hat{\sigma}_{hk\mu}\|_{l^2(H)} \le C.$$

Hence, there exist subsequences of μ and $\hat{\sigma}_{hk\mu}$ such that for $\mu \to 0$,

$$\hat{\sigma}_{hk\mu} \rightharpoonup \hat{\sigma}_{hk} \quad \text{(weakly) in the space } l^2(H). \tag{3.24}$$

Lemma 2.4 implies that we can also write

$$\partial\hat{\sigma}_{hk\mu} \rightharpoonup \hat{s}_{hk} \quad \text{(weakly) in } l^2(H). \tag{3.25}$$

It is not difficult to verify that $\hat{s}_{hk} = \partial\hat{\sigma}_{hk}$. We will show that $\hat{\sigma}_{hk}$ is a solution of problem (3.6). Since J_μ is convex, we have

$$J_\mu(\hat{\tau}^m) \geq J_\mu(\hat{\sigma}^m_{hk\mu}) + (J'_\mu(\hat{\sigma}^m_{hk\mu}), \hat{\tau}^m - \hat{\sigma}^m_{hk\mu})_0.$$

If $\hat{\tau}^m \in E_h(t_m) \times V_h$, then $\hat{\tau}_0 \equiv \hat{\tau}^m - \hat{\sigma}^m_{hk\mu} \in E^0_h \times V_h$, and we can apply (3.13), thus deriving (omitting the indices $hk\mu$ again)

$$(J'_\mu(\hat{\sigma}^m), \hat{\tau}^m - \hat{\sigma}^m)_0 = -\{\partial\hat{\sigma}^m, \hat{\tau}^m - \hat{\sigma}^m\}.$$

In this way we obtain the inequality

$$\{\partial\hat{\sigma}^m, \hat{\tau}^m - \hat{\sigma}^m\} + J_\mu(\hat{\tau}^m) - J_\mu(\hat{\sigma}^m) \geq 0.$$

Put $\hat{\tau}^m \in K_h(t_m)$. Then $J_\mu(\hat{\tau}^m) = 0$ and

$$\{\partial\hat{\sigma}^m, \hat{\tau}^m - \hat{\sigma}^m\} \geq 0 \quad \forall \hat{\tau}^m \in K_h(t_m).$$

Using (3.24) and (3.25) once more, we conclude for $\mu \to 0$

$$0 \leq \limsup \left[-\sum_{m=1}^{M} k\{\partial\hat{\sigma}^m, \hat{\sigma}^m\} + \sum_{m=1}^{M} k\{\partial\hat{\sigma}^m, \hat{\tau}^m\} \right]$$

$$= \limsup \left[\frac{1}{2}|||\hat{\sigma}^0|||^2 - \frac{1}{2}|||\hat{\sigma}^M|||^2 - \frac{1}{2}\sum_{m=1}^{M} |||\hat{\sigma}^m - \hat{\sigma}^{m-1}|||^2 + \sum_{m=1}^{M} k\{\partial\hat{\sigma}^m, \hat{\tau}^m\} \right]$$

$$\leq \frac{1}{2}|||\hat{\sigma}^0|||^2 - \frac{1}{2}|||\hat{\sigma}^M|||^2 - \frac{1}{2}\sum_{m=1}^{M} |||\hat{\sigma}^m - \hat{\sigma}^{m-1}|||^2$$

$$+ \sum_{m=1}^{M} k\{\partial\hat{\sigma}^m, \hat{\tau}^m\} = \sum_{m=1}^{M} k\{\partial\hat{\sigma}^m, \hat{\tau}^m - \hat{\sigma}^m\}.$$

Choosing $\hat{\tau}^m = \hat{\sigma}^m_{hk}$ for $m < M$, we obtain

$$\{\partial\hat{\sigma}^M_{hk}, \hat{\tau} - \hat{\sigma}^M_{hk}\} \geq 0 \quad \forall \hat{\tau} \in K_h(t_M), \quad M = 1, \ldots, N.$$

It remains to show that $\hat{\sigma}_{hk}^m \in K_h(t_m) = (E_h(t_m) \times V_h) \cap P$. Recall that $\sigma_{hk\mu}^m \subset E_h(t_m)$, $\sigma_{hk\mu}^m = \chi^m + \bar{\sigma}^m$. As $E_h(t_m)$ is closed and convex in S, it is weakly closed in S as well. However, it follows from (3.24) that $\sigma_{hk\mu}^m \rightharpoonup \sigma_{hk}^m$ weakly in S, hence, $\sigma_{hk}^m \in E_h(t_m)$.

In order to verify that $\hat{\sigma}_{hk}^m \in P$ we apply (ii) from lemma 2.3. Hence,

$$C \geq kJ_\mu(\hat{\sigma}_{hk\mu}^m) = \frac{k}{2\mu}\|\hat{\sigma}_{hk\mu}^m - \pi\hat{\sigma}_{hk\mu}^m\|^2,$$

$$\|\hat{\sigma}_{hk}^m - \pi\hat{\sigma}_{hk}^m\|^2 \leq \liminf_{\mu\to 0} \|\hat{\sigma}_{hk\mu}^m - \pi\hat{\sigma}_{hk\mu}^m\| \leq \liminf_{\mu\to 0} \frac{2C\mu}{k} = 0,$$

which implies that $\hat{\sigma}_{hk}^m \in P$.

Finally, using (3.25) and lemma 2.4 we can write

$$\sum_{m=1}^{N} k\|\partial\hat{\sigma}_{hk}^m\|^2 \leq \sum_{m=1}^{N} k \liminf_{\mu\to 0} \|\partial\hat{\sigma}_{hk\mu}^m\|^2$$

$$\leq \liminf_{\mu\to 0} \|\partial\hat{\sigma}_{hk\mu}\|^2_{l^2(H)} \leq C. \qquad \square$$

Theorem 2.3. *Denote*

$$\epsilon(h,k) = \inf_{\hat{\tau}\in\mathcal{K}} \|\hat{\sigma} - \hat{\tau}\|_{l^2(H)},$$

where

$$\mathcal{K} = \{\hat{\tau} = (\hat{\tau}^1, \ldots, \hat{\tau}^N) \mid \hat{\tau}^m \in K_h(t_m) \quad \forall m\}.$$

Assume that (3.7) *is fulfilled. Then there exist positive constants C and k_0 such that for $k \leq k_0$,*

$$\max_{1\leq m\leq N} \|\hat{\sigma}^m - \hat{\sigma}_{hk}^m\| \leq C(\sqrt{\epsilon(h,k)} + \sqrt{k}). \tag{3.26}$$

Proof. Analogous to that of theorem 1.3. $\square$

3.2.22. A Priori Error Estimates for the Plane Problem. Let us now consider $n = 2$; that is, the plane problem, and evaluate the quantity $\epsilon(h,k)$ provided piecewise linear triangular elements are used. However, to this end we must assume a certain regularity of the weak solution $\hat{\sigma}$ of problem (3.5).

Let the reference body forces F^0 be *constant* and the reference load g^0 *piecewise linear* on Γ_σ. We will consider a *regular* system $\{\mathcal{T}_h\}$, $0 < h \leq h_0$, of triangulations of the domain Ω (i.e., there is $\theta_0 > 0$ such that no angle in the triangulations $\mathcal{T}_h$ is less than ω_0); let h denote the maximal length of side in $\mathcal{T}_h$.

We will use finite elements for the equilibrium model of stress which have been defined by linear polynomials on subtriangles $K_i \subset K$, $i = 1, 2, 3$, and which generate subspaces $N_h(\Omega) \subset S$ (see section 2.4.11).

In the papers by C. Johnson and Mercier (1978) and Hlaváček (1979), approximation properties of the spaces $N_h(\Omega)$ were studied. Here we will use the following result (see Hlaváček (1979), theorem 2.3):

Let $\tau \in S \cap [C^2(\bar{\Omega})]^4$. Then there is a linear mapping

$$r_h : E(0,0) \cap [C^2(\bar{\Omega})]^4 \to N_h(\Omega)$$

such that on each triangle $K \in N_h(\Omega)$ satisfies

$$\max_{i=1,2,3} \|\tau - (r_h\tau)^i\|_{[C(K_i)]^4} \leq C h_K^2 \|\tau\|_{[C^2(K)]^4}, \tag{3.27}$$

where $(r_h\tau)^i$ is the restriction of $r_h\tau$ on K_i, h_K is the maximal side of K, and C is independent of τ, h_K.

Let us define the spaces of finite elements

$$E_h^0 = N_h(\Omega) \cap E(0,0) = \{\tau \in N_h(\Omega) \mid \tau \cdot \nu = 0 \text{ on } \Gamma_\sigma\},$$

$$V_h = \{\beta \in L^2(\Omega) \mid \beta|_{K_i} \in P_1(K_i) \quad \forall K_i \subset K \in \mathcal{T}_h\}.$$

Under the assumptions imposed on F^0, g^0 we can find a triangulation $\mathcal{T}_{h_0}$ and an auxiliary function χ^0 that is piecewise linear with respect to $\mathcal{T}_{h_0}$ (see remark 2.2). Then $\chi(t_m) = \gamma(t_m)\chi^0$ is also piecewise linear. In the following, we will assume that the system of triangulations $\{\mathcal{T}_h\}$ results from refining the original triangulation $\mathcal{T}_{h_0}$.

Theorem 2.4. *Let the solution $\hat{\sigma} \equiv (\sigma, \alpha)$ be such that the following inequalities hold for $\sigma_0 = \sigma - \chi$ and α in each triangle $K^0 \in \mathcal{T}_{h_0}$:*

$$\sup_{t\in I} \|\sigma_0(t)\|_{[C^2(K^0)]^4} < \infty,$$

$$\sup_{t\in I} \|\alpha(t)\|_{2,K_i^0} < \infty, \quad i = 1, 2, 3.$$

Then

$$\epsilon(h,k) \leq Ch^2, \tag{3.28}$$

where $C = C(\sigma_0, \alpha)$ is independent of h, k.

Proof. Recall that $\hat{\tau}^m = (\tau^m, \beta^m)$, $\tau^m = \chi^m + \tau_0^m$, $\sigma^m = \chi^m + \sigma_0^m$, where $\tau_0^m \in E_h^0$ and $\sigma_0^m \in E(0,0)$. Hence, we can write for all $m = 1, \ldots, N$ (omitting the index m):

$$\|\hat{\sigma} - \hat{\tau}\|^2 = \|\sigma_0 - \tau_0\|_S^2 + \|\alpha - \beta\|_{0,\Omega}^2. \tag{3.29}$$

Put $\tau_0 = r_h\sigma_0$. Then the definition of the mapping r_h (see Hlaváček (1979)) implies that $r_h\sigma_0 \cdot \nu = 0$ on Γ_σ, hence $r_h\sigma_0 \in E_h^0$.

Consider an arbitrary triangle $K_i \subset K$, $K \in \mathcal{T}_h$ and denote its vertices by a_j. Then, provided $\beta \in P_1(K_i)$ and

$$\beta(a_j) \geq f(\chi + r_h\sigma_0)(a_j) \equiv d_j, \quad j = 1, 2, 3, \tag{3.30}$$

we have

$$\beta \geq f(\chi + r_h\sigma_0) \text{ in } K_i.$$

This is a consequence of the linearity of $\chi + r_h\sigma_0$, β and of the convexity of the function f.

Let $\Pi\alpha \in P_1(K_i)$ denote the Lagrangian interpolation of the function α on K_i. Define $r \in P_1(K_i)$, $\beta_h \in P_1(K_i)$ by the relations

$$r(a_j) = [d_j - \alpha(a_j)]^+, \quad \beta_h = \Pi\alpha + r. \tag{3.31}$$

Then, evidently $\beta_h(a_j) \geq d_j$, $j = 1, 2, 3$.

The following estimate is well known:

$$\|\alpha - \Pi\alpha\|_{0,K_i} \leq Ch^2|\alpha|_{2,K_i}, \tag{3.32}$$

where the right-hand side is the seminorm of the second derivative (see Ciarlet (1978)).

By using assumptions (3.1), (3.2) and estimate (3.27), we obtain

$$|d_j - f(\chi + \sigma_0)(a_j)| \leq \|f(\chi + r_h\sigma_0) - f(\chi + \sigma_0)\|_{C(K_i)}$$

$$\leq C\|\sigma_0 - r_h\sigma_0\|_{[C(K_i)]^4} \leq Ch^2\|\sigma_0\|_{[C^2(K^0)]^4} \equiv \epsilon_1(h).$$

This immediately yields

$$\|r\|_{0,K_i} \leq Ch\,\epsilon_1(h). \tag{3.33}$$

Now (3.32) and (3.33) successively imply

$$\|\alpha - \beta_h\|_{0,\Omega}^2 \leq Ch^4 \sum_{K^0 \in \mathcal{T}_{h_0}} \left(\sum_{i=1}^{3} |\alpha|_{2,K_i^0}^2 + \|\sigma_0\|_{[C^2(K^0)]^4}^2\right). \tag{3.34}$$

It follows from (3.27) that

$$\|\sigma_0 - r_h\sigma_0\|_S^2 \leq Ch^4 \sum_{K^0 \in \mathcal{T}_{h_0}} \|\sigma_0\|_{[C^2(K^0)]^4}^2. \tag{3.35}$$

By substituting (3.34) and (3.35) into (3.29), we derive

$$\|\hat{\sigma}^m - \hat{\tau}^m\|^2 \le Ch^4 \sum_{K^0 \in \mathcal{T}_{h_0}} \left(\|\sigma_0^m\|^2_{[C^2(K^0)]^4} + \sum_{i=1}^{3} |\alpha^m|^2_{2,K_i^0} \right)$$

$$\le Ch^4 \sum_{K^0 \in \mathcal{T}_{h_0}} \left(\sup_{t \in I} \|\sigma_0(t)\|^2_{[C^2(K^0)]^4} + \sum_{i=1}^{3} \sup_{t \in I} |\alpha(t)|^2_{2,K_i^0} \right)$$

$$= h^4 C_1(\sigma_0, \alpha).$$

Now it is easy to establish the estimate

$$\|\hat{\sigma} - \hat{\tau}\|^2_{l^2(H)} = \sum_{m=1}^{N} k \|\hat{\sigma}^m - \hat{\tau}^m\|^2 \le C_1(\sigma_0, \alpha) h^4 T,$$

which completes the proof of the theorem. □

Remark 2.3. Given a three-dimensional problem we can use a four-faced element consisting of four tetrahedra, which is analogous to the triangular block-element. Then, estimates of the type (3.27), (3.32) are valid (see Křížek (1982), Ciarlet (1978), and proceeding in the same way as in the proof of the previous theorem we arrive at an analogous assertion.

Corollary of Theorem 2.4. *Let the assumptions of theorem 2.4 be fulfilled. Then there are constants C and k_0 such that for all $k \le k_0$ and $h \le h_0$,*

$$\max_{1 \le m \le N} \|\hat{\sigma}^m - \hat{\sigma}^m_{hk}\| \le C(h + \sqrt{k}).$$

Proof. The corollary is an immediate consequence of theorem 2.3 and 2.4. □

Remark 2.4. Algorithm for the Solution of the Approximate Problem (3.6). Defining E_h^0 and V_h as above, we obtain for $\tau_0^m \in E_h^0$:

$$(\chi^m + \tau_0^m, \beta^m) \in P \iff \beta^m(a_j) \ge f(\chi^m + \tau_0^m)(a_j) \tag{3.36}$$

at all vertices $a_j \in K_i \subset K$ of all triangles $K \in \mathcal{T}_h$.

Hence, we again have nonlinear constraints for the parameters β^m and τ_0^m. (In the case of the Mises yield function these constraints are quadratic.) At each time level it is thus necessary to minimise the quadratic functional (3.6′) with nonlinear constraints (3.36) and with linear constraints—equations—which guarantee the continuity of the stress vector on the boundaries between the triangles. For this purpose, we choose a suitable algorithm of nonlinear programming.

3.2.23. Convergence in the Case of Nonregular Solution. First, let us again assume $n = 2$, the plane problem. Let us keep the assumptions on F^0 and g^0 from section 3.2.22, so that the functions χ^0 and $\chi(t_m)$ are piecewise linear with respect to the triangulation $\mathcal{T}_{h_0}$.

Theorem 2.5. *Let us assume*

(i) *if* $\Gamma = \Gamma_\sigma$, *then* Ω *is a starlike domain;*[2]

(ii) *if* $\Gamma = \bar{\Gamma}_u \cup \bar{\Gamma}_\sigma$, *then there is a point* $A \in \mathbf{R}^2$ *such that, provided the origin of the coordinate system is at* A, *we obtain for* $\lambda = 1 + \epsilon$ *and for all sufficiently small positive* ϵ *that*

$$\text{either } \lambda\bar{\Gamma}_\sigma \subset \mathbf{R}^2 - \bar{\Omega} \text{ or } \lambda\bar{\Gamma}_\sigma \subset \Omega.$$

Here, $\lambda\bar{\Gamma}_\sigma$ *stands for the image of* $\bar{\Gamma}_\sigma$ *under the dilatation mapping* $x \mapsto \lambda x$. *Let the system* $\{\mathcal{T}_h\}$ *of triangulation result by refining the original triangulation* $\mathcal{T}_{h_0}$. *Then, for every fixed* $k < T$,

$$\lim_{h \to 0} \epsilon(h, k) = 0. \tag{3.37}$$

The proof will involve the following theorem:

Theorem 2.6. *Let conditions* (i), (ii) *from theorem* 2.5 *be fulfilled. Then, the set*

$$E(0,0) \cap [C^\infty(\bar{\Omega})]^4$$

is dense in $E(0,0)$ *(with respect to the norm in the space* S*).*

Proof. Given in detail in Hlaváček (1979). □

Proof of Theorem 2.5. Consider a time moment t_m and omit the index m for simplicity. Theorem 2.6 on density implies that there exists

$$\sigma_{0\epsilon} \in E(0,0) \cap [C^\infty(\bar{\Omega})]^4, \quad \|\sigma_0 - \sigma_{0\epsilon}\| \le \epsilon. \tag{3.38}$$

Regularizing α we obtain

$$\alpha_\epsilon \in C^\infty(\bar{\Omega}), \quad \|\alpha - \alpha_\epsilon\|_{0,\Omega} \le \epsilon. \tag{3.39}$$

Let us define functions

$$\mathcal{F} = f(\chi + \sigma_0) - \alpha, \quad \mathcal{F}_\epsilon = f(\chi + \sigma_{0\epsilon}) - \alpha_\epsilon.$$

---

[2]In the case $\Gamma = \Gamma_\sigma$, the assumption that Ω is starlike can be omitted provided we apply the Airy stress function in the proof of the density from theorem 2.6. The reason why we introduce condition (i) even here is that the application of the stress functions does not suit the case $n = 3$. Nonetheless, the technique of proof from Hlaváček (1979) works in $\mathbf{R}^3$ under condition (i).

Evidently $\mathcal{F} \leq 0$ a.e. in Ω, but $\mathcal{F}_\epsilon$ generally does not satisfy such an inequality. Recall that $\chi \in [P_1(K^0)]^4$ for all triangles $K^0 \in \mathcal{T}_{h_0}$. Choose $\hat{\tau} = (\chi + \tau_0, \beta_h)$, where $\tau_0 = r_h\sigma_{0\epsilon}$, $\beta_h = \Pi_h\alpha_\epsilon + \rho$. Here $\Pi_h\alpha_\epsilon$ and ρ are defined locally in each $K_i \subset K \subset K^0$ in the following way: $\Pi_h\alpha_\epsilon$ is the linear Lagrangian interpolation $\Pi_{K_i}\alpha_\epsilon$ on K_i,

$$\rho \in P_1(K_i), \quad \rho(a_j) = [d_j - \alpha_\epsilon(a_j)]^+, \quad j = 1, 2, 3,$$

where a_j are the vertices of K_i and $d_j = f(\chi + r_h\sigma_{0\epsilon})(a_j)$. It is easily seen that $\chi + r_h\sigma_{0\epsilon} \in E(t_m)$ and

$$f(\chi + r_h\sigma_{0\epsilon}) - \beta_h \leq 0 \quad \text{a.e. on } \Omega.$$

Now we need an estimate for ρ in $L^2(\Omega)$. For $j = 1, 2, 3$ we can write

$$0 \leq \rho(a_j) = [d_j - \alpha_\epsilon(a_j)]^+ \leq |d_j - f(\chi + \sigma_{0\epsilon})(a_j)| + \mathcal{F}_\epsilon^+(a_j),$$

since

$$-\alpha_\epsilon(a_j) \leq \mathcal{F}_\epsilon^+(a_j) - f(\chi + \sigma_{0\epsilon})(a_j).$$

Further,

$$|d_j - f(\chi + \sigma_{0\epsilon})(a_j)| \leq Ch^2\|\sigma_{0\epsilon}\|_{[C^2(\bar{\Omega})]^4} \equiv \epsilon_1(h, \sigma_{0\epsilon}),$$

hence,

$$\rho \leq \epsilon_1(h, \sigma_{0\epsilon}) + \Pi_{K_i}\mathcal{F}_\epsilon^+ \quad \text{on } K_i,$$

$$\|\rho\|_{0,K_i}^2 \leq 2\epsilon_1^2 \operatorname{mes} K_i + 2\|\Pi_{K_i}\mathcal{F}_\epsilon^+\|_{0,K_i}^2. \quad \square \tag{3.40}$$

Lemma 2.5. *For each triangle $K^0 \in \mathcal{T}_{h_0}$ the estimate*

$$\|\mathcal{F}_\epsilon^+\|_{0,K^0} \leq C\epsilon$$

holds.

Proof. By virtue of (3.1) and (3.2) we obtain

$$|f(\chi + \sigma_0) - f(\chi + \sigma_{0\epsilon})| \leq C\|\sigma_0 - \sigma_{0\epsilon}\|_{R_\sigma}$$

$$\|f(\chi + \sigma_0) - f(\chi + \sigma_{0\epsilon})\|_{0,K^0}^2 \leq C\|\sigma_0 - \sigma_{0,\epsilon}\|_S^2 \leq C\epsilon^2,$$

hence, we have the estimates

$$\|\mathcal{F} - \mathcal{F}_\epsilon\|_{0,K^0} \leq \|f(\chi + \sigma_0) - f(\chi + \sigma_{0\epsilon})\|_{0,K^0} + \|\alpha - \alpha_\epsilon\|_{0,K^0} \leq C\epsilon.$$

Further, denoting $\Omega_1 = \operatorname{supp} \mathcal{F}_\epsilon^+ \cap K^0$, we have

$$\|\mathcal{F}_\epsilon^+\|_{0,K^0}^2 \leq \int_{\Omega_1} (\mathcal{F}_\epsilon^+ - \mathcal{F})^2 dx = \|\mathcal{F}_\epsilon - \mathcal{F}\|_{0,\Omega_1}^2 \leq C\epsilon^2. \quad \square$$

Lemma 2.6. *Let us define $\Pi_h \mathcal{F}_\epsilon^+$ locally on each $K_i \subset K \subset K^0$ as the linear Lagrangian interpolation $\Pi_{K_i} \mathcal{F}_\epsilon^+$. Then*

$$\|\mathcal{F}_\epsilon^+ - \Pi_h \mathcal{F}_\epsilon^+\|_{0,K^0} \le \epsilon_2(h), \quad \lim_{h\to 0} \epsilon_2(h) = 0 \tag{3.41}$$

holds for each triangle $K^0 \in \mathcal{T}_{h_0}$.

Proof. Since $\mathcal{F}_\epsilon \in C(K^0)$, we have $\mathcal{F}_\epsilon^+ \in C(K^0)$ as well. For every $\eta > 0$ there is a polynomial p such that

$$\|\mathcal{F}_\epsilon^+ - p\|_{C(K^0)} \le \eta.$$

Further,

$$\|\Pi_h p - \Pi_h \mathcal{F}_\epsilon^+\|_{C(K_i)} \le \|p - \mathcal{F}_\epsilon^+\|_{C(K_i)} \le \|p - \mathcal{F}_\epsilon^+\|_{C(K^0)},$$

$$\|p - \Pi_h p\|_{C(K_i)} \le C_0 h^2 \|p\|_{C^2(K^0)},$$

for every $K_i \subset K \subset K^0$. Hence, we can write

$$\|\mathcal{F}_\epsilon^+ - \Pi_h \mathcal{F}_\epsilon^+\|_{C(K^0)} \le \|\mathcal{F}_\epsilon^+ - p\|_{C(K^0)} + \|p - \Pi_h p\|_{C(K^0)}$$

$$+ \|\Pi_h p - \Pi_h \mathcal{F}_\epsilon^+\|_{C(K^0)} \le 2\eta + C_0 h^2 \|p\|_{C^2(K^0)} = \delta(h, K^0)$$

and this implies (3.41) if we put

$$\epsilon_2(h) = \max_{K^0 \in \mathcal{T}_{h_0}} [\delta(h, K^0)(\text{mes } K^0)^{1/2}]. \qquad \square$$

Proof. Now we will complete the proof of theorem 2.5. Lemmas 2.5, 2.6 and the inequality (3.40) yield

$$\|\rho\|_{0,K^0}^2 \le 2\epsilon_1^2 \,\text{mes}\, K^0 + 4(\|\mathcal{F}_\epsilon^+\|_{0,K^0}^2 + \|\Pi_h \mathcal{F}_\epsilon^+ - \mathcal{F}_\epsilon^+\|_{0,K^0}^2)$$

$$\le 2\epsilon_1^2 \,\text{mes}\, K^0 + C(\epsilon^2 + \epsilon_2^2).$$

Finally, we obtain

$$\|\hat{\sigma} - \hat{\tau}\|^2 = \|\sigma_0 - r_h \sigma_{0\epsilon}\|_S^2 + \|\alpha - \beta_h\|_{0,\Omega}^2 \le 2(\|\sigma_0 - \sigma_{0\epsilon}\|_S^2$$

$$+ \|\sigma_{0\epsilon} - r_h \sigma_{0\epsilon}\|_S^2) + 3(\|\alpha - \alpha_\epsilon\|_{0,\Omega}^2 + \|\alpha_\epsilon - \Pi_h \alpha_\epsilon\|_{0,\Omega}^2 + \|\rho\|_{0,\Omega}^2)$$

$$\le 2(\epsilon^2 + C h^4 \|\sigma_{0\epsilon}\|_{[C^2(\bar{\Omega})]^4}^2) + C(\epsilon^2 + h^4 |\alpha_\epsilon|_{2,\Omega}^2$$

$$+ h^4 \|\sigma_{0\epsilon}\|_{[C^2(\bar{\Omega})]^4}^2 + \epsilon^2 + \epsilon_2^2(h)).$$

Hence,

$$\lim_{h\to 0} \|\hat{\sigma}^m - \hat{\tau}\| = 0, \quad m = 1, \dots, N.$$

Thus, for $k = T/N$ fixed we obtain the assertion (3.37). □

Remark 2.5. In three-dimensional problems, the analog of theorem 2.5 can be established by applying four-faced elements for the equilibrium model, consisting of four subtetrahedra (see Křížek (1982)). □

In conclusion, we would like to point out that there are a number of other methods for approximate solution of problems for elasto-plastic bodies. Only a few of them have been subjected to a theoretical analysis of the convergence problem (see Gröger (1979), C. Johnson (1977), Nguyen Quoc Son (1977), and Moreau (1974)).

References

Aubin, J. P. (1972). *Approximation of Elliptic Boundary Value Problems.* New York: Wiley-Interscience.

Babuška, I., M. Práger, and E. Vitásek. (1966). *Numerical Processes in Differential Equations.* Prague: SNTL.

Brézis, H. (1973). *Operateurs maximaux monotones.* Amsterdam: North-Holland.

Brezzi, F., W. W. Hager, and P. A. Raviart. (1979). Error estimates for the finite element solution of variational inequalities. *Numer. Math.* **28**, 431–443.

Céa, J. (1971). *Optimisation, théorie et algorithmes.* Paris: Dunod.

Chan, S. H., and I. S. Tuba. (1971). A finite element method for contact problems of solid bodies. *Int. J. Mech. Sci.* **13**, 615–639.

Ciarlet, P. G. (1978). *The Finite Element Method for Elliptic Problems.* Amsterdam: North-Holland.

Conry, T. F., and A. Seireg. (1971). A mathematical programming method for design of elastic bodies in contact. *J. A. M. ASME* **2**, 387–392.

Duvaut, G. (1976). Problémes de contact entre corps solides deformables. Applications of methods of functional analysis. *In: Lecture Notes in Math.* ed. P. Germain and B. Nayroles, 317–327. Berlin: Springer-Verlag.

Duvaut, G., and J. L. Lions. (1972). *Les inéquations en mécanique et en physique.* Paris: Dunod.

Ekeland, I., and R. Temam. (1974). *Analyse convexe et problémes variationnels.* Paris: Dunod.

Falk, R. S. (1974). Error estimates for the approximation of a class of variational inequalities. *Math. Comp.* **28**, 963–971.

Fichera, G. (1964). Problemi elastostatici con vincoli unilaterali; il problema di Signorini con ambigue condizioni al contorno. *Mem. Accad. Naz. Lincei* **8(7)**, 91–140.

Fichera, G. (1972). *Boundary value problems of elasticity with unilateral constraints. In:* Encyclopedia of Physics, ed. S. Flügge, vol. VIa/2, Berlin: Springer-Verlag.

Fraeijs de Veubeke, B., and M. Hogge. (1972). Dual analysis of heat conduction problems by finite elements. *Int. J. Numer. Meth. Eng.* **5**, 65–82.

Francavilla, A., and O. C. Zienkiewicz. (1975). A note on numerical computations of elastic contact problems. *Int. J. Numer. Meth. Eng.* **9**, 913–924.

Fredriksson, B. (1976). Finite element solution of surface nonlinearities in structural mechanics. *Comp. and Struct.* **6**, 281–290.

Fučík, S., and A. Kufner. (1980). *Nonlinear Differential Equations.* Amsterdam: Elsevier.

Glowinski, R., J. L. Lions, and R. Trémolières. (1976). *Analyse numérique des inéquations variationelles.* Paris: Dunod. (English translation, 1981. Amsterdam: North-Holland.)

Grisvard, P., and G. Iooss. (1976). Problémes aux limites unilatéraux dans les domaines non régulieres. *Publ. Seminaires Math.* Rennes: Université de Rennes.

Gröger, K. (1977). *Evolution equations in the theory of plasticity.* Proceedings of the fifth summer school on nonlinear operators, Berlin.

Gröger, K. (1978). Zur Theorie des quasi-statischen Verhaltens von Elastisch-Plastischen Körpern. *ZAMM* **58**, 81–88.

Gröger, K. (1978). Zur Theorie des dynamischen Verhaltens von Elastisch-Plastischen Körpern. *ZAMM* **58**, 483–487.

Gröger, K. (1979). Initial value problems for elastoplastic and elastoviscoplastic systems. *In: Nonlinear Analysis, Function Spaces, Applications, Proceedings,* ed. S. Fučík and A. Kufner. Leipzig: Teubner Texte zur Mathematik.

Gröger, K., and J. Nečas. (1979). On a class of nonlinear initial value problems in Hilbert spaces. *Math. Nachr.* **98**, 21–81.

Gröger, K., J. Nečas, and L. Trávníček. (1979). Dynamic deformation processes of elasto-plastic systems. *ZAMM* **59**, 567–572.

Halphen, B., and Q. S. Nguyen. (1975). Sur les matériaux standarts généralisés. *J. Mécanique* **14**, 39–63.

Haslinger, J. (1977). Finite element analysis for unilateral problems with obstacles on the boundary. *Apl. Mat.* **22**, 180–188.

Haslinger, J. (1979). Dual finite element analysis for an inequality of the 2nd order. *Apl. Mat.* **24**, 118–132.

Haslinger, J. (1979). Finite element analysis of the Signorini problem. *Comm. Math. Univ. Carol.* **20**, **1**, 1–17.

Haslinger, J. (1981). Mixed formulation of variational inequalities and its approximation. *Apl. Mat.* **26**, 462–475.

Haslinger, J., and I. Hlaváček. (1976). Convergence of a finite element method based on the dual variational formulation. *Apl. Mat.* **21**, 43–65.

Haslinger, J., and I. Hlaváček. (1980). Contact between elastic bodies. I. Continuous problems. *Apl. Mat.* **25**, 324–347.

Haslinger, J., and I. Hlaváček. (1981). Contact between elastic bodies. II. Finite element analysis. *Apl. Mat.* **26**, 263–290.

Haslinger, J., and I. Hlaváček. (1981). Contact between elastic bodies. III. Dual finite element analysis. *Apl. Mat.* **26**, 321–344.

Haslinger, J., and I. Hlaváček. (1982). Approximation of the Signorini problem with friction by a mixed finite element method. *J. Math. Anal. Appl.* **86**, 99–122.

Haslinger, J., and J. Lovíšek. (1980). Mixed variational formulation of unilateral problems. *CMUC* **21**, 231–246.

Haslinger, J., and P. D. Panagiotopoulos. (1984). Approximation of contact problems with friction by reciprocal variational formulations. *Proc. Roy. Soc. Edinburgh,* **98A**, 365–383.

Haslinger, J., and M. Tvrdý. (1983). Approximation and numerical realization of contact problems with friction. *Apl. Mat.* **28**, 55–71.

Hertz, H. (1896). *Miscellaneous Papers.* London: MacMillan.

Hlaváček, I. (1977). Dual finite element analysis for unilateral boundary value problems. *Apl. Mat.* **22**, 14–51.

Hlaváček, I. (1977). Dual finite element analysis for elliptic problems with obstacles on the boundary, I. *Apl. Mat.* **22**, 244–255.

Hlaváček, I. (1978). Dual finite element analysis for semi-coercive unilateral boundary value problems. *Apl. Mat.* **23**, 52–71.

Hlaváček, I. (1979). Convergence of an equilibrium finite element model for plane elastostatics. *Apl. Mat.* **24**, 427–457.

Hlaváček, I. (1980). A finite element solution for plasticity with strain-hardening. *R. A. I. R. O. Analyse numérique* **14**, 347–368.

Hlaváček, I. (1980). Convergence of dual finite element approximations for unilateral boundary value problems. *Apl. Mat.* **25**, 375–386.

Hlaváček, I., and J. Lovíšek. (1977). A finite element analysis for the Signorini problem in plane elastostatics. *Apl. Mat.* **22**, 215–228.

Hlaváček, I., and J. Lovíšek. (1980). Finite element analysis of the Signorini problem in semi-coercive cases. *Apl. Mat.* **25**, 273–285.

Hodge, P. G. (1959). *Plastic Analysis of Structures*. New York: McGraw-Hill.

Hühnlich, R. (1979). Quasistatische Anfangswertprobleme für elastisch-plastische Materialen mit linearer Verfestigung. *Rep. Akad. Wiss. DDR, ZIMM, R-02/79*, Berlin.

Jakovlev, G. N. (1961). Boundary properties of functions of class $W_p^{(1)}$ on the domains with angular points. *D.A.N. USSR*, **140**, 73–76.

Janovský, V., and P. Procházka. (1980). Contact problem of two elastic bodies. I–III. *Apl. Mat.* **25**, 87–146.

Johnson, C. (1976). Existence theorems for plasticity problems. *J. Math. Pure Appl.* **55**, 431–444.

Johnson, C. (1976). On finite element methods for plasticity problems. *Numer. Math.* **26**, 79–84.

Johnson, C. (1977). A mixed finite element method for plasticity problems with hardening. *SIAM J. Numer. Anal.* **14**, 575–583.

Johnson, C. (1978). On plasticity with hardening. *J. Math. Anal. Appl.* **62**, 325–336.

Johnson, C., and B. Mercier. (1978). Some equilibrium finite element methods for two-dimensional elasticity problems. *Numer. Math.* **30**, 103–116.

Kachanov, L. M. (1974). *Fundamentals of the Theory of Plasticity*. Moscow: Mir.

Kikuchi, N., and J. T. Oden. (July, 1979). Contact problems in elasticity. TICOM Report 79-8.

Kinderlehrer, D., and G. Stampacchia. (1980). *An Introduction to Variational Inequalities and their Applications*. New York: Academic Press.

Koiter, W. T. (1960). General theorems for elastic-plastic solids. *In: Progress in Solid Mechanics*, vol. 1, chapt. IV, 167–221. Amsterdam: North-Holland.

Křížek, M. (1982). An equilibrium finite element method in three-dimensional elasticity. *Apl. Mat.* **28**, 46–75.

Kufner, A., S. Fučík, and O. John. (1977). *Function Spaces*. Prague: Academia.

Lions, J. L., and G. Stampacchia. (1967). Variational inequalities. *Comm. Pure Appl. Math.* **20**, 493–519.

Moreau, J. J. (1974). On Unilateral Constraints. Friction and Plasticity. In: *New Variational Techniques in Mathematical Physics*. CIME, II ciclo, ed. G. Capriz and G. Stampacchia. 175–322.

Mosco, U., and G. Strang. (1974). One-sided approximations and variational inequalities. *Bull. Am. Math. Soc.* **80**, 308–312.

Nečas, J. (1967). *Les methodes directes en theorie des équations elliptiques*. Prague: Academia.

Nečas, J. (1975). On regularity of solutions to nonlinear variational inequalities for second order elliptic systems. *Rend. di Matematica,* **2**, vol. 8, ser. VI, 481–498.

Nečas, J., and I. Hlaváček. (1981). *Mathematical Theory of Elastic and Elasto-Plastic Bodies: An Introduction*. Amsterdam: Elsevier.

Nečas, J., J. Jarušek, and J. Haslinger. (1980). On the solution of the variational inequality to the Signorini problem with small friction. *Boll. Unione Mat. Ital.* **5**, 17–B, 796–811.

Nečas, J., and L. Trávníček. (1978). Variational inequalities of elastoplasticity with internal state variables. *Abh. Akad. Wiss. DDR, Abt. Mathematik,* GN, 195–204.

Nečas, J., and L. Trávníček. (1980). Evolutionary variational inequalities and applications in plasticity. *Apl. Mat.* **25**, 241–256.

Nguyen, Q. S. (1973). Materiaux elastoplastiques écrouissables. *Arch. Mech. Stos.* **25**, 695–702.

Nguyen, Q. S. (1977). On the elastic-plastic initial boundary value problem and its numerical integration. *Int. J. Numer. Math. Eng.* **11**, 817–832.

Nitsche, J. (1971). Über ein Variationsprinzip zur Lösung von Dirichlet Problem bei Verwendung von Teilräumen, die keinen Randbedingungen unterworfen sind. *Abh. Mat. Sem. Univ. Hamburg,* **36**, 9–15.

Panagiotopoulos, P. D. (1975). A nonlinear programming approach to the unilateral contact and friction-boundary value problem in the theory of elasticity. *Ing. Archiv.* **44**, 421–432.

Pšeničnyj, B. N., and J. M. Danilin. (1975). *Čislennyje metody v extremal'nych zadačach*. Moskva: Nauka.

Rektorys, K. (1977). *Variational Methods*. Dordrecht-Boston: Reidel Co.

Signorini, A. (1959). Questioni di elasticità non linearizzata e semi-linearizzata. *Rend. di Matem.* **18**, 1–45.

Trávníček, L. (1976). *Existence and uniqueness to the boundary value problem for elasto-plastic materials*. Thesis.

Washizu, K. (1968). *Variational Methods in Elasticity and Plasticity*. New York: Pergamon Press.

Zoutendijk, G. (1960). *Methods of Feasible Directions*. Amsterdam: Elsevier.

Zoutendijk, G. (1966). Nonlinear programming. *SIAM J. Control,* **4**, 194–210.

Index

Applied Mathematical Sciences

cont. from page ii

44. Pazy: **Semigroups of Linear Operators and Applications to Partial Differential Equations.**
45. Glashoff/Gustafson: **Linear Optimization and Approximation: An Introduction to the Theoretical Analysis and Numerical Treatment of Semi-Infinite Programs.**
46. Wilcox: **Scattering Theory for Diffraction Gratings.**
47. Hale et al.: **An Introduction to Infinite Dimensional Dynamical Systems—Geometric Theory.**
48. Murray: **Asymptotic Analysis.**
49. Ladyzhenskaya: **The Boundary-Value Problems of Mathematical Physics.**
50. Wilcox: **Sound Propagation in Stratified Fluids.**
51. Golubitsky/Schaeffer: **Bifurcation and Groups in Bifurcation Theory, Vol. I.**
52. Chipot: **Variational Inequalities and Flow in Porous Media.**
53. Majda: **Compressible Fluid Flow and Systems of Conservation Laws in Severa Space Variables.**
54. Wasow: **Linear Turning Point Theory.**
55. Yosida: **Operational Calculus: A Theory of Hyperfunctions.**
56. Chang/Howes: **Nonlinear Singular Perturbation Phenomena: Theory and Applications.**
57. Reinhardt: **Analysis of Approximation Methods for Differential and Integral Equations.**
58. Dwoyer/Hussaini/Voigt (eds.): **Theoretical Approaches to Turbulence.**
59. Sanders/Verhulst: **Averaging Methods in Nonlinear Dynamical Systems.**
60. Ghil/Childress: **Topics in Geophysical Dynamics: Atmospheric Dynamics, Dynamo Theory and Climate Dynamics.**
61. Sattinger/Weaver: **Lie Groups and Algebras with Applications to Physics, Geometry, and Mechanics.**
62. LaSalle: **The Stability and Control of Discrete Processes.**
63. Grasman: **Asymptotic Methods of Relaxation Oscillations and Applications.**
64. Hsu: **Cell-to-Cell Mapping: A Method of Global Analysis for Nonlinear Systems.**
65. Rand/Armbruster: **Perturbation Methods, Bifurcation Theory and Computer Algebra.**
66. Hlaváček/Haslinger/Nečas/Lovíšek: **Solution of Variational Inequalities in Mechanics.**
67. Cercignani: **The Boltzmann Equation and Its Applications.**
68. Temam: **Infinite Dimensional Dynamical System in Mechanics and Physics.**
69. Golubitsky/Stewart/Schaeffer: **Singularities and Groups in Bifurcation Theory, Vol. II.**